Intelligent Systems Reference Library

Volume 277

The aim of this series is to publish a Reference Library, including novel advances and developments in all aspects of Intelligent Systems in an easily accessible and well structured form. The series includes reference works, handbooks, compendia, textbooks, well-structured monographs, dictionaries, and encyclopedias. It contains well integrated knowledge and current information in the field of Intelligent Systems. The series covers the theory, applications, and design methods of Intelligent Systems. Virtually all disciplines such as engineering, computer science, avionics, business, e-commerce, environment, healthcare, physics and life science are included. The list of topics spans all the areas of modern intelligent systems such as: Ambient intelligence, Computational intelligence, Social intelligence, Computational neuroscience, Artificial life, Virtual society, Cognitive systems, DNA and immunity-based systems, e-Learning and teaching, Human-centred computing and Machine ethics, Intelligent control, Intelligent data analysis, Knowledge-based paradigms, Knowledge management, Intelligent agents, Intelligent decision making, Intelligent network security, Interactive entertainment, Learning paradigms, Recommender systems, Robotics and Mechatronics including human-machine teaming, Self-organizing and adaptive systems, Soft computing including Neural systems, Fuzzy systems, Evolutionary computing and the Fusion of these paradigms, Perception and Vision, Web intelligence and Multimedia.

Indexed by SCOPUS, DBLP, zbMATH, SCImago.

All books published in the series are submitted for consideration in Web of Science.

Pallavi Jain · Sapna Raghav · Anil Kumar Jangir
Editors

Towards Green Chemical Processes: Strategies and Innovations

Springer

Editors
Pallavi Jain
Department of Chemistry
Faculty of Engineering and Technology
SRM Institute of Science and Technology
Delhi NCR Campus
Ghaziabad, Uttar Pradesh, India

Sapna Raghav
Department of Chemistry
Shri Jagdishprasad Jhabarmal Tibrewala University
Jhunjhunu, Rajasthan, India

Anil Kumar Jangir
Department of Chemistry
Shri Jagdishprasad Jhabarmal Tibrewala University
Jhunjhunu, Rajasthan, India

ISSN 1868-4394 ISSN 1868-4408 (electronic)
Intelligent Systems Reference Library
ISBN 978-3-031-94089-7 ISBN 978-3-031-94090-3 (eBook)
https://doi.org/10.1007/978-3-031-94090-3

This Springer imprint is published by the registered company Springer Nature Switzerland AG
The registered company address is: Gewerbestrasse 11, 6330 Cham, Switzerland

Contents

Chapter 1
Green Chemical Processes: An Introduction

Pallavi Jain, Sapna Raghav, and Anil Kumar Jangir

Abstract Green chemistry is known as sustainable chemistry mainly focused on making chemical processes and products that minimises hazardous substances. This chapter mainly focuses on the principles and significance of green chemistry in modern industry and day to day life. Beginning from the environmental movement of the 1960s and 1970s green chemistry has come into mainstream practice as a sustainable technology. Twelve principles of green chemistry were introduced by Paul Anastas and John Warner gives a framework for making safer, more efficient and environmentally friendly chemical processes. The chapter also highlights the importance of green chemistry in addressing global challenges such as climate change, resource depletion and pollution by focusing its importance in providing innovation, regulatory compliance and economic benefits. By using green chemistry principles industries can reduce their environmental footprint, increase efficiency and contribute to a sustainable future goal.

Keywords Green chemistry · Twelve principles · Chemical processes · Sustainable future

1.1 Introduction

Green chemistry also known as sustainable chemistry has main focuses on the design, development and implementation of chemical method that minimise the utilisation and generation of harmful substances [1]. The primary aim of green chemistry is to minimise the environmental effect of chemical methods and keeping economic

P. Jain (✉)
Department of Chemistry, Faculty of Engineering and Technology, SRM Institute of Science and Technology, NCR Campus, Delhi-NCR Campus, Uttar Pradesh, Modinagar, Ghaziabad, India
e-mail: pallavij@srmist.edu.in; palli24@gmail.com

S. Raghav · A. Kumar Jangir
Department of Chemistry, Shri Jagdishprasad Jhabarmal Tibrewala University, Vidyanagri, Jhunjhunu, Rajasthan, India

P. Jain et al. (eds.), *Towards Green Chemical Processes: Strategies and Innovations*, Intelligent Systems Reference Library 277,
https://doi.org/10.1007/978-3-031-94090-3_1

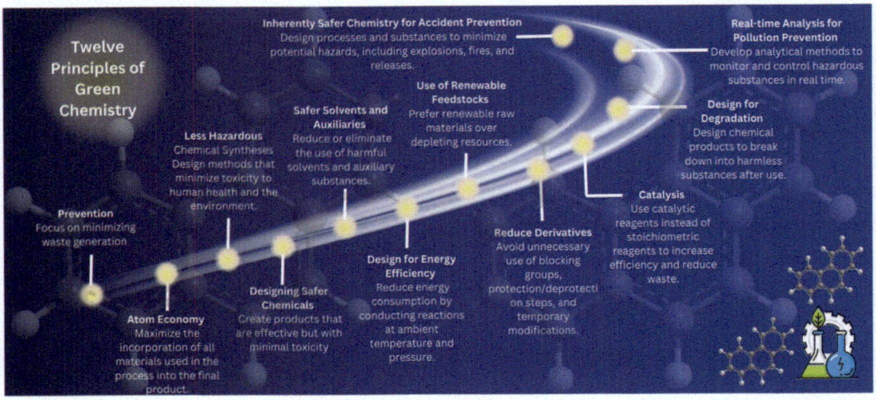

Fig. 1.1 Twelve principles of green chemistry

viability and efficiency. This can be achieved by following a set of principles that help us to the development of safer chemicals, energy efficient processes and sustainable practices [2]. The concept of green chemistry was introduced in the 1990s with the publication of the Twelve principles of green chemistry (Fig. 1.1) by Paul Anastas and John Warner [3]. These principles has main focus on the waste prevention, atom economy, the use of renewable feedstocks and the making of biodegradable products [4]. Figure 1.1 highlights the twelve principles of green chemistry.

1.1.1 Historical Development and Evolution

The research article published by Rachel Carson Silent Spring in 1962 shows the harmful impact of pesticides and other synthetic chemicals [5]. In thar decades regulatory frameworks such as the U.S. Clean Air Act (1970) and Clean Water Act (1972) were made to control pollution and protect natural resources [6, 7]. However these regulations has main focus on the end of pipe solutions such as waste treatment and dumping instead of addressing the root causes of pollution [8]. This approach made the realization that there is a requirement of some more strategy that integrates environmental considerations into the design of chemical processes and products from the outset [9]. In 1991 the U.S. Environmental Protection Agency (EPA) promoted the Green Chemistry Program to promote research and innovation in this area [10]. The program gives financial assistance for innovative projects and also gives the Presidential Green Chemistry Challenge Awards to honour outstanding performance in the field [11]. Today green chemistry is a world-wide movement with new ideas, initiatives and collaborations of countries and industries.

1.1.2 Importance and Relevance in Modern Industry

The need and urgency of green chemistry in industry cannot be ignored. As the population continues to grow globally and natural resources are getting reduced hence there is a need for a more sustainable method of production and consumption. The chemical industry has an important role in these changes because it is both a consumer of resources and a major source of pollution. Traditional chemical processes depend on non renewable raw materials such as petroleum and generate large amounts of waste [12]. These processes are not environment friendly and also they are economically inefficient because they result in the loss of valuable materials and energy. Green chemistry gives a way to tackle these challenges by making the development processes more efficient and less harmful to the environment. One of the major positive points of green chemistry is that it gives innovative ideas and provides business opportunities. Companies that follow the green chemistry principles can minimize the environmental effect by reducing operating costs and get a competitive advantage in the marketplace [13]. For example the development of bio based plastics and renewable energy sources has provided a new market and created new jobs in the green economy. Green chemistry is important for meeting requirements and for achieving sustainability goals.

1.1.3 The Role of Green Chemistry in Addressing Global Challenges

Green chemistry has an important role in solving some of the most concerning global challenges of our time such as climate change, resource depletion and pollution. The chemical industry has a major contribution to greenhouse gas emissions so there is a need for processes that will help in reducing carbon footprint. For example the development of carbon capture technologies that convert CO_2 into different products is a needed area of research in green chemistry [14]. Resource depletion is another major challenge because the availability of non-renewable resources such as petroleum and rare earth metals is limited. Green chemistry promotes the use of renewable resources such as biomass and the development of technologies to recover valuable materials from waste streams [15]. These approaches not only minimise the need for new resources but also help to the circular economy where materials are kept in use for as long as possible. Pollution causes from hazardous chemicals have a serious threat to human health and the environment. Green chemistry aims to minimise the use of toxic substances in chemical processes and products thereby minimizing the risk of exposure and contamination. The making of non toxic pesticides and biodegradable plastics has the potential to reduce the environmental impact of agriculture and consumer goods. By making safe and more sustainable products green chemistry can improve health and quality of life. In the 21st century green chemistry will play an important role in making a sustainable future. The journey toward sustainability is a joint effort and green chemistry offers a way to get this vision.

1.2 Environmental and Economic Drivers

1.2.1 Environmental Impact of Traditional Chemical Processes

Traditional chemical processes are being used for industrial development and making the production of materials, fuels, pharmaceuticals and consumer goods. However these processes were done at a some environmental cost [16]. The main dependence on non-renewable resources such as petroleum and natural gas as sources has caused the depletion of resources and led to environmental degradation. The generation of harmful waste, greenhouse gas emissions and the release of toxic by-products into air, water and soil have caused serious concerns about the sustainability of traditional chemical practices [17]. One of the most serious environmental issues with traditional chemical processes is pollution. The manufacturing of plastics mainly depends on petrochemical sources that generate large amounts of waste and gas emissions. The improper disposal of plastic waste has caused the accumulation of microplastics in oceans that harms marine life and also entered the food chain [18]. Synthesis of agrochemicals such as pesticides and fertilizers involves the utilization of toxic reagents and solvent that contaminate soil and water systems which directly or indirectly affect ecosystems and human health. Combustion of fossil fuels for energy in chemical manufacturing also releases CO_2 and other greenhouse gases into the atmosphere that results in increased global warming [19]. Furthermore certain chemical reactions produce major greenhouse gases such as N_2O and CH_4 that have a higher global warming capacity than CO_2 [20]. The combined effect of these emissions is a major reason for climate change. The environmental effect of traditional chemical processes is not limited to pollution and climate change. The extraction and processing of raw materials such as mining for metals and drilling for oil led to habitat loss, biodiversity loss and the damage of ecosystems. The mining of rare earth metals that are needed for electronics and renewable energy technologies but this can cause soil erosion, water contamination and the displacement of local communities. Among these challenges there is an urgent need of transition to more sustainable chemical practices that minimises the environmental effect along with meeting the demands of an increasing population. Green chemistry provides a way to achieve this by the way of green chemical processes.

1.2.2 Regulatory Frameworks and Policies

The environmental effect of traditional chemical processes has made the governments and international organizations to frame a regulation and policies that has main aim to reduce pollution, follow sustainability and protect human health. One of the most regulatory frameworks is the U.S. Clean Air Act (1970) that was made to control air pollution and set limits on the emission of harmful substances. This rules and act has been successful in reducing the release of pollutants such as SO_2, NO and volatile organic compounds (VOCs) from chemical plants and other industries [21]. Clean Water Act (1972) controls the discharge of pollutants into water bodies and make sure that chemical processes do not pollute drinking water sources [22]. In Europe the Registration, Evaluation, Authorisation and Restriction of Chemicals (REACH 2007) rules shows a good approach for chemical safety [23]. REACH put eyes on the companies and made them to register and provide safety data of all chemicals made or imported in the European Union promote the alternative of harmful substances with safer alternatives. This rules has a major role for the development and adoption of green chemistry practices in the European chemical industry. At the global level the Stockholm Convention on Persistent Organic Pollutants (POPs 2001) has motive to control the manufacturing and employment of harmful chemicals that continues in the environment and bioaccumulate in living organisms [24]. The rule has control the use various harmful chemicals such as polychlorinated biphenyls (PCBs) and various pesticides. Along with regulatory frameworks, governments have also introduced various policies and incentives to promote green chemistry [25]. U.S. Green Chemistry Program started by the Environmental Protection Agency (EPA) in 1991 gives financial support for research and development in green chemistry and finds new solutions by the Presidential Green Chemistry Challenge Awards [26]. The European Union Horizon 2020 program has provided major financial support for projects based on sustainable chemistry and the circular economy [27]. Principle of green chemistry and its application was given in Table 1.1. These regulations and policies not only solve the environmental impact of chemical processes but also provide economic opportunities for companies that follow green chemistry.

1.2.3 Economic Benefits of Green Chemistry

Environment benefited by green chemistry are documented in respect of economic advantages. Green chemistry gives way to minimise costs, increase efficiency and create new market opportunities. One of the most economic benefits of green chemistry is waste reduction. Traditional chemical processes mainly creates large amounts of waste and to treat and dispose that waste needed huge amount of cost. Making a new processes that can minimise waste, reduce the operating cost and increases the efficacy of company. Following catalytic processes that use less concentration of reagents and generate negligible by products can help substantial cost savings.

Table 1.1 Twelve principles of green chemistry and their applications

Principle	Description	Example applications
1. Waste prevention	Minimize waste production at the source rather than managing it later	Using solvent-free methods in pharmaceutical manufacturing
2. Atom efficiency	Ensure that most of the starting materials become part of the final product	Employing catalytic hydrogenation instead of excess reagent-based reductions
3. Safer chemical synthesis	Develop methods that reduce or eliminate hazardous substances	Using water as a reaction medium instead of harmful organic solvents
4. Designing eco-friendly chemicals	Create compounds that are effective while minimizing toxicity	Producing biodegradable polymers for sustainable packaging
5. Use of benign solvents and auxiliaries	Prefer environmentally friendly or no solvents when feasible	Applying supercritical CO_2 in extractions rather than toxic chlorinated solvents
6. Energy-efficient processes	Reduce energy consumption in chemical reactions	Implementing microwave-assisted synthesis to accelerate reactions
7. Renewable resource utilization	Prioritize raw materials derived from renewable sources	Manufacturing bio-based polymers from agricultural residues
8. Minimizing unnecessary modifications	Reduce extra steps like blocking or protecting groups in synthesis	Directly converting carboxylic acids into amides without using acid chlorides
9. Use of catalysts	Prefer catalytic systems over reactions requiring large amounts of reagents	Employing metal–organic frameworks (MOFs) as catalysts in organic reactions
10. Designed for decomposition	Ensure that chemical products break down into harmless components	Creating compostable bioplastics that degrade naturally
11. Continuous monitoring for waste reduction	Develop techniques to track reactions in real time to minimize pollution	Using in situ spectroscopy to monitor and control reaction processes
12. Safer chemical handling	Design processes and materials that lower the risks of accidents	Replacing highly reactive or explosive substances with safer alternatives

Many green chemistry process such as use of renewable energy sources and optimization of reaction conditions can reduce energy consumption and lower utility bills [28]. The market for bio based plastics that can made from renewable resources and are biodegradable is increasing as for both consumers and businesses [29]. The manufacturing of non toxic pesticides and eco friendly products has provided a new way for revenue for companies in the both agrochemical and consumer goods sectors. Consumers, investors and stakeholders are mainly focusing on sustainability and companies that are following green chemistry. By limiting the use of harmful substances and following safer processes companies can limit the risk of accidents.

1.2.4 Sustainability and Corporate Social Responsibility

Following green chemistry is linked with the principles of sustainability and corporate social responsibility (CSR). Sustainability is composed of the responsible use of resources and keeping protecting the environment for future generations. By making processes that are efficient, non-toxic and renewable helps companies to get their sustainability target and contribute to sustainable development [30]. CSR is for the company to have commitment to work in an ethical and sustainable manner and also look into the interests of stakeholders such as employees, customers, communities and the environment. Companies that follow green chemistry help to minimize the environmental pollution and contribute to the work against climate change aligning with the United Nations Sustainable Development Goals (SDGs) mainly Goal 12 (Responsible Consumption and Production) and Goal 13 (Climate Action) [31]. Making sustainable products and following processes industries can also help in solving some social challenges such as access to clean water, affordable healthcare and safe working conditions. Green Chemistry & Commerce Council (GC3) and the International Sustainable Chemistry Collaborative Centre (ISC3) come together from across the world to promote innovation and practices in green chemistry [32]. The environmental and economic drivers of green chemistry are interconnected. The environmental challenges posed by traditional chemical processes composed with the regulatory and economic incentives for sustainability have made a compelling case for the following of green chemistry. As the world continues to grapple with the challenges of resource depletion, pollution and climate change, green chemistry provides a pathway to a more sustainable and prosperous future.

1.3 Fundamental Concepts in Green Chemical Processes

Green chemistry is a transformative approach to chemical design, production and its application that aims to reduce environmental impact with maximum efficiency and sustainability. Green chemistry has a set of fundamental principles that guide how chemical processes are conceptualized and executed. These principles atom economy, renewable feedstocks, energy efficiency and waste minimization form the foundation of sustainable chemical engineering. This section focuses on each of these concepts exploring their scientific basis, practical applications and significance in advancing green chemistry.

1.3.1 Atom Economy and Reaction Efficiency

Atom economy is a cornerstone of green chemistry focusing the efficient use of atoms in chemical reactions to reduce waste. Introduced by Barry Trost in 1991 atom economy evaluates the proportion of reactant atoms that are incorporated into the final product [33]. A reaction with 100% atom economy means that all atoms from the reactants are utilized in the product leaving no waste. High atomic economy is particularly important in large scale industrial processes where even small inefficiencies can cause significant waste generation. In Diels–Alder reaction a cycloaddition process is celebrated for its high atom economy as all atoms from the diene and dienophile are used into the cyclic product [34]. In contrast traditional synthetic routes such as the Wittig reaction mainly produce stoichiometric amounts of by-products like triphenylphosphine oxide that results in lower atom economy. Reaction efficiency complements the atom economy by ensuring that chemical transformations proceed with high yields and selectivity. Efficient reactions reduce the need for excess reagents, energy-intensive purification steps and hazardous solvents. Catalytic processes such as hydrogenation or oxidation mainly achieve high efficiency by enabling reactions under milder conditions and reducing the formation of unwanted by-products. Together atom economy and reaction efficiency drive the development of cleaner, more sustainable chemical processes.

1.3.2 Renewable Feedstocks and Resources

The shift from fossil resources to renewable feedstocks is a defining feature of green chemistry. Renewable feedstocks are derived from biomass, agricultural waste or other replenishable sources providing a sustainable alternative to petroleum based raw materials [35]. This transition not only reduces dependence on non-renewable resources but also reduces the carbon footprint of chemical production [36]. Biomass a renewable feedstock can be converted into valuable chemicals through processes like fermentation, pyrolysis and enzymatic catalysis. Lignocellulosic biomass composed of cellulose, hemicellulose and lignin can be depolymerized into sugars and aromatic compounds that can serve as building blocks for biofuels, bioplastics and pharmaceuticals [37]. The making of polylactic acid (PLA) from corn starch shows how renewable resources can replace the product of petroleum derived plastics [38].

1.3.3 Energy Efficiency and Reduction of Energy Consumption

Energy efficiency is an important factor in green chemistry because in chemical processes energy is the main factor. Traditional methods such as high temperature reactions and distillation require huge amounts of energy that is a major contributor to greenhouse gas emissions and environmental degradation. Green chemistry's main motto is to address this challenge by using energy efficient technologies and methods. Use of catalyst to lower activation energies and make the reaction proceed under milder conditions. Enzymatic catalysis works at normal temperatures and pressures by lowering energy requirements in comparison to conventional methods. Photocatalysis uses light energy to proceed chemical changes giving a renewable and energy efficient method in comparison to thermal processes. Techniques such as microwave assisted synthesis and continuous flow reactors make the reaction faster with minimum energy input. Microwave heating directly transfers energy to reactions by lowering heat loss and improving reaction efficiency. Renewable energy sources such as solar, wind and hydropower further increase the sustainability of chemical processes.

1.3.4 Waste Minimization and Pollution Prevention

Waste minimization and pollution prevention are central to green chemistry and aim to eliminate waste at the source rather than managing. This approach reduces the environmental and economic burdens associated with waste disposal and remediation. One key strategy is the design of inherently safer processes that produce minimal waste. Solvent free reactions such as mechanochemical synthesis eliminate the need for hazardous solvents and reduce waste generation [39]. Mechanochemistry that depends on mechanical force to drive chemical reactions has been successfully applied to the synthesis of metal organic frameworks (MOFs), pharmaceuticals and organic compounds often produce water as the only by-product [40]. The use of solvents is another important aspect of waste minimization. Traditional solvents such as dichloromethane and benzene are toxic and harmful. Green alternatives including water, supercritical CO_2 and ionic liquids provide safer and more sustainable options [41]. Supercritical CO_2 is non-toxic, non-flammable and easily recyclable making it an ideal solvent for extraction and purification processes [42]. Advanced analytical techniques such as in situ spectroscopy and process control systems enable real time adjustments to reaction parameters minimizing the formation of by products and improving overall efficiency.

For Pollution control there is also a need to design safer chemicals and materials. The development of biodegradable polymers and less harmful flame retardants helps to reduce the environmental effect of products throughout their lifecycle. By combining both waste management and pollution control into chemical design green chemistry helps towards a sustainable approach for chemical synthesis. The main concepts of green chemistry are atom economy, renewable resources, energy efficiency and waste management gives a detailed outline for making sustainable chemical methods. These principles help in the control of the environmental pollution of chemical manufacturing as well as increase the efficiency and safety. Smart and effective use of resources, adoption of renewable material and limiting the use of energy gives the way for a more sustainable future.

1.4 Green Synthesis and Catalysis

Catalysis plays an important role in green chemistry by increasing the efficiency and sustainability of chemical processes [43]. One of the most interesting areas in this field is biocatalysts in which enzymes obtained from living organisms were used to carry organic compounds reactions. This process is very helpful in the pharmaceutical industries because it follows and comes under the principles of green chemistry. Pharmaceutical sector uses bio catalysis in the discovery and synthesis of small drug molecules. It is important for the enantioselective synthesis of active pharmaceutical ingredients (APIs), many of which are found as enantiomers [44]. Enzyme is used for the synthesis of specific enantiomers with high selectivity and minimise the need for excess purification steps. Compared to traditional chemical methods, bio catalysis has many advantages such as less waste generation, improved atom economy, less environmental pollution and the use of safer solvents, mainly water based. These enzyme based reactions occur at mild conditions, enhancing energy efficiency and reducing the need for functional group protection and deprotection. The enzyme based hydrolysis of penicillin G to make 6-aminopenicillanic acid is a well known biocatalytic process [45]. Using water at 37 °C with ammonia for pH adjustment process this approach is safer and more sustainable in comparison to traditional chemical methods which depend on harmful reagents and extreme conditions. Bio catalysis also speeds up the one pot reactions where various steps occur stepwise under the same conditions and increasing the efficiency and sustainability. Nowadays research has focused on making catalysts through effective methods such as green synthesis that minimise waste and harmful solvents. A remarkable research in this area is the solvent free synthesis of porous nanomaterials like SBA-15, MCM-41 and MOFs. These materials give high surface areas, adjustable acidity and stability makes them ideal for catalytis [46]. Fe_2O_3/SBA-15 nanomaterials made by a planetary ball mill method give magnetically separable catalysts with high surface areas [47]. Similarly the green synthesis of 2D materials such as GO and MXenes has shown potential in electrocatalysis and energy harvesting. Another sustainable approach involves using plant extracts for the green synthesis of metal and metal oxide nanoparticles [48]. These extracts are rich in phytochemicals that act as natural reducing agents and stabilizers leading to controlled nanoparticle size and shape. Ag and Au nanoparticles synthesized through this method have good catalytic and antibacterial properties. Organic based nanocomposites derived from renewable materials like starch, chitosan, lignin and cellulose have also been used for their catalytic potential. These biopolymer based catalysts offer high surface areas and active sites that makes them suitable for various applications. Pd nanoparticles immobilized on a starch halloysite hybrid system have shown high catalytic efficiency in coupling reactions. Catalysis also plays a vital role in using renewable feedstocks transforming biomass derived platform chemicals into valuable products. Cobalt oxide nanoparticles encapsulated within mesoporous materials have been used to oxidize 5-hydroxymethylfurfural (HMF) into 2, 5-furandicarboxylic acid (FDCA) with high selectivity and conversion rates [49]. These advancements underscore how catalytic processes can contribute to a circular economy where waste is minimized and resources are efficiently utilized.

Nano catalysis further enhances the field of green chemistry by leveraging the high surface area and unique properties of nanoparticles to improve reaction rates and selectivity. Nano catalysts bridge the gap between homogeneous and heterogeneous catalysis by combining high activity with ease of recovery and reusability. Solvent free synthesis of nanomaterials like SBA-15 and MOFs increases their catalytic applications while plant extract based synthesis reduces the need for toxic reagents [50]. When coupled with microwave assisted heating nano catalysis achieves even greater energy efficiency and faster reaction rates. Ruthenium hydroxide nano catalysts for nitrile hydration and glutathione based nano organ catalysts for heterocycle synthesis both offer high activity and easy recovery. Supported metal nanoparticles also play an important role in green chemistry. These catalysts often anchored onto mesoporous materials provide higher activity and selectivity because of their high surface area and well ordered pore structures. A key application is the selective oxidation of CO in hydrogen for fuel cell purification. Pt nanoparticles supported on mesoporous silica such as FSM-16 have high activity and selectivity for CO oxidation even at low temperatures because of the strong interaction between the Pt nanoparticles and the silica support [51]. Supported metal catalysts facilitate biomass conversion such as the hydrogenolysis of cellulose into sugar alcohols like sorbitol and mannitol using Pt/γ-Al_2O_3 catalysts [52]. These bifunctional catalysts combine acid and metal sites enabling efficient hydrolysis and reduction reactions. Another exciting development in green catalysis is the use of organ catalysts like proline and its derivatives for multi component reactions in aqueous media. Proline is a naturally occurring, non-toxic, inexpensive and available in both enantiomeric forms of amino acid. It can act as a Brønsted acid or base and form iminium or enamine intermediates that makes it highly versatile for catalytic applications. Proline catalysed reactions in water provide a green pathway for synthesizing heterocycles that are important in pharmaceuticals and agrochemicals. These reactions including the synthesis of pyrroles, isoxazolidines, pyridines and quinazolines proceed efficiently under mild conditions, reducing waste and eliminating the need for intermediate purification steps [53]. Catalysis in green chemistry comprises a broad spectrum of sustainable innovations from bio catalysis and nano catalysis to supported metal catalysts and organo-catalysts. These advancements not only increase reaction efficiency and selectivity but also reduce environmental impact by reducing hazardous reagents, waste and energy consumption [54]. By green synthesis techniques, renewable feedstocks and innovative catalytic materials researchers continue to develop sustainable solutions that align with the principles of green chemistry providing the way for a cleaner and more efficient future in chemical manufacturing.

1.5 Green Solvents and Reaction Media

Different types of green solvent and its application was shown in Fig. 1.2.

1.5.1 Deep Eutectic Solvents (DESs)

Green solvents and reaction media are important for sustainable chemistry aiming to reduce environmental pollution and enhance process efficiency. DESs have emerged as promising green solvents because of their unique properties such as low toxicity, biodegradability and ease of preparation. DESs are formed by mixing hydrogen bond donors and acceptors and have melting points much lower than their individual components making them versatile alternatives to traditional VOCs [55]. They showed remarkable effectiveness in polar organometallic chemistry enabling the use of highly reactive S-block elements like Grignard and organolithium reagents under milder conditions and in the presence of air. DESs also facilitate metal catalyzed reactions such as Suzuki Miyaura and Heck couplings in milder conditions having efficient recycling of catalysts and solvents [56]. In bio catalysis DESs increases the stability, selectivity and performance of enzymes and whole cell microorganisms leading to more efficient bio transformations [57]. These advantages make DESs an

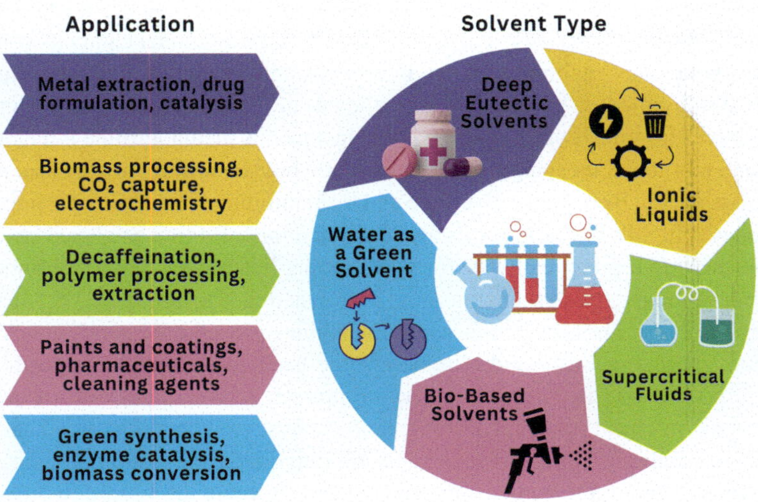

Fig. 1.2 Types of solvent and its application

important component in green chemistry giving eco-friendly solutions for different chemical processes [58]. Green solvents and reaction media are an important portion of sustainable chemistry that aim to minimise environmental impact and increase efficiency. DESs are formed by mixing a hydrogen bond donor and acceptor. They make possible polar organometallic reactions, facilitate metal catalysed couplings and bio catalysis by increasing enzyme stability and selectivity [59]. DESs give a pathway for a sustainable alternative to traditional solvents by supporting green chemistry goals by decreasing hazardous waste and improving reaction conditions.

1.5.2 *Ionic Liquid (ILs)*

ILs are a class of green solvents that have significant attention because of their unique properties and versatile applications. ILs remain liquid at temperatures below 100 °C because of the large size and asymmetry of their ions that prevent them from forming a solid lattice. This characteristic of ILs makes them highly tunable and allow chemists to design specific properties by changing the cation and anion components. One of the most important features of ILs is their negligible vapor pressure that significantly decreases the risk of VOC emissions and makes them environmentally friendly. ILs have high thermal stability and low flammability further increases their safety profile. These properties align well with the principles of green chemistry that focuses the reduction of hazardous substances and the let to the development of safer chemicals and processes [60]. ILs have applications across different fields such as solvent extraction, CO_2 capture, battery technology and catalysis. CO_2 capture ILs provide advantages over traditional amine-based solvents because of their lower volatility and higher thermal stability that reduce energy consumption and operational costs. In catalysis ILs can stabilize reactive intermediates and increase the selectivity and efficiency of catalytic processes. This has been shown in various studies where ILs have been used to increase the performance of transition metal catalysts and biocatalysts [61]. ILs are treated as green solvents but their environmental impact depends on their synthesis, use and disposal. Some ILs can be toxic and their biodegradability differs widely. Therefore it is important to design ILs having low toxicity and high biodegradability to fully realize their potential as sustainable solvents. However ongoing research is needed to address their environmental impact and ensure their safe and sustainable use.

1.5.3 *Bio-based Solvents*

Bio based solvents are a rapidly developing area in green chemistry providing sustainable alternatives to traditional petrochemical derived solvents. Obtained from renewable resources such as agricultural crops, lignocellulosic biomass and organic waste materials these solvents follow green chemistry principles by decreasing environmental impact, enhancing biodegradability and minimizing toxicity. They show properties such as non toxicity, biodegradability and low volatility that makes them suitable for different industrial and laboratory applications. The synthesis of bio based solvents mainly involves the fermentation of biomass or the chemical transformation of biomass derived intermediates. Bioethanol obtained through the fermentation of sugars from crops like corn and sugarcane can be used as a solvent or starting material for other chemicals. Glycerol is obtained as a by-product of biodiesel manufacturing and is employed as a solvent or it can be also converted into propylene glycol. Lactic acid and its esters such as ethyl lactate are obtained from carbohydrate fermentation and are used in food as additives. Comparison of green solvents and their applications were given Table 1.2. These solvents have environmental benefits by minimising use of fossil fuels, lowering greenhouse gas emissions and enhancing biodegradability. Advances in biotechnology and process continue to improve the efficiency and sustainability of bio based solvent production.

Table 1.2 Comparison of green solvents and their applications

Solvent type	Examples	Advantages	Applications
Deep eutectic solvents	Choline chloride and urea, Choline chloride and glycerol	Non-toxic, biodegradable, tuneable properties	Metal extraction, drug formulation, catalysis
Ionic liquids	1-Butyl-3-methylimidazolium chloride, 1-Ethyl-3-methylimidazolium acetate	Low volatility, high thermal stability, recyclable	Biomass processing, CO_2 capture, electrochemistry
Supercritical fluids	Supercritical CO_2, Supercritical water	Low toxicity, high diffusivity, tunable properties	Decaffeination, polymer processing, extraction
Bio-based solvents	Ethyl lactate, Limonene, Glycerol	Renewable, biodegradable, low toxicity	Paints and coatings, pharmaceuticals, cleaning agents
Water as a green solvent	Pure water, Subcritical water	Abundant, non-toxic, environmentally friendly	Green synthesis, enzyme catalysis, biomass conversion

1.6 Industrial Applications of Green Chemistry

1.6.1 Pharmaceutical Industry

Green chemistry is used in the pharmaceutical industry for the development of safer sustainable processes. The industry follows the green chemistry principles because it helps in the reduction of environmental pollution, increases efficiency and improves safety. One of the notable practices is the use of bio based solvents which are derived from renewable resources and provide lower toxicity and better biodegradability in comparison to traditional solvents. Ethyl lactate derived from lactic acid is used as a green solvent in various pharmaceutical processes because of its low toxicity and high biodegradability. Pharmaceutical companies have developed solvent selection guidelines to identify safer and more sustainable solvents. These guidelines rank solvents based on their environmental, health and safety profiles encouraging the use of preferred solvents [62]. Pfizer solvent selection guide categorizes solvents into preferred usable and undesirable categories promoting the use of greener alternatives [63]. The use of efficient catalysts and the development of catalytic processes that minimize waste and energy consumption are the main focus of green chemistry in the pharmaceutical industry. Enzymatic catalysis provides a sustainable alternative to traditional chemical catalysts by providing high selectivity and work under mild conditions. Several case studies show the successful implementation of green chemistry in the pharmaceutical industry. The development of a green synthesis route for the anti-inflammatory drug ibuprofen to reduce waste by 80% and avoid the use of hazardous reagents [64]. The pharmaceutical industry adoption of green chemistry principles has led to significant advancements in sustainability, safety and efficiency. By the use of safer solvents optimizing processes and employing innovative catalytic methods the industry continues to reduce its environmental impact and improve the sustainability of its products.

1.6.2 Agrochemicals and Pesticides

The industrial applications of green chemistry in agrochemicals and pesticides focusing on biopesticides. Biopesticides derived from natural sources provide a sustainable alternative to conventional synthetic pesticides by following the green chemistry principles. These principles include waste prevention, atom economy, less hazardous chemical syntheses and the use of renewable feedstocks [65]. Biopesticides are categorized into microbial biopesticides, biochemical biopesticides and plant incorporated protectants (PIPs) [66]. They have various advantages such as environmental safety, reduced pest resistance, low toxicity and compatibility with combined pest management (IPM) [67]. However some challenges such high costs, limited market supply and slower action in comparison to synthetic pesticides that hinder their widespread adoption. Despite these challenges biopesticides contribute significantly to sustainable agriculture and the achievement of several Sustainable Development Goals (SDGs) including zero hunger, good health and well-being, responsible consumption and production and life on land [68]. By promoting environmental stewardship, economic productivity and social acceptability biopesticides represent a critical driver of sustainable development in agriculture the industrial applications of green chemistry in the context of polymers and materials science highlighting the importance of sustainable practices in these fields. In polymers and materials science these principles are applied to develop environmentally friendly materials with reduced ecological footprints.

1.6.3 Polymers and Materials Science

Polymers which are long chains of repeating molecular units are used in various industries such as packaging, construction, automotive and electronics. Traditional polymer production involves the use of toxic chemicals and generates a good amount of waste. Green chemistry addresses these issues by the use of renewable feedstocks, safer chemical synthesis and processes that minimize waste and energy consumption. One thing to be focused on is the development of biopolymers that are derived from natural sources such as plants, algae and microorganisms. Biopolymers have many advantages over conventional polymers such as biodegradability, reduced reliance on fossil fuels and lower environmental impact. Examples of biopolymer such as polylactic acid, polyhydroxyalkanoates and starch based polymers [69]. These materials are used in applications in biodegradable packaging to medical implants. Another important thing of green chemistry in materials science is the design of safer and more sustainable materials. This involves the use of non toxic additives, catalysts and solvents that can be easily recycled or repurposed. The use of green solvents in polymer synthesis can reduce the environmental and health risks associated with traditional solvents [70]. Nanotechnology also has a significant role in advancing green chemistry in materials science. Nanomaterials have properties because of their small size due to which it can increase the performance of polymers and other materials that reduce the need of harmful chemicals. Nanocellulose extracted from plant fibres can be used to reinforce biopolymers for improving their mechanical properties and extending its applications. The use of green chemistry principles in polymers and materials science contributes to the achievement of several Sustainable Development Goals (SDGs) such as responsible consumption and production (SDG 12), climate action (SDG 13) and life on land (SDG 15) [71]. Table 1.3 highlights the benefits of green chemistry, emphasizing both environmental sustainability and economic viability. By using sustainable practices green chemistry helps to reduce the environmental impact of material production and supports the transition to a circular economy. The application of green chemistry principles in polymers and materials science is needed for making sustainable materials that completes the needs of modern society by reducing environmental harm.

Table 1.3 Environmental and economic benefits of green chemistry

Aspect	Environmental benefits	Economic benefits
Waste reduction	Minimizes hazardous waste and toxic by-products, reducing environmental contamination	Lowers waste disposal costs and regulatory compliance expenses
Energy efficiency	Reduces energy consumption, lowering greenhouse gas emissions	Decreases operational costs by optimizing energy use
Use of renewable feedstocks	Reduces dependence on fossil fuels and non-renewable resources	Creates new market opportunities for bio-based industries
Safer chemical synthesis	Lowers risks of toxicity, spills, and exposure to hazardous chemicals	Reduces liability costs and enhances workplace safety
Green solvents and processes	Eliminates or reduces use of volatile organic compounds (VOCs), improving air and water quality	Lowers costs of solvent purchase, recycling, and disposal
Improved product lifecycle	Enhances biodegradability, reducing long-term environmental impact	Increases product value by appealing to eco-conscious consumers
Regulatory compliance	Helps meet stricter environmental regulations and sustainability goals	Avoids fines, legal liabilities, and enhances brand reputation

1.7 Challenges and Future Directions

Despite the significant advancements in green chemistry several challenges remain. One of the primary hurdles is the high cost of making and using green chemical processes that can be a barrier for small and medium sized enterprises. The scalability of green chemistry from laboratory to industrial scale have technical and economic challenges. There is also a need for awareness and education about green chemistry principles among chemists, engineers and policymakers to make it usable everywhere.

Future directions in green chemistry include the continued development of renewable feedstocks and the optimization of catalytic processes to increase efficiency and reduce waste. Advances in biotechnology such as synthetic biology and metabolic engineering have ability for making new biocatalysts and bio based chemicals. The combination of digital technologies such as artificial intelligence and machine learning can further enhance chemical processes and improve sustainability. The development of green solvents and reaction media such as DESs and ILs will play an important role in reducing the environmental impact of chemical toxicity. Collaboration between academia, industry and government is needed to overcome these challenges and increase the transition to sustainable chemical practices.

1.8 Conclusion

Green chemistry represents a transformative approach to chemical design and manufacturing focusing sustainability, efficiency and safety. Following the Twelve Principles of Green Chemistry industries can develop processes that reduce waste, use renewable feedstocks and reduce energy consumption. The use of green chemistry practices is necessary for addressing present global challenges such as climate change, resource depletion and pollution. As the field continues to grow it provides great opportunities for innovation, economic growth and improved public health. The combination of green chemistry into industrial practices not only increases environmental safety but also helps to achieve the corporate social responsibility and regulatory requirements. Green chemistry provides a pathway to a more sustainable and prosperous future where chemical processes contribute positively to society and the environment.

References

1. Andraos, J., Matlack, A.S.: Introduction to green chemistry, CRC press (2022)
2. Ardila-Fierro, K.J., Hernández, J.G.: Sustainability assessment of mechanochemistry by using the twelve principles of green chemistry. Chemsuschem **14**, 2145–2162 (2021)
3. Anastas, P.T., Warner, J.C.: Green chemistry: theory and practice, Oxford university press (2000)
4. Aksoy, A.C.A.: Green chemistry and science education. Curr. Stud. Educ. Discip. **1** (2024)
5. Lear, L.J.: Rachel Carson's silent spring. Environ. Hist. Rev. **17**, 23–48 (1993)
6. Hines, N.W.: History of the 1972 Clean water act: the story behind how the 1972 act became the capstone on a decade of extraordinary environmental reform. Geo. Wash. J. Energy Envtl. L. **4**, 80 (2013)
7. Fowler, L.: Assessing the framework of policy outcomes: the case of the US clean air act and clean water act. J. Environ. Assess. Policy Manag. **16**, 1450034 (2014)
8. Schoenbrod, D.: Goal statutes or rules statutes: the case of the clean air act. UClA L. REv. **30**, 740 (1982)
9. Anastas, P.T., Warner, J.C.: Principles of green chemistry. Green Chem. Theory Pract. **29**, 14821–14842 (1998)
10. U.S.E.P.A.O. of W. Enforcement, U.S.E.P.A.O. of W. Regulations, Standards, Technical support document for water quality-based toxics control, US Environmental Protection Agency (1991)
11. Lipshutz, B.H.: On the role of surfactants: rethinking "aqueous" chemistry. Green Chem. **26**, 739–752 (2024)
12. Gabrielli, P., Rosa, L., Gazzani, M., Meys, R., Bardow, A., Mazzotti, M., Sansavini, G.: Net-zero emissions chemical industry in a world of limited resources. One Earth **6**, 682–704 (2023). https://doi.org/10.1016/j.oneear.2023.05.006
13. Zhu, Y., Zhang, H., Siddik, A.B., Zheng, Y., Sobhani, F.A.: Understanding corporate green competitive advantage through green technology adoption and green dynamic capabilities: does green product innovation matter? Systems **11** (2023). https://doi.org/10.3390/systems11090461.
14. Tawalbeh, M., Javed, R.M.N., Al-Othman, A., Almomani, F., Ajith, S.: Unlocking the potential of CO_2 hydrogenation into valuable products using noble metal catalysts: a comprehensive review. Environ. Technol. Innov. **31**, 103217 (2023). https://doi.org/10.1016/j.eti.2023.103217

15. Kataya, G., Cornu, D., Bechelany, M., Hijazi, A., Issa, M.: Biomass waste conversion technologies and its application for sustainable environmental development—a review. Agronomy **13** (2023). https://doi.org/10.3390/agronomy13112833
16. Xia, R., Overa, S., Jiao, F.: Emerging electrochemical processes to decarbonize the chemical industry. JACS Au. **2**, 1054–1070 (2022). https://doi.org/10.1021/jacsau.2c00138
17. Hajam, Y.A., Kumar, R., Kumar, A.: Environmental waste management strategies and vermi transformation for sustainable development. Environ Chall. **13**, 100747 (2023). https://doi.org/10.1016/j.envc.2023.100747
18. Thushari, G.G.N., Senevirathna, J.D.M.: Plastic pollution in the marine environment. Heliyon **6**, e04709 (2020). https://doi.org/10.1016/j.heliyon.2020.e04709
19. Alli, Y.A., Bamisaye, A., Bamidele, M.O., Etafo, N.O., Chkirida, S., Lawal, A., Hammed, V.O., Akinfenwa, A.S., Hanson, E., Nwakile, C., Kazeem, K.O., Ayanwunmi, R.J., Ige, A.S., Parga Torres, J.R., Al Nageim, H.: Transforming waste to wealth: harnessing carbon dioxide for sustainable solutions. Results Surf. Interfaces **17**, 100321 (2024). https://doi.org/10.1016/j.rsurfi.2024.100321
20. Mar, K.A., Unger, C., Walderdorff, L., Butler, T.: Beyond CO_2 equivalence: the impacts of methane on climate, ecosystems, and health. Environ. Sci. Policy. **134**, 127–136 (2022). https://doi.org/10.1016/j.envsci.2022.03.027
21. Meo, S.A., Salih, M.A., Alkhalifah, J.M., Alsomali, A.H., Almushawah, A.A.: Environmental pollutants particulate matter (PM2.5, PM10), carbon monoxide (CO), nitrogen dioxide (NO_2), sulfur dioxide (SO_2), and ozone (O_3) impact on lung functions. J. King Saud Univ.—Sci. **36**, 103280 (2024). https://doi.org/10.1016/j.jksus.2024.103280
22. Hall Jr, R.M.: The clean water act of 1977. Nat. Resour. Lawyer. 343–372 (1978)
23. Lilienblum, W., Dekant, W., Foth, H., Gebel, T., Hengstler, J.G., Kahl, R., Kramer, P.-J., Schweinfurth, H., Wollin, K.-M.: Alternative methods to safety studies in experimental animals: role in the risk assessment of chemicals under the new European chemicals legislation (REACH). Arch. Toxicol. **82**, 211–236 (2008)
24. Lallas, P.L.: The stockholm convention on persistent organic pollutants. Am. J. Int. Law. **95**, 692–708 (2001)
25. Kulsoom, M., Shukla, V., Kumar, N.: International legislation for containment of persistent organic pollutants and hazardous chemicals. Persistent Org. Pollut. Environ. Orig. Role, CRC Press Boca Raton, pp. 319–338 (2021)
26. Wilson, M.P., Schwarzman, M.R.: Toward a new US chemicals policy: rebuilding the foundation to advance new science, green chemistry, and environmental health. Environ. Health Perspect. **117**, 1202–1209 (2009)
27. Mhatre, P., Panchal, R., Singh, A., Bibyan, S.: A systematic literature review on the circular economy initiatives in the European Union. Sustain. Prod. Consum. **26**, 187–202 (2021)
28. Shah, K.: Green chemistry for environmental sustainability-prevention-assurance-sustainability (PAS) approach: prevention-assurance-sustainability (PAS) approach. BoD–Books on Demand (2024)
29. Szekely, G.: The 12 principles of green membrane materials and processes for realizing the United Nations' sustainable development goals. RSC Sustain. **2**, 871–880 (2024)
30. Voukkali, I., Papamichael, I., Loizia, P., Economou, F., Stylianou, M., Naddeo, V., Zorpas, A.A.: Fashioning the future: green chemistry and engineering innovations in biofashion. Chem. Eng. J. **497**, 155039 (2024)
31. Luo, Y.: The political effects of sustainable development goals on sustainable consumption and production efforts of united nations environment programme (2024)
32. Friege, H., Friege, N.T.A.D., Partners, V., Heidbüchel, E., Lahl, B.Z.: Indicators for sustainable management of chemicals (n.d.)
33. Trost, B.M.: The atom economy—a search for synthetic efficiency. Science (80) **254**, 1471–1477 (1991)
34. Zuin, V.G., Eilks, I., Elschami, M., Kümmerer, K.: Education in green chemistry and in sustainable chemistry: perspectives towards sustainability. Green Chem. **23**, 1594–1608 (2021)

35. Gallezot, P.: Process options for converting renewable feedstocks to bioproducts. Green Chem. **9**, 295–302 (2007)
36. Clark, J.H., Budarin, V., Deswarte, F.E.I., Hardy, J.J.E., Kerton, F.M., Hunt, A.J., Luque, R., Macquarrie, D.J., Milkowski, K., Rodriguez, A.: Green chemistry and the biorefinery: a partnership for a sustainable future. Green Chem. **8**, 853–860 (2006)
37. Sheldon, R.A.: Green and sustainable manufacture of chemicals from biomass: state of the art. Green Chem. **16**, 950–963 (2014)
38. Raj, T., Chandrasekhar, K., Kumar, A.N., Kim, S.-H.: Lignocellulosic biomass as renewable feedstock for biodegradable and recyclable plastics production: a sustainable approach. Renew. Sustain. Energy Rev. **158**, 112130 (2022)
39. Pattanayak, P., Saha, S., Chatterjee, T., Ranu, B.C.: Sustainable and solvent-free synthesis of molecules of pharmaceutical importance by ball milling. Chem. Commun. **61**, 247–265 (2025)
40. Reynes, J.F., Leon, F., García, F.: Mechanochemistry for organic and inorganic synthesis. ACS Org. Inorg. Au. **4**, 432–470 (2024). https://doi.org/10.1021/acsorginorgau.4c00001
41. Mainkar, P., Ray, A., Chandrasekhar, S.: SOLVENTS: from past to present. ACS Omega **9**, 7271–7276 (2024). https://doi.org/10.1021/acsomega.3c07508
42. López-Hortas, L., Rodríguez, P., Díaz-Reinoso, B., Gaspar, M.C., de Sousa, H.C., Braga, M.E.M., Domínguez, H.: Supercritical fluid extraction as a suitable technology to recover bioactive compounds from flowers. J. Supercrit. Fluids. **188**, 105652 (2022). https://doi.org/10.1016/j.supflu.2022.105652
43. Fu, G., Jiang, X., Ding, L., Tao, L., Chen, Y., Tang, Y., Zhou, Y., Wei, S., Lin, J., Lu, T.: Green synthesis and catalytic properties of polyallylamine functionalized tetrahedral palladium nanocrystals. Appl. Catal. B Environ. **138–139**, 167–174 (2013). https://doi.org/10.1016/j.apcatb.2013.02.052
44. Kinner, A., Nerke, P., Siedentop, R., Steinmetz, T., Classen, T., Rosenthal, K., Nett, M., Pietruszka, J., Lütz, S.: Recent advances in biocatalysis for drug synthesis. Biomedicines **10** (2022). https://doi.org/10.3390/biomedicines10050964
45. Grulich, M., Štěpánek, V., Kyslík, P.: Perspectives and industrial potential of PGA selectivity and promiscuity. Biotechnol. Adv. **31**, 1458–1472 (2013). https://doi.org/10.1016/j.biotechadv.2013.07.005
46. Mane, P.V., Rego, R.M., Yap, P.L., Losic, D., Kurkuri, M.D.: Unveiling cutting-edge advances in high surface area porous materials for the efficient removal of toxic metal ions from water. Prog. Mater. Sci. **146**, 101314 (2024). https://doi.org/10.1016/j.pmatsci.2024.101314
47. Saberi, F., Rodríguez-Padrón, D., Doustkhah, E., Ostovar, S., Franco, A., Shaterian, H.R., Luque, R.: Mechanochemically modified aluminosilicates for efficient oxidation of vanillyl alcohol. Catal. Commun. **118**, 65–69 (2019). https://doi.org/10.1016/j.catcom.2018.09.017
48. Kogularasu, S., Chen, Y.-L., Lee, Y.-Y., Huang, B.-W., Murugesan, P., Chang-Chien, G.-P., Govindasamy, M.: Two-dimensional MXenes for toxin management: advanced electrocatalytic detection, degradation, and adsorption dynamics. Environ. Chem. Ecotoxicol. **6**, 390–407 (2024). https://doi.org/10.1016/j.enceco.2024.08.005
49. Sajid, M., Zhao, X., Liu, D.: Production of 2, 5-furandicarboxylic acid (FDCA) from 5-hydroxymethylfurfural (HMF): recent progress focusing on the chemical-catalytic routes. Green Chem. **20**, 5427–5453 (2018)
50. Chaudhari, D.S., Upadhyay, R.P., Shinde, G.Y., Gawande, M.B., Filip, J., Varma, R.S., Zboril, R.: A review on sustainable iron oxide nanoparticles: synthesis and application in organic catalysis and environmental remediation. Green Chem. (2024)
51. Kumar, K.M., Kumar, M.K., Kumar, N.S., Kumar, L.N.S.: Engineering the Invisible: synthesis, characterization, and applications of nanoporous materials. Innov. Mater. Ind. Appl. Synth. Charact. Eval. 271–328 (2025)
52. Fukuoka, A., Dhepe, P.L.: Catalytic conversion of cellulose into sugar alcohols. Angew. Chemie Int. Ed. **45**, 5161–5163 (2006). https://doi.org/10.1002/anie.200601921
53. Rajesh, S.M., Bala, B.D., Perumal, S., Menéndez, J.C.: l-Proline-catalysed sequential four-component "on water" protocol for the synthesis of structurally complex heterocyclic ortho-quinones. Green Chem. **13**, 3248–3254 (2011). https://doi.org/10.1039/C1GC15794A

54. Varma, R.S.: Greener and sustainable trends in synthesis of organics and nanomaterials. ACS Sustain. Chem. Eng. **4**, 5866–5878 (2016). https://doi.org/10.1021/acssuschemeng.6b01623
55. Alqahtani, A.S.: Indisputable roles of different ionic liquids, deep eutectic solvents and nanomaterials in green chemistry for sustainable organic synthesis. J. Mol. Liq. **399**, 124469 (2024). https://doi.org/10.1016/j.molliq.2024.124469
56. Rodríguez-Álvarez, M.J., Ríos-Lombardía, N., García-Garrido, S.E., Concellón, C., del Amo, V., Capriati, V., García-Álvarez, J.: Recent advancements in the utilization of s-block organometallic reagents in organic synthesis with sustainable solvents. Molecules **29** (2024). https://doi.org/10.3390/molecules29071422
57. Panić, M., Bubalo, M.C., Redovniković, I.R: Designing a biocatalytic process involving deep eutectic solvents. J. Chem. Technol. Biotechnol. **96**, 14–30 (2021). https://doi.org/10.1002/jctb.6545
58. Baby, E.K., Savitha, R., Kinsella, G.K., Nolan, K., Ryan, B.J., Henehan, G.T.M.: Influence of deep eutectic solvents on redox biocatalysis involving alcohol dehydrogenases, Heliyon **10**, e32550 (2024). https://doi.org/10.1016/j.heliyon.2024.e32550
59. Perna, F.M., Vitale, P., Capriati, V.: Deep eutectic solvents and their applications as green solvents. Curr. Opin. Green Sustain. Chem. **21**, 27–33 (2020)
60. Salas, R., Villa, R., Velasco, F., Cirujano, F.G., Nieto, S., Martin, N., Garcia-Verdugo, E., Dupont, J., Lozano, P.: Ionic liquids in polymer technology. Green Chem. **27**, 1620–1651 (2025). https://doi.org/10.1039/D4GC05445H
61. Wu, Y., Xu, J., Mumford, K., Stevens, G.W., Fei, W., Wang, Y.: Recent advances in carbon dioxide capture and utilization with amines and ionic liquids. Green Chem. Eng. **1**, 16–32 (2020). https://doi.org/10.1016/j.gce.2020.09.005
62. Gutiérrez, G.O.: Solvent assessment and selection in terms of sustainability, performance, and cost: design of a solvent database in a pharmaceutical environment (2023)
63. Bozza, D.: Greening the downstream of biopharmaceuticals through process intensification in the research on innovative eco-compatible solvent (2024)
64. Mureşan, A.C.: Ibuprofen: original versus green synthesis, ann. "Dunarea Jos" Univ. Galati. Fascicle IX, Metall. Mater. Sci. **41** (2018). https://doi.org/10.35219/mms.2018.3.05
65. Fenibo, E.O., Ijoma, G.N., Nurmahomed, W., Matambo, T.: The potential and green chemistry attributes of biopesticides for sustainable agriculture. Sustainability **14** (2022). https://doi.org/10.3390/su142114417
66. Villarreal, G.U., Campos, E.V.R., de Oliveira, J.L., Fraceto, L.F.: Chapter 3—Development and commercialization of pheromone-based biopesticides: a global perspective. In: Koul, O. (ed.) Development and Commercialization of Biopesticides, Academic Press, pp. 37–56 (2023). https://doi.org/10.1016/B978-0-323-95290-3.00008-X
67. Dara, S.K.: The new integrated pest management paradigm for the modern age. J. Integr. Pest Manag. **10**, 12 (2019)
68. Mawcha, K.T., Malinga, L., Muir, D., Ge, J., Ndolo, D.: Recent advances in biopesticide research and development with a focus on microbials. F1000Research **13**, 1071 (2024)
69. Mtibe, A., Motloung, M.P., Bandyopadhyay, J., Ray, S.S.: Synthetic biopolymers and their composites: advantages and limitations—an overview. Macromol. Rapid Commun. **42**, 2100130 (2021)
70. Farrán, A., Cai, C., Sandoval, M., Xu, Y., Liu, J., Hernáiz, M.J., Linhardt, R.J.: Green solvents in carbohydrate chemistry: from raw materials to fine chemicals. Chem. Rev. **115**, 6811–6853 (2015)
71. Bontempi, E., Sorrentino, G.P., Zanoletti, A., Alessandri, I., Depero, L.E., Caneschi, A.: Sustainable materials and their contribution to the sustainable development goals (SDGs): a critical review based on an Italian example. Molecules **26**, 1407 (2021)

Chapter 2
Scalable Solutions in Green Chemistry for Industrial Manufacturing

Jasvinder Kaur, Neha Choudhary, and Shivanshi Tandon

Abstract Green Chemistry is the utilization of a set of principles that reduces or eliminates the use of generation of hazardous substances in the design, manufacturing and application of chemical products. The industrial sector plays a critical role in driving economic development but often at the expense of environmental sustainability. Green chemistry offers innovative pathways to address these challenges, focusing on the design of chemical processes and products that minimize environmental impact. Scalable Solutions refers to approaches or strategies that can be expanded or adapted to increase impact without a corresponding increase in cost or resources. Hereby scalable solutions in green chemistry tailored for industrial manufacturing, emphasizing their potential for widespread adoption. An economic and ecological assessment that incorporates quantitative analysis, including life cycle assessments and cost–benefit analyses, provides a richly textured view of the pros and cons involved. Policy relevance, industry-academic partnerships, and international partnership also play an important role in enabling scalable solutions. This final chapter takes up future directions and opportunities for growth by exploring the integration of artificial intelligence, bioengineering, and synthetic biology into the green chemistry domain. It includes potential applications in emerging industries such as energy storage and water purification. By taking a look at technical and systemic barriers to scale, this chapter should provide a strategic framework for industrial-scale adoption of green chemistry in manufacturing to catalyse a step toward a more sustainable and resilient industrial future.

Keywords Green chemistry · Scalability · Bioengineering sustainable solutions · Life cycle assessment

J. Kaur (✉) · N. Choudhary · S. Tandon
Department of Chemistry, School of Sciences, IFTM University, Moradabad, UP, India
e-mail: jasvinderkaur2911@gmail.com

P. Jain et al. (eds.), *Towards Green Chemical Processes: Strategies and Innovations*, Intelligent Systems Reference Library 277,
https://doi.org/10.1007/978-3-031-94090-3_2

2.1 Introduction

The EPA defines green chemistry or sustainable chemistry as "the production of chemical products and processes that reduce or eliminate the use or production of hazardous substances." Green chemistry is a branch of chemistry that focuses on the development of chemical products and processes that reduce or eliminate the use and production of hazardous substances. It seeks to develop safer and more sustainable ways of producing, using, and disposing of drugs [1]. Green chemistry is a marriage of environmental protection, sustainable development, and economic efficiency, driven by the need to reduce pollution, conserve resources, and protect human rights. For the past years, people have depended on doctors and chemists to develop better herbal methods, and this is most likely to continue in the coming years [2].

Green chemistry (sustainable chemistry) involves designing chemical products and processes that make them safer to use and dispose of, use fewer hazardous materials, minimize risk to people and the environment, maximize efficiency in all aspects of designing chemical products, and provide sound science to help achieve a sound economy. Such practice is part of the 12 Principles of Green Chemistry outlined by Paul Anastas and John Warner. Such principles focus on preventing waste, having an atom-economy-based principle, use of renewable feed stocks, and products designed to break down harmlessly after their purpose is served [3].

Green chemistry was born in the early 1990s from such endeavours as the U.S. Pollution Prevention Act of 1990 and textbook work like Green Chemistry: Theory and Practice. It has impacted areas such as chemical manufacturing, energy production, and water treatment that provide solutions with a lower environmental footprint [4]. Green chemistry supports the Sustainable Development Goals through resource use efficiency and emissions of SDG 12: Responsible Consumption and Production, SDG 13: Climate Action, and SDG 7: Affordable and Clean Energy [5]. Green chemistry has a huge potential in changing the world's real-life scenarios. Atom-economical syntheses and enzymatic processes in the pharmaceutical sector increased efficiency and minimized toxic byproducts. In materials science, the advent of biodegradable plastics and green solvents have replaced traditional materials that were damaging the environment. These achievements only reinforce the role that green chemistry will play in setting the tone for sustainable industrial operations and contributing to a circular economy at the global level [6].

Industrial manufacturing processes, in general, have focused more on productivity and cost efficiency and less so concerning environmental sustainability. Processes are highly energy-intensive and have a significant dependency on fossil fuels. This creates carbon footprints associated with the activities to be very high. Steel, cement, and chemicals are some excellent examples of sectors that contribute greatly to greenhouse emissions [7]. A main characteristic of dependency on non-renewable resources—petroleum, coal, and natural gas—would make the system prone to resource depletion and price. Moreover, this process produces hazardous byproducts, including heavy metals, Volatile Organic Compounds, and greenhouse gases that would contribute to the pollution and degradation of the environment, which are mainly long-lasting in their ecological effects [8]. These practices affect environmental, economic, and social dimensions. In the environmental dimension, they destroy ecosystems, contribute to climate change, and accelerate deforestation. In the economic dimension, industries experience increasing costs related to resource scarcity, pollution control, and possible regulatory penalties. In the social dimension, communities suffer adverse health effects from exposure to industrial toxins and displacement resulting from pollution and industrial expansion [9].

Challenges in adopting green methodologies. Transitioning to sustainable technologies is costly, and there are limited financial incentives. Scalability of green innovations is also a problem. Industries face the challenge of integrating new technologies without disrupting existing operations [10]. Regulatory barriers include the lack of globally harmonized standards and region-specific regulations, which complicate implementation, especially for multinational companies. All this requires collaboration on the part of policymakers, industries, and researchers for developing economically viable, technologically feasible, and globally standardized green manufacturing solutions [11].

One of the key characteristics of green chemistry is scalability, the area that makes the difference between laboratory-scale promising innovation and impact at the industrial scale. Laboratory processes are often optimized for small-scale experiments and then encounter real challenges in being scaled up to industrial levels. Reaction kinetics, heat transfer, and equipment design all require re-engineering: investing a great deal of money and time, which deters industries from adopting [12]. However, large-scale implementation is critical to the realization of the whole environmental and economic benefit of green chemistry. It helps reduce carbon emissions, limits harmful waste, and maximizes resource efficiency, boosting advancement towards global climate goals. Economically, large-scale green processes bring down the cost of production, open new markets for sustainable products, and also create growth. Among these beneficial aspects, technological constraints, lack of standardization, and regulatory barriers have stood in the way of many green applications. These hurdles need firm policy frameworks with subsidies and mandates for green practices as well as international cooperation to share resources, expertise, and standards. Scalling over these barriers ensures that green chemistry is part of sustainable industrial development [13].

2.2 Principles of Green Chemistry in Industrial Applications

Green chemistry is a transformational approach focused on designing chemical processes and products so that they have minimal or no hazardous substances and waste and improve sustainability. In industrial applications, this leads to the formulation of cleaner, safer, and more efficient manufacturing processes. 12 Principles of Green Chemistry, developed by chemist Paul Anastas and his colleagues in the 1990s, are what really found the foundation of green chemistry in industrial settings, see Table 2.1. These principles are adapted towards making industrial processes more environmentally friendly while still maintaining productivity and cost-effectiveness. Below is a detailed look at how these principles apply to industrial applications.

2.3 Challenges in Scaling Green Chemistry

Scaling green chemistry will be a real challenge since such a transition of small laboratory processes to industrial-sized applications while staying sustainable, efficient, and safe is quite tricky [26]. Some of the major challenges with scaling green chemistry include:

2.3.1 Technical Challenges (Reaction Optimization and Catalyst Efficiency)

Optimizing reactions in green chemistry, therefore, means answering a number of technical questions designed to produce efficiency and scalability. The main issues are substrate variability, as different substrates behave inconsistently; reaction condition sensitivities, mainly in the form of temperature, pressure, and pH, which significantly affect yield and product purity; and time inefficiency, particularly in slow reactions or requiring high activation energies. Scaling up from laboratory to industrial settings adds further complexity due to differences in heat transfer, mixing, and reactant concentrations [27].

Another major challenge in the catalyst lies in its efficiency. Issues associated with catalyst fouling, sintering, or poisoning affect the catalyst; selectivity toward unwanted reactions affects the catalyst and also stability towards harsh conditions make it difficult. Additionally, minimal recycling and reuse of catalysts increase operational expenses and sustainability [28]. These will be overcome through research in new innovative solutions. The advanced catalysts, such as bifunctional catalysts, nanocatalysts, and enzyme mimics, ensure better performance. Machine learning tools and computational methods will be applied in optimizing the reaction conditions and finding optimal catalysts by exhaustive search. Techniques in process intensification such as microreactors and flow chemistry allow more control in reactions and the behavior of catalysts. Lastly, the pursuit of sustainable catalysis, where bio-based and eco-friendly catalysts are involved, advances efficiency as well as eco-compatibility [29].

Table 2.1 The principles of green chemistry in industrial applications

Principle	Description	Industrial applications
Waste prevention [14]	Minimize waste generation by designing chemical processes that produce little to no waste	Implementation of closed-loop systems, solvent recovery in manufacturing, waste minimization in pharmaceuticals
Atom economy [15]	Maximize the incorporation of all materials used in the process into the final product	Design of more efficient catalytic processes, and atom-efficient synthetic routes for polymers and fine chemicals
Less hazardous chemical syntheses [16]	Choose synthetic methods that use and generate substances with little or no toxicity	Safer reagents in the production of chemicals, replacing hazardous solvents in pharmaceuticals and agrochemicals
Designing safer chemicals [17]	Design chemical products that are less toxic and hazardous, ensuring the safety of consumers and the environment	Development of biodegradable, non-toxic surfactants and solvents in cleaning, detergents, and cosmetics
Safer solvents and auxiliaries [18]	Avoid the use of hazardous solvents and reagents in chemical processes	Transition to water-based or supercritical CO_2 solvents for extraction and purification in pharmaceuticals and agrochemicals
Energy efficiency [19]	Optimize energy use by designing processes that require less energy, ideally using renewable sources	Use of low-temperature processes, green catalysts for energy-efficient reactions, and integration of renewable energy in production
Renewable feedstocks [20]	Use renewable raw materials over depleting, non-renewable ones	Biocatalysis for the production of biofuels, sustainable sourcing of natural ingredients in pharmaceuticals and food production
Reduce derivatives [21]	Minimize the need for chemical derivatization, which often generates waste	Development of direct synthesis methods to reduce additional chemical steps in the manufacture of drugs, plastics, and specialty chemicals

(continued)

Table 2.1 (continued)

Principle	Description	Industrial applications
Catalysis [22]	Use catalysts to increase the efficiency of reactions, thus reducing energy and raw material use	Use of homogeneous and heterogeneous catalysts in reactions such as polymerization, hydrogenation, and pharmaceutical synthesis
Design for degradation [23]	Design chemicals to degrade into harmless substances after use, preventing environmental contamination	Development of biodegradable plastics, and eco-friendly agricultural chemicals that break down safely
Real-time analysis for pollution prevention [24]	Develop methods for real-time monitoring of chemical processes to avoid the formation of hazardous substances and ensure efficient reactions	Implementation of sensors and real-time monitoring in chemical reactors to optimize yields and minimize byproducts
Inherently safer chemistry for accident prevention [25]	Design chemical processes to minimize the risk of accidents, including explosions, fires, and releases of toxic materials	Design of safer storage and transport methods for chemicals, use of inert atmospheres for hazardous reactions in large-scale industries

2.3.2 *Economic Barriers (Initial Investment Costs and Market Readiness)*

Technical challenges related to reaction optimization and catalyst efficiency are key hurdles in scaling green chemistry processes. These challenges involve achieving the desired outcomes (such as high yield, selectivity, and sustainability) while reducing energy consumption, minimizing waste, and improving process efficiency [30]. Below are some specific technical challenges in green chemistry related to these areas, represented in flow chart Fig. 2.1.

2.3.3 *Regulatory and Policy Constraints*

One of the major barriers to scaling green chemistry innovations, focusing on regulatory and policy issues. Regulatory frameworks often lack clarity for new materials, chemicals, and methods used in green chemistry, such as bio-based chemicals and novel catalysts. This will cause uncertainty in approval processes, legal risks, and delays as well as high costs of certification and commercialization. The regional

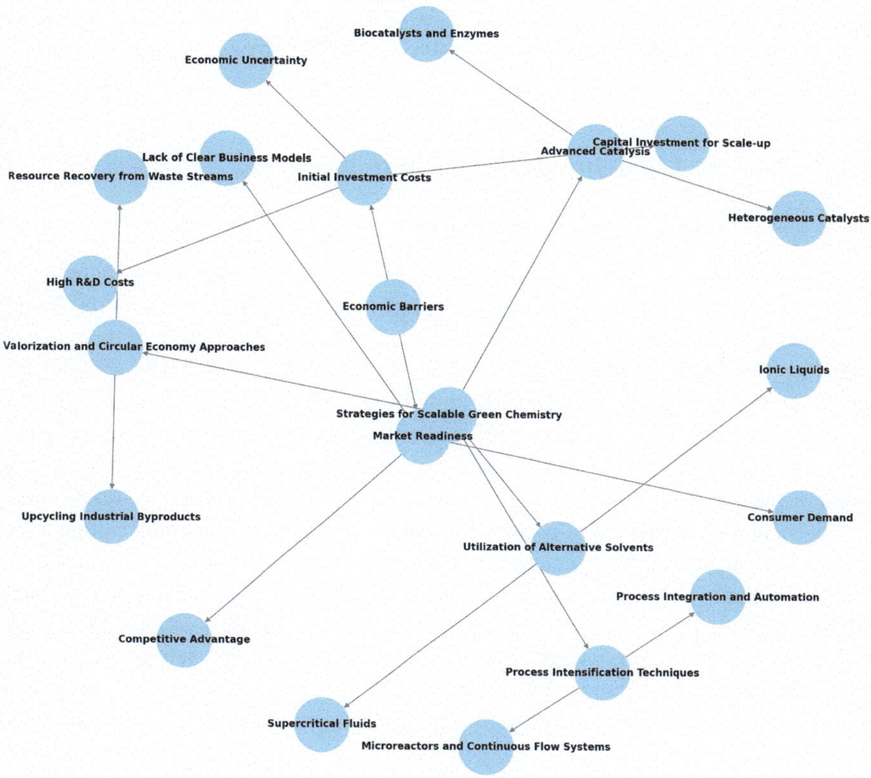

Fig. 2.1 The economic barriers and strategies for scalable green chemistry

diversity in regulatory requirements further complicates compliance and raises the cost of scale globally [31].

Policy imperatives are inconsistent governmental support towards fossil fuel-related industries, which are often subsidized more than the green chemistry. Long-term policy stability is crucial in adopting green chemistry as the uncertainty over incentives would certainly deter investment. The carbon pricing mechanisms in the form of taxes or cap-and-trade systems either promote or discourage the development of green innovations, based on their design. Misaligned policies risk increasing the cost of adoption and slow the pace of developing green chemistry technologies [32].

2.4 Strategies for Scalable Green Chemistry

The development of strategies to make green chemistry scalable requires answering the questions associated with increased productivity, environmental benefits, and commercial feasibility as green chemistry processes transition from the laboratory stage to industrial practice [33]. The following strategy can be very helpful in paving the way to widespread adoption and scaling of green chemistry:

2.4.1 Process Intensification Techniques

Process Intensification (PI) techniques are instrumental in optimizing the chemical process as it is toward efficiency and in sustainability, keeping green chemistry's door open wider. Microreactors and continuous flow systems bring about the intensification of scalable green chemistry application. Microreactors are mainly used for producing fine chemicals/pharmaceutical products, catalysis, and polymeric materials involving high purity material with low to negligible solvent volumes. Continuous Flow Systems allow for uninterrupted and controlled reactions, thus beneficial for sustainable synthesis, biocatalysis, and energy-efficient green hydrogen production [34].

Green chemistry processes can also be optimized with the use of Process integration and automation of processes. Process integration is when production steps were combined to save energy, materials, waste and pollution, and also reduced costs and increased the flexibility of production. Automation is primarily done to facilitate the control of production, which improves the quality and consistency of output as well as more resource utilization. Both methods highly facilitate scaling up the green chemistry inventions from the laboratory to applications at the industrial level [35].

2.4.2 Advanced Catalysis

Advanced catalysis is the backbone of green chemistry, making chemical processes more sustainable by increasing efficiency, selectivity, and energy utilization. Biocatalysts, which include mainly enzymes, are natural catalysts that speed up chemical reactions under mild conditions, reducing the need for harsh reagents, solvents, and energy. Due to high specificity, they are very good for sustainable applications such as pharmaceutical synthesis, biofuels production, green synthesis of fine chemicals, and environmental remediation with less waste production and energy consumption possible [36]. The role of heterogeneous catalysts is played in large-scale industrial processes. These catalysts play a vital role in sustainable fuel production, green chemical synthesis, petroleum refining, and environmental applications including pollution degradation and emission reductions. Both heterogeneous catalysts and biocatalysts are essential to making the various chemical processes more sustainable: biocatalysts have the merit of precision in mild conditions but heterogeneous catalysts drive scalable efficient industry-based reactions [37].

2.4.3 Utilization of Alternative Solvents

By the use of alternative solvents that include supercritical fluids (SCFs) and ionic liquids (ILs), the realms of green chemistry actually get expanded with relevance to their sustainability, environmental impact, and increased efficiency in the chemical processes. In fact, the SCFs are different from gases and liquids because they are above their critical temperature and pressure, which makes it very advantageous to use in a number of applications. Among the numerous applications of SCF, supercritical fluid extraction (SFE) has proven to be beneficial because bioactive compounds can be extracted from plants or other organisms without using solvent systems toxic to human health and the environment [38]. They are also used as crucial agents in polymer synthesis and green chemical synthesis and in synthesis nanomaterials by providing lower energy consumption, higher reaction rates, and more selective chemical transformations. Ionic liquids are salts that remain liquid at lower temperatures; they share many attractive features of the traditional organic solvents. ILs exhibit high thermal stability and low vapor pressure, making them suitable for a range of green chemistry applications in catalysis, extraction and separation, and carbon dioxide capture for enhanced reaction efficiency with the possibility of selective dissolution of compounds [39]. In addition, ILs play a key role in the production of biofuels and act as green solvents, lowering hazardous emissions commonly associated with VOCs [40]. SCFs and ILs together ensure innovation in green chemistry, providing sustainable solutions to industrial processes.

2.4.4 Waste Valorization and Circular Economy Approaches

Strategies to advance green chemistry involve waste valorization and circular economy approaches aimed at reducing impacts on the environment and promoting efficiency in resource usage. Waste valorization is generally the process through which waste materials or by-products are converted into valuable products or resources, whereas circular economy approaches seek to make systems where less waste is created and materials continue to be recycled, reused, or upcycled. These principles do not only minimize the need to extract raw materials and lower the cost of disposal but also provide economic sustainability. Upcycling industrial byproducts is transforming waste into higher-value products, which reduces environmental pollution and saves raw materials. Economic benefits include creating new sources of revenue, energy saving, and resource use optimization [41]. The recovery of resources from waste streams through the extraction of valuable materials or energy from industrial, agricultural, or municipal waste helps to support a sustainable circular economy in terms of the minimization of waste and the reduced dependence on virgin resources. All these approaches are therefore crucial in setting up an environmentally friendly as well as an economically beneficial closed-loop system [42].

2.5 Case Studies of Scalable Green Chemistry Solutions

Green chemistry solution is about the cure for environmental degradation due to chemical processes without sacrificing or even improving the efficacy of these processes themselves. The following case studies show scaled implementations of green chemistry in various industries:

2.5.1 Example 1: Greener Synthesis of Pharmaceuticals

The Pharmaceuticals, in general, have been known for their high-waste production processes and intensive use of resources, and so far, it is one of the main thrust areas of green chemistry: environmental impact reduction, efficiency enhancement, and safety enhancement. Generally, traditional synthesis methods of pharmaceuticals produce much solvent and chemical waste because multi-step reactions involving hazardous reagents are used. Processes can be made more efficient and sustainable with green chemistry principles like atom economy, safer solvents, and renewable feedstocks [43].

Significantly, the eco-friendlier synthesis of ibuprofen. Then, development in the resource-consuming traditional process is low atom economy and a huge amount of waste through toxic reagents and solvents. An example is Boots Company afterward with Hoechst Celanese (BHC) in developing a three-step catalytic synthetic process that involved catalytic hydrogenation, recyclable palladium catalyst, and elimination of toxic chemicals. An overall reduction of waste to below 70%, improved yield, increased scalability, cost cutting, and safety of the worker are among the achievements this innovation enables. All these successes strongly signify the tremendous potential that green chemistry has in promising very substantial environmental and economic benefits in pharmaceuticals production [44].

2.5.2 Example 2: Bioplastic Manufacturing

Made of renewable feedstock sources, plant-based material-bioplastics offer the chance for the implementation of alternative substitutes to traditional plastics; such products result in fewer environmentally adverse effects, with functional application preserved. With their production process relying on fossil fuels, thus increasing the incidence of pollution as well as leading to a low level of biodegradability, a massive amount of production needs to meet this gap as alternatives are eco-friendly and also sustainable [45].

Example of PLA Bioplastics: Polylactic Acid (PLA) is the outstanding example for mass production of bioplastics. It has provided evidence of the realization of green chemistry through the manufacturing process, such as that beginning with the renewable feedstocks taken from corn starch or sugarcane, fermenting glucose into lactic acid, and afterwards polymerizing into PLA. In material science, it shows sustainability together with innovation [46]. An example of scaling up green chemistry to address environmental challenges of traditional plastics is bioplastic production. In fact, as a material science revolution, bioplastics are projected to lead into a more sustainable and a circular economy in the future [47].

2.5.3 *Example 3: Sustainable Agrochemical Production*

Although agrochemicals like fertilizers and pesticides are acknowledged as prerequisites in enhancing crop yield and protecting from pest attacks, the unsustainable energy-intensive production processes along with toxic wastes and nonrenewable resources have rather limited adoption and application of such products. This case study examines a green chemistry innovation that is targeted on solving these problems through innovative methods like the use of bio-catalysis, replacing traditional catalysts with enzymes; renewable feedstocks derived from the agricultural waste; process intensification using continuous flow reactor systems; and waste minimization through closed-loop systems. Eco-friendly methodology scalable for production that minimizes greenhouse gas emissions and toxic by-products to reduce costs and meet regulatory specifications could provide a template for sustainable agrochemical manufacture [48].

Examples of Biopesticides from Neem Oil: Neem-based biopesticides are an environment-friendly substitute of synthetic pesticides; all these biopesticides are basically ingredient of *Azadirachta indica*. These products were obtained by some companies that are commercially selling neem extracts for worldwide agribusiness based on the evidence of scale in pest control and reduction of soil as well as water pollution [49].

2.5.4 *Textile Industry: Supercritical CO_2 Dyeing Technology*

Conventional textile dyeing processes consume lots of water, and toxic effluent is released during the process. Supercritical CO_2 dyeing technology saves water and diminishes the harmful chemicals' requirement, ensuring sustainability and scalability for companies such as Dye Coo [50].

2.5.5 *Energy Sector: Bioethanol Production*

The large-scale production of bioethanol from lignocellulosic biomass has been a potential source of renewable and cleaner energy. Companies like DuPont and POET-DSM have optimized enzymatic hydrolysis and fermentation processes, which significantly reduced greenhouse gas emissions compared to fossil fuels [51].

These case studies really highlight the scope of green chemistry in transforming industries by bringing about environmental sustainability together with economic viability.

2.6 Economic and Environmental Impact Assessment

Economic and environmental impact assessments play an important role in the assessment of green chemistry solutions in terms of their viability and sustainability. Assessments give quantitative and qualitative insight into the cost-effectiveness, resource efficiency, and ecological benefits of introducing innovative technologies in industrial manufacturing [52].

2.6.1 *Life Cycle Assessment (LCA)*

A Life Cycle Assessment is one of the key tools in green chemistry to estimate the environmental impact of a product or process through its lifecycle from raw material extraction to end-of-life disposal. LCA provides insight into energy consumption, waste generation, and greenhouse gas emissions that help industries head in the direction of more sustainable practices [53].

2.6.2 *Cost–Benefit Analysis of Green Chemistry Implementations*

The cost–benefit analysis of implementing green chemistry sustains its economic viability, return on investment, and incentives that encourage one to gain competitive advantage through sustainable practices. While short-term benefits include gains in raw material usage, energy efficiency, and the reduction of waste in operations, long-term benefits include compliance with relevant regulations and access to markets. The return on investment assessment is a financial comparison of returns generated from using green chemistry solutions against traditional solutions. It further goes to include governmental and institutional incentives, subsidies, and support mechanisms that assist in offsetting the initial investment required for a transition to green practices.

The above analysis brings to light not only the financial but also environmental benefits of using green chemistry [54].

2.6.3 Metrics for Sustainability and Scalability

Sustainability and scalability metrics are the key focus of environmental, economic, and scalability performance. Environmental metrics will reduce the carbon footprint and hazardous waste and improve energy efficiency and resource conservation. Economic metrics will consider cost-effectiveness and market competitiveness for green chemistry solutions, in addition to the contribution of green chemistry solutions toward job creation and industrial growth. Indicators for scalability illustrate modularity and versatility in the production processes of green chemistry to achieve applicability on the industrial level. Moreover, its capacity for scaled-up sustainable growth is achievable due to this indication. As such, both parameters guarantee the innovation of green chemistry as scalable yet effective on industrial levels [55]. This validation of the sustainability and scalability of green chemistry solutions can be done using economic and environmental impact assessments such as LCA and cost–benefit analysis. Clear metrics for such transition into eco-friendly practices help stimulate economic growth and reinforce regulation.

2.7 Role of Policy and Collaboration in Driving Scalability

Policies and cooperation go a long way in driving scalable sustainable and innovative technologies. Solid policies offer an incentive for going green by means of subsidies, tax benefits, and regulatory incentives to promote use. Standardization guidelines and conformity requirements also pave the way to equal ground. Collaboration among stakeholder groups that include governments, industries, academics, and other nongovernmental organizations offers opportunities for shared knowledge, resource sharing, and expertise. Public–private partnerships and international cooperation help to amplify the impact of green innovations by pooling investments and dealing with cross-border challenges. Both policy and collaboration ensure that scalability is not just technically feasible but also economically and socially sustainable [56].

2.7.1 Government Incentives and Regulations

Scalability of green chemistry solutions is largely supported by friendly governmental policies and teamwork. Incentives by governments such as subsidy, tax credit, and R&D grants create a relief effect in terms of finances and promote innovations in industries-in particular, SMEs-by taking on the 'eco-friendly technologies'. Taxation benefits boost investments in green infrastructure while public–private partnerships expedite the actualization of green methodologies in the industries. Strong regulatory framework, which consists of obligatory reduction targets of emission, waste minimization protocols, and standardized guidelines, helps achieve environmental compliance. Success stories emanating from strong policy-driven initiatives across nations depict both economic and ecological benefits associated with such measures [57]. The primary challenges for scaling up, however, lie in policy gaps and enforcement. Addressing these issues through adaptive policies and enhanced collaboration between governments, industries, and researchers will be key in allowing general adoption of green chemistry to transform industrial manufacturing into a sustainable and resilient enterprise.

2.7.2 Industry-Academia Partnerships

Industry-academia partnerships help bridge the gap between theoretical research and industrial application, which are necessary for innovation and rapid development of green chemistry solutions [58]. Such collaborations allow academic rigor and industrial practicality to be meshed together so that the innovations are not only scientifically sound but also operationally feasible. Benefits include accelerated innovation, talent building through hands-on student training, and resource sharing to optimize the efficiency of the research. More collaborative models, like research consortia, technology transfer programs, and joint ventures, are necessary to scale up sustainable technologies into economic and environmental benefits, such as reduced costs and emissions [59]. However, issues in communication gaps and priority differences are seen. More global partnerships that improve communication and ensure support from the government would be crucial for bridging these gaps, making cost-effective and sustainable manufacturing practices that can integrate green chemistry into industrial processes [60].

2.7.3 International Collaborations and Standards

Green chemistry refers to a transformative approach in the design of chemical products and processes such that hazardous substances are reduced or eliminated. As industrial manufacturing strictly embraces the principles of green chemistry, the challenge remains scalability. What follows is discussing the critical role of policy frameworks and international collaborations that drive scalability and set the basis for a universal standard in green chemistry [61].

2.7.3.1 The Need for Policy Intervention

Policy intervention, therefore, has to be directed at addressing the specific challenges of industrial adoption and scale-up of green chemistry solutions. Subsidies and incentives to companies will ensure that financial stress is relieved to companies switching towards sustainable practices so that environmental friendliness is economical. The main role of the regulatory mandate, however, comes in terms of setting standards or requirements for nontoxic renewable resources and other similar parameters of adoption. R&D grants for the development of new green chemistry technologies that are scalable and cost-effective will fuel innovation. Public–private partnerships will allow pooling of resources and expertise to encourage collaboration between stakeholders to speed up the implementation of sustainable practices. Such a regulatory frame supports both encouraging and investment within green industries on a level ground toward achieving environmental stability along with promoting the pace of growth as stated [62].

2.7.3.2 International Collaborations: A Pathway to Scalability

Global collaboration is a very critical foundation toward scaling green chemistry by harmonizing efforts and providing shared expertise. Knowledge exchange platforms, such as those inspired by the United Nations' Sustainable Development Goals, offer a universal framework in which to align green chemistry initiatives across nations toward collective progress. The co-development of scalable solutions through cross-border R&D collaborations among academic institutions, industries, and governments can facilitate the utilization of diverse expertise and resources to overcome global sustainability challenges. Standardized metrics are at the heart of these efforts: they ensure that universally accepted methods for evaluating impacts on the environment and economy prevail, and thereby promote transparency and consistency. Responsible raw material sourcing and environmentally conscious manufacturing processes also find a footing in integrating sustainability into global supply chains across borders. This represents critical relationships through international partnership and could bring to bear full scope potential through impactful and scalable green chemistry around the globe [63].

2.7.3.3 Standards for Universal Adoption

Standards are the framework of universal adoption in green chemistry, ensuring consistency, reliability, and scalability in diverse industries. ISO standards related to environmental management and sustainability help provide a worldwide framework for adopting eco-friendly practices. Robust material safety and performance benchmarks help evaluate the viability of new materials and processes, and thus, can align innovation with sustainability [64]. LCA standards have universal guidelines in measuring environmental impacts throughout the life cycle of a product, from design to disposal. This creates openness and accountability. Industry and consumers, through certification programs such as green labels and eco-certifications, have the power to make informed choices. This will influence market demand for solutions that are considered sustainable. Combining these standards would create a consistent basis for the practice of green chemistry, its integration into industrial manufacturing, and global supply chains [65].

2.7.3.4 Case Studies in Policy and Collaboration

Real-world examples clearly show the potential of policy frameworks and collaborative effort in scaling up green chemistry. The European Green Deal is one of the all-encompassing strategies for attaining climate neutrality while promoting sustainable industries, an example of strong policy in encouraging green innovation. Global Green Chemistry Initiative (GGCI) is the power of international collaboration and knowledge exchange in bringing sustainable practices across borders. Bilateral agreements among nations underscore further that collective investment will have a key role in the promotion and adoption of green technologies [66]. These, however, do not come without the hurdles of economic disparity, heterogeneity in regulatory aspects, and technological disparities which impede global convergence on standards and practices for green chemistry. The hurdles are also avenues for innovation, specifically the integration of policy frameworks which could converge local, national, and international efforts. Accommodating emergent technologies and discoveries will require a higher level of global collaboration as well as more adaptive and dynamic standards. Thus, the route to scaling up green chemistry is via policy and collaboration as dual pillars toward sustainable industrial manufacturing, wherein economic viability matches environmental responsibility [67].

2.8 Future Trends and Opportunities

Emerging Technologies in Green Chemistry (Artificial Intelligence and Machine Learning for Process Design, Bioengineering and Synthetic Biology), and Potential Areas for Expansion (Energy Storage and Battery Manufacturing, and Water Purification Technologies).

2.8.1 Emerging Technologies in Green Chemistry

The field of green chemistry is advancing very fast with emerging innovations in diverse technological, biological, and environmental fields. While sustainability is taking the center stage in industrial processes, several of these emerging technologies and potential areas of expansion create massive opportunities for growth in the area

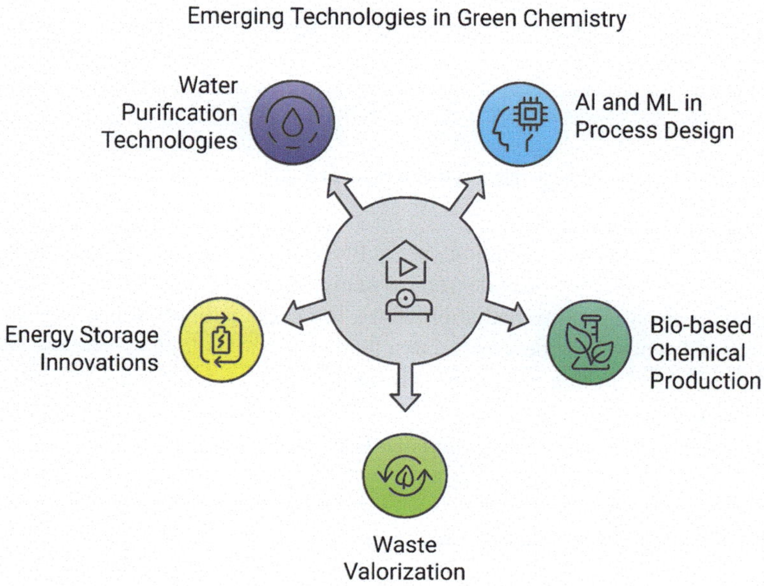

Fig. 2.2 Emerging technologies in green chemistry

of green chemistry. A few of the most important trends and areas promising future developments, which can also be observed in Fig. 2.2. The integration of AI and ML in green chemistry is changing the industrial landscape of manufacturing by offering innovative solutions for sustainable process design. These technologies promote the optimization and advancement of chemical processes as aligned with the principles of green chemistry to achieve efficiency, waste reduction, and bettering environmental standing [68].

Artificial Intelligence and Machine Learning are disruptive technologies which promote the growth of green chemistry by advancing the design of the process and its sustainability. Optimization of processes in manufacturing through AI can analyze complex data to reveal hidden patterns improving yields in reactions, reducing energy usage, and reducing the usage of toxic solvents. Predictive modeling allows ML to predict the behavior of materials and reactions, making it less reliant on resource-intensive trial-and-error experiments. AI-driven pathway design for the eco-friendly synthetic route identifies renewable feedstocks, thereby reducing the generation of byproducts. Catalyst discovery and optimization are accelerated in ML, thus identifying and fine-tuning catalysts to increase reaction efficiency and selectivity and reduce energy usage [69]. AI-powered LCA tools offer all-encompassing assessments of product environmental impacts, allowing manufacturers to devise strategies on how to bring down carbon footprint. In process monitoring and control, real time is achieved that dynamically adjusts parameters of reactions during the process by optimizing resource consumption and avoiding security or environmental dangers.

In sum, AI/ML is pivotal in advancing the cause of green chemistry by driving eco-friendly, efficient, scaleable, and globally set sustainability targets [70].

2.8.2 *Bioengineering and Synthetic Biology*

Bioengineering and synthetic biology are at the center of the future green chemistry, because, as expected, they provide transformative methods of producing sustainable chemicals, environmental remediation, and a life cycle outlook, especially through the design and synthesis of new materials, chemicals, and biological processes within biological systems. In this way, they will further support a transition towards more sustainable industrial practices [71], Fig. 2.3. The major trends and opportunities for green chemistry in bioengineering and synthetic biology are as follows:

2.8.2.1 Bio-Based Chemical Production

Bioengineering and synthetic biology are bringing chemicals into sustainability by using the power of biological systems. Microbial Production of Chemicals uses microbes such as bacteria, fungi, and yeast in chemical production to produce chemicals once obtained from fossil fuels. Some of these bioprocesses are based on renewable feedstocks, such as agricultural waste or CO_2 or biomass, rather than

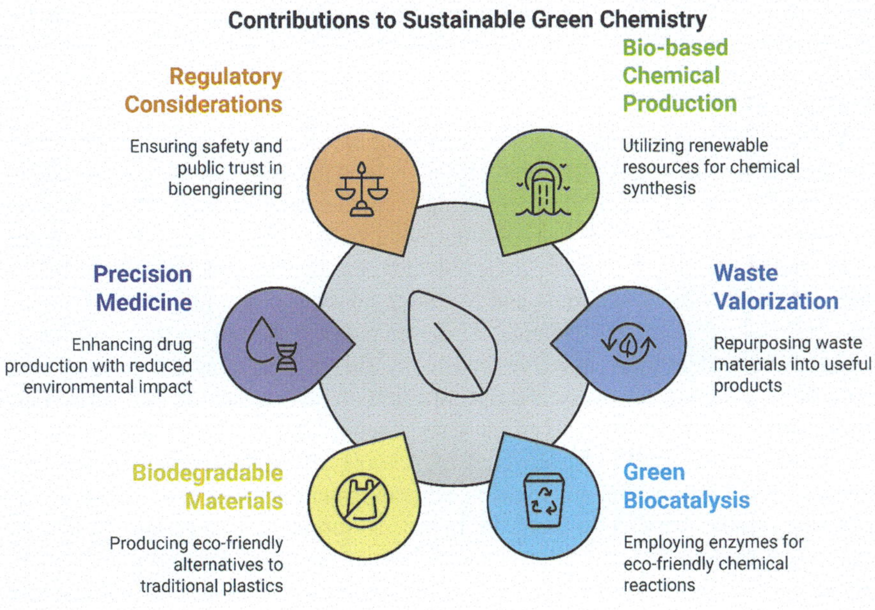

Fig. 2.3 Contributions to sustainable green chemistry

petroleum inputs and have thus lowered the environmental impacts [72]. Biofuel, bioplastic (polylactic acid), and specialty chemicals like lactic acid and succinic acid production through microorganisms can serve as sustainable alternatives to the petrochemical routes. Meanwhile, Enzyme Engineering for Green Chemistry is enhanced in industrial processes through optimization of enzymes for a specific application. Engineered enzymes are used for selective chemical reactions under mild conditions with the goal of energy and waste consumption. These enzymes are important in the production of fine chemicals, pharmaceuticals, and biodegradable plastics as well as in the development of enzyme-based waste treatment systems that effectively degrade toxic substances. Combined, these innovations represent the potential of bio-based chemical production to transform industries in a way that fosters environmental sustainability [73].

2.8.2.2 Waste Valorization and Circular Economy

Bioengineering and synthetic biology are opportunities that will revolutionize sustainable development by valorizing waste and the circular economy. Biological conversion of waste into valuable products is one focus area, utilizing industrial or agricultural waste to produce biofuels, biodegradable plastics, and essential chemicals. Bio-systems employ, from microbial fermentation to enzymatic processes within biorefineries, which convert waste streams of food scraps, agricultural residues, and even direct emission of CO_2 to non-renewable resources, thereby putting a close to the loop of waste [74]. Carbon Capture and Utilization (CCU) is another facet of the power of bioengineering as microorganisms are engineered to use atmospheric CO_2 as the feedstock for production of desirable products. For example, such engineered algae or bacteria may convert CO_2 into biofuels, plastics and other high-value chemicals, therefore solving the climate change crisis and paving the way for a sustainable industrial pathway. They also reduce waste and promote a regenerative model of resource use, adhering to the tenets of a circular economy [75].

2.8.2.3 Green Biocatalysis and Process Optimization

Green biocatalysis is transforming the landscape of chemical synthesis by providing a more environmentally friendly alternative to traditional methods. Thanks to recent developments in synthetic biology, it is now feasible to create biocatalysts—that is, enzymes that catalyse chemical processes under mild circumstances, therefore drastically lowering energy consumption and avoiding extreme conditions including high temperatures or pressures. Biocatalysts are appropriate for sustainable industrial use as, unlike conventional chemical catalysts, they generate extremely low harmful byproducts. Drug synthesis, biofuels, specialised chemicals, and targeted synthesis of complicated compounds all find use for biocatalysis [76]. Synthetic biology also allows metabolic pathways tailored in microorganisms for better sustainability. Microbes can be genetically modified to generate strains with low raw material and waste inputs that might generate high yield target molecules. This strategy marks a major stride towards a greener and more sustainable industrial future since it not only maximises production processes but also greatly lowers the environmental impact compared to conventional chemical manufacturing methods [77].

2.8.2.4 Biodegradable Materials and Green Polymers

Biodegradable materials and green polymers represent a transformative shift in addressing the environmental challenges posed by traditional plastics. Using bioengineered microbes and renewable feedstocks, synthetic biology advances allow the synthesis of sustainable bioplastics including polyhydroxyalkanoates (PHA) and polylactic acid (PLA). These new materials are eco-friendly alternatives to the traditional fossil fuel-based polymers. Unlike usual plastics, the threat posed by bioplastics is being easily dismantled in natural conditions; thereby, it reduces plastic waste and environmental impacts in the long run. The inclusion of these biodegradable polymers in the production and packaging industry, with such green solutions, is thus near to opening a circular economy and consequently marked resource and environmental accountability [78].

2.8.2.5 Precision Medicine and Sustainable Healthcare

Bioengineering and synthetic biology are revolutionizing precision medicine and sustainability in healthcare. Engineered microorganisms and biosynthetic pathways can be used to produce pharmaceuticals with less resource usage and much lower environmental impact. The technology is regarded as fundamental to sustainable healthcare, reducing waste and the carbon footprint of pharmaceutical production. For example, routinely manufactured biopharmaceuticals including vaccinations, growth hormones, and insulin in microbial systems currently use recombinant DNA technology. This change not only lessens dependency on animal-based production systems but also improves scalability and efficiency, so complementing the objectives of precision medicine by allowing the customised manufacture of medications to satisfy specific patient demands while protecting environmental resources [79].

2.8.2.6 Future Challenges and Regulatory Considerations

There has been significant development in bioengineering and synthetic biology, which further includes their integration with environmentally friendly applications of chemicals. The challenges with such developments also include regulatory consideration. One such challenge is regulating modified organisms, mainly GMOs (genetically modified organisms) in industrial and environmental applications. Regulatory systems must adapt to tackle confinement, controlled release, and the ethical ramifications of large-scale deployment of bioengineered organisms. Guaranteeing safety through thorough risk evaluations and stringent supervision will be essential. Moreover, public acceptance and ethical concerns are significant obstacles to the use of these technologies. Public anxieties about GMOs and synthetic biology demand honest communication, rigorous safety assessments, and evidence-based policy-making in order to build public confidence. Proactive regulation, stakeholder involvement, and fidelity to ethical principles will be paramount for the proper and sustainable use of bioengineering innovation in green chemistry [80].

In conclusion, bioengineering and synthetic biology have significant prospects for enhancing green chemistry through the sustainable production of chemicals, minerals, and energy, while concurrently minimising waste, energy usage, and environmental repercussions. These technologies are positioned to be pivotal in tackling global issues of sustainability and resource scarcity, rendering them essential to the future of green chemistry.

2.8.3 Potential Areas for Expansion

Green chemistry has significant potential to transform areas such as energy storage and water purification, providing sustainable solutions to urgent environmental issues. Green chemistry can enhance the development of sustainable battery materials in energy storage and battery manufacturing by identifying alternatives to toxic and limited resources such as lithium and cobalt. Effective minimising of the carbon footprint connected with battery production depends on environmentally friendly production techniques including the use of renewable energy and the reduction of toxic chemicals. Green chemistry strongly supports projects on the circular economy and recycling by improving battery recycling methods and facilitating more effective, ecologically sustainable activities. Innovations such as organic flow and sodium-ion batteries, which are alternate energy storage alternatives, represent promising ways for creating greener and more affordable technologies [81].

Green chemistry provides sustainable alternatives to conventional water treatment methods that depend on harsh chemicals and significant energy consumption in water purification technologies. Biodegradable natural filtration media in the form of biomass-based activated carbon, on the other hand, proved useful in removal capabilities and lowering adverse environmental impact through engineered nanomaterials, for instance. Desalination technologies through direct use of sun power and energy efficiency also produce alternative sources of pure water derived from seawater. Wastewater treatment and resource recovery systems where nutrients such as nitrogen and phosphorus are reused as fertilisers are aligned to the circular economy principles. Development of catalysts and membranes to improve efficiency of water treatment decreases energy usage as well as less chemical use in the process. Such areas emphasize green chemistry's integral role in averting resource exhaustion and pollution; hence, aiding in achieving goals for global sustainability [82].

2.9 Conclusion

Scaling green chemistry solutions in industrial manufacturing is a critical step in achieving sustainable development objectives. The concepts, problems, and ways of implementing green chemistry industrially have been explained while underlining the need to integrate advanced technologies like artificial intelligence, bioengineering, and circular economy practices. Such approaches facilitate environmental sustainability while also bringing about economic benefits through cost reduction and higher output of resources. This change needs joint efforts from the governments, corporations, and academic circles. Effective policies combined with global partnerships can make the implementation of green chemistry by breaking regulatory and economic barriers more feasible. The case studies are examples illustrating concrete benefits to the methods under consideration, from environment-friendly pharma synthesis through eco-friendly agrochemical production up to bioplastics. As the field continues to grow, future applications in energy storage, water purification, and bio-based manufacturing will increase the impact of green chemistry. By emphasising scalability, innovation, and collaboration, industries can transform their processes to conform to a sustainable and resilient industrial framework. The path to universal adoption presents both a problem and an opportunity to transform the industrial landscape for future generations.

References

1. Singh, N.B.: Green synthesis of nanomaterials. Handbook of Microbial Nanotechnology, Elsevier, pp. 225–54 (2022). https://doi.org/10.1016/B978-0-12-823426-6.00007-3
2. Al-Shatti, B.J., Alsairafi, Z., Al-Tannak, N.F.: Green chemistry and its implementation in pharmaceutical analysis. Rev. Anal. Chem. **42** (2023). https://doi.org/10.1515/revac-2023-0069
3. de Marco, B.A., Rechelo, B.S., Tótoli, E.G., Kogawa, A.C., Salgado, H.R.N.: Evolution of green chemistry and its multidimensional impacts: a review. Saudi Pharm. J. **27**, 1–8 (2019). https://doi.org/10.1016/j.jsps.2018.07.011
4. Zuin, V.G., Eilks, I., Elschami, M., Kümmerer, K.: Education in green chemistry and in sustainable chemistry: perspectives towards sustainability. Green Chem. **23**, 1594–1608 (2021). https://doi.org/10.1039/D0GC03313H
5. Oliveira, J.R.P., Tusset, A.M., Andrade, D.I., Balthazar, J.M., Pagani, R.N., Lenzi, G.G.: Action plans study: principles of green chemistry, sustainable development, and smart cities. Sustainability **16**, 8041 (2024). https://doi.org/10.3390/su16188041
6. Ratti, R.: Industrial applications of green chemistry: status, challenges and prospects. SN Appl. Sci. **2**, 263 (2020). https://doi.org/10.1007/s42452-020-2019-6

7. Gabrielli, P., Rosa, L., Gazzani, M., Meys, R., Bardow, A., Mazzotti, M., et al.: Net-zero emissions chemical industry in a world of limited resources. One Earth **6**, 682–704 (2023). https://doi.org/10.1016/j.oneear.2023.05.006
8. Neupane, D.: Biofuels from renewable sources, a potential option for biodiesel production. Bioengineering **10**, 29 (2022). https://doi.org/10.3390/bioengineering10010029
9. Butzer, K.W.: Environmental history in the Mediterranean world: cross-disciplinary investigation of cause-and-effect for degradation and soil erosion. J. Archaeol. Sci.Archaeol. Sci. **32**, 1773–1800 (2005). https://doi.org/10.1016/J.JAS.2005.06.001
10. Onifade, M., Zvarivadza, T., Adebisi, J.A., Said, K.O., Dayo-Olupona, O., Lawal, A.I., et al.: Advancing toward sustainability: the emergence of green mining technologies and practices. Green Smart Min. Eng. **1**, 157–174 (2024). https://doi.org/10.1016/J.GSME.2024.05.005
11. Allan, J., Belz, S., Hoeveler, A., Hugas, M., Okuda, H., Patri, A., et al.: Regulatory landscape of nanotechnology and nanoplastics from a global perspective. Regul. Toxicol. Pharmacol.. Toxicol. Pharmacol. **122**, 104885 (2021). https://doi.org/10.1016/J.YRTPH.2021.104885
12. Rogers, L., Jensen, K.F.: Continuous manufacturing—the green chemistry promise? Green Chem. **21**, 3481–3498 (2019). https://doi.org/10.1039/C9GC00773C
13. Söderholm, P.: The green economy transition: the challenges of technological change for sustainability. Sustain. Earth **3**, 1–11 (2020). https://doi.org/10.1186/S42055-020-00029-Y
14. Padmavathy, A., Anbarashan, M.: Potential of solid waste prevention and minimization strategies. In: Cost Effective Technologies for Solid Waste and Wastewater Treatment, pp. 219–28 (2022). https://doi.org/10.1016/B978-0-12-822933-0.00019-X
15. Song, Q.-W., He, L.-N.: Atom economy. Green Chem. Chem. Eng. 3–22 (2019). https://doi.org/10.1007/978-1-4939-9060-3_1001
16. Kharissova, O.V., Kharisov, B.I., González, C.M.O., Méndez, Y.P., López, I.: Greener synthesis of chemical compounds and materials. R. Soc. Open Sci. **6**, 191378 (2019). https://doi.org/10.1098/RSOS.191378
17. Parker, O.: Designing safer chemicals: reducing toxicity and environmental impact. To Chem. J. **11**, 10–10 (2024)
18. Doble, M., Kruthiventi, A.K.: Alternate solvents. Green Chem. Eng. 93–104 (2007). https://doi.org/10.1016/B978-012372532-5/50006-7
19. Jayabal, R.: Towards a carbon-free society: innovations in green energy for a sustainable future. Results Eng. **24**, 103121 (2024). https://doi.org/10.1016/J.RINENG.2024.103121
20. Anil, N., Rao, P.K., Sarkar, A., Kubavat, J., Vadivel, S., Manwar, N.R., et al.: Advancements in sustainable biodiesel production: a comprehensive review of bio-waste derived catalysts. Energy Convers. Manag. **318**, 118884 (2024). https://doi.org/10.1016/J.ENCONMAN.2024.118884
21. Jiang, A., Li, J., Shah, K.J., You, Z., Jiang, A., Li, J., et al.: Perspective chapter: implementing green chemistry principles for pollution control to achieve environmental sustainability—a review. In: Green Chemistry for Environmental Sustainability-Prevention-Assurance-Sustainability (P-A-S) Approach (2023). https://doi.org/10.5772/INTECHOPEN.1003627
22. Isahak, W.N.R.W., Al-Amiery, A.: Catalysts driving efficiency and innovation in thermal reactions: a comprehensive review. Green Technol. Sustain. **2**, 100078 (2024). https://doi.org/10.1016/J.GRETS.2024.100078
23. Afshar, S.V., Boldrin, A., Astrup, T.F., Daugaard, A.E., Hartmann, N.B.: Degradation of biodegradable plastics in waste management systems and the open environment: a critical review. J. Clean. Prod. **434**, 140000 (2024). https://doi.org/10.1016/J.JCLEPRO.2023.140000
24. Sharma, M., Mahajan, P., Alsubaie, A.S., Khanna, V., Chahal, S., Thakur, A., et al.: Next-generation nanomaterials-based biosensors: real-time biosensing devices for detecting emerging environmental pollutants. Mater. Today Sustain. **29**, 101068 (2025). https://doi.org/10.1016/J.MTSUST.2024.101068
25. Andrade-Guel, M., Cabello-Alvarado, C., Caicedo, C., Melo, L., Ávila-Orta, C.: Inherent safer chemistry for accident prevention. Green Chem. Appl. 244–75 (2020). https://doi.org/10.1201/9780429291166-14

26. O'Neil, N.J., Scott, S., Relph, R., Ponnusamy, E.: Approaches to incorporating green chemistry and safety into laboratory culture. J. Chem. Educ. **98**, 84–91 (2021). https://doi.org/10.1021/ACS.JCHEMED.0C00134/ASSET/IMAGES/LARGE/ED0C00134_0002.JPEG
27. Taylor, C.J., Pomberger, A., Felton, K.C., Grainger, R., Barecka, M., Chamberlain, T.W., et al.: A Brief Introduction to chemical reaction optimization. Chem. Rev. **123**, 3089 (2023). https://doi.org/10.1021/ACS.CHEMREV.2C00798
28. Sobczak, J., Wysocka, I., Murgrabia, S., Rogala, A.: A review on deactivation and regeneration of catalysts for dimethyl ether synthesis. Energies **15**, 5420 (2022). https://doi.org/10.3390/EN15155420.
29. Moses, O.A., Chen, W., Adam, M.L., Wang, Z., Liu, K., Shao, J., et al.: Integration of data-intensive, machine learning and robotic experimental approaches for accelerated discovery of catalysts in renewable energy-related reactions. Mater. Rep. Energy **1**, 100049 (2021). https://doi.org/10.1016/J.MATRE.2021.100049
30. Osman, A.I., Ayati, A., Krivoshapkin, P., Tanhaei, B., Farghali, M., Yap, P.S., et al.: Coordination-driven innovations in low-energy catalytic processes: advancing sustainability in chemical production. Coord. Chem. Rev.. Chem. Rev. **514**, 215900 (2024). https://doi.org/10.1016/J.CCR.2024.215900
31. Ramchuran, S.O., O'Brien, F., Dube, N., Ramdas, V.: An overview of green processes and technologies, biobased chemicals and products for industrial applications. Curr. Opin. Green Sustain. Chem. **41**, 100832 (2023). https://doi.org/10.1016/J.COGSC.2023.100832
32. Agrawal, R., Agrawal, S., Samadhiya, A., Kumar, A., Luthra, S., Jain, V.: Adoption of green finance and green innovation for achieving circularity: an exploratory review and future directions. Geosci. Front.. Front. **15**, 101669 (2024). https://doi.org/10.1016/J.GSF.2023.101669
33. Constable, D.J.C.: Green and sustainable chemistry—the case for a systems-based, interdisciplinary approach. IScience **24**, 103489 (2021). https://doi.org/10.1016/J.ISCI.2021.103489
34. Kong, Z.Y., Sánchez-Ramírez, E., Sim, J.Y., Sunarso, J., Segovia-Hernández, J.G.: The importance of process intensification in undergraduate chemical engineering education. Digit. Chem. Eng. **11**, 100152 (2024). https://doi.org/10.1016/J.DCHE.2024.100152
35. Idoko, F.A., Ezeamii, G.C., Ojochogwu, O.J.: Green chemistry in manufacturing: innovations in reducing environmental impact **23**, 2826–41 (2024). https://doi.org/10.30574/WJARR.2024.23.3.2938. https://wjarr.com/sites/default/files/WJARR-2024-2938.pdf
36. Lozano, P., García-Verdugo, E.: From green to circular chemistry paved by biocatalysis. Green Chem. **25**, 7041–7057 (2023). https://doi.org/10.1039/D3GC01878D
37. Farouk, S.M., Tayeb, A.M., Abdel-Hamid, S.M.S., Osman, R.M.: Recent advances in transesterification for sustainable biodiesel production, challenges, and prospects: a comprehensive review. Environ. Sci. Pollut. Res. **31**, 12722–47 (2024). https://doi.org/10.1007/S11356-024-32027-4
38. Salas, R., Villa, R., Velasco, F., Cirujano, F.G., Nieto, S., Martin, N., et al.: Ionic liquids in polymer technology. Green Chem. (2025). https://doi.org/10.1039/D4GC05445H
39. Sowbhagyam, Dr D.V.: Ionic liquids as green solvents: a comprehensive review. Int. Res. J. Adv. Eng. Hub (IRJAEH) **2**, 220–4 (2024). https://doi.org/10.47392/IRJAEH.2024.0035
40. Alam, M.M., Greco, A., Rajabimashhadi, Z., Esposito, C.C.: Efficient and environmentally friendly techniques for extracting lignin from lignocellulose biomass and subsequent uses: a review. Clean. Mater. **13**, 100253 (2024). https://doi.org/10.1016/J.CLEMA.2024.100253
41. Vianna, L.M.C., de Oliveira, L., Mühl, D.D.: Waste valorization in agribusiness value chains. Waste Manag. Bull. **1**, 195–204 (2024). https://doi.org/10.1016/J.WMB.2023.10.009
42. Arun, M., Barik, D., Chandran, S.S.R.: Exploration of material recovery framework from waste—a revolutionary move towards clean environment. Chem. Eng. J. Adv. **18**, 100589 (2024). https://doi.org/10.1016/J.CEJA.2024.100589
43. Kharissova, O.V., Kharisov, B.I., González, C.M.O., Méndez, Y.P., López, I.: Greener synthesis of chemical compounds and materials. R. Soc. Open Sci. **6** (2019). https://doi.org/10.1098/RSOS.191378

44. Ha, M.W., Paek, S.M.: Recent advances in the synthesis of ibuprofen and naproxen. Molecules **26**, 4792 (2021). https://doi.org/10.3390/MOLECULES26164792
45. Yadav, K., Nikalje, G.C.: Comprehensive analysis of bioplastics: life cycle assessment, waste management, biodiversity impact, and sustainable mitigation strategies. PeerJ **12**, e18013 (2024). https://doi.org/10.7717/PEERJ.18013/SUPP-1
46. Bukharbayeva, F., Zharmagambetova, A., Talgatov, E., Auyezkhanova, A., Akhmetova, S., Jumekeyeva, A., et al.: The synthesis of green palladium catalysts stabilized by chitosan for hydrogenation. Molecules **29**, 4584 (2024). https://doi.org/10.3390/MOLECULES29194584
47. Shah, K.U., Gangadeen, I.: Integrating bioplastics into the US plastics supply chain: towards a policy research agenda for the bioplastic transition. Front. Environ. Sci. **11**, 1245846 (2023). https://doi.org/10.3389/FENVS.2023.1245846/BIBTEX
48. Yadav, N., Garg, V.K., Chhillar, A.K., Rana, J.S.: Recent advances in nanotechnology for the improvement of conventional agricultural systems: a review. Plant Nano Biol. **4**, 100032 (2023). https://doi.org/10.1016/J.PLANA.2023.100032
49. Chaudhary, S., Kanwar, R.K., Sehgal, A., Cahill, D.M., Barrow, C.J., Sehgal, R., et al.: Progress on azadirachta indica based biopesticides in replacing synthetic toxic pesticides. Front. Plant Sci. **8**, 610 (2017). https://doi.org/10.3389/FPLS.2017.00610
50. Goñi, M.L., Gañán, N.A., Martini, R.E.: Supercritical CO_2-assisted dyeing and functionalization of polymeric materials: a review of recent advances (2015–2020). J. CO_2 Util. **54**, 101760 (2021). https://doi.org/10.1016/J.JCOU.2021.101760
51. Afedzi, A.E.K., Parakulsuksatid, P.: Recent advances in process modifications of simultaneous saccharification and fermentation (SSF) of lignocellulosic biomass for bioethanol production. Biocatal. Agric. Biotechnol.. Agric. Biotechnol. **54**, 102961 (2023). https://doi.org/10.1016/J.BCAB.2023.102961
52. Jiang, Y., Ding, X., Zhang, J.: Toward environmental efficiency: analyzing the impact of green innovation initiatives in enterprises. Manag. Decis. Econ.. Decis. Econ. (2024). https://doi.org/10.1002/MDE.4429
53. Li, H., Saboori, A., Cao, X.: Information synthesis and preliminary case study for life cycle assessment of reflective coatings for cool pavements. Int. J. Transp. Sci. Technol. **5**, 38–46 (2016). https://doi.org/10.1016/j.ijtst.2016.06.005
54. Chojnacka, K.: Sustainable chemistry in adaptive agriculture: a review. Curr. Opin. Green Sustain. Chem. **46**, 100898 (2024). https://doi.org/10.1016/J.COGSC.2024.100898
55. Mugoni, E., Kanyepe, J., Tukuta, M.: Sustainable supply chain management practices (SSCMPS) and environmental performance: a systematic review. Sustain. Technol. Entrep. **3**, 100050 (2024). https://doi.org/10.1016/J.STAE.2023.100050
56. Okwandu, A.C., Esho, A.O.-O., Iluyomade, T.D., Olatunde, T.M.: The role of policy and regulation in promoting green buildings **22**, 139–50 (2024). https://doi.org/10.30574/WJARR.2024.22.1.1047. https://wjarr.com/sites/default/files/WJARR-2024-1047.pdf
57. Berikhanovna, C.M., Bauirzhanovna, B.A., Kudaibergenovna, N.G., Gulbagda, B., Serikovna, Y.G.: The influence of green credit policy on green innovation and transformation and upgradation as a function of corporate diversification: the case of Kazakhstan. Economies **11**, 210 (2023). https://doi.org/10.3390/ECONOMIES11080210
58. Esangbedo, C.O., Zhang, J., Esangbedo, M.O., Kone, S.D., Xu, L.: The role of industry-academia collaboration in enhancing educational opportunities and outcomes under the digital driven industry 4.0. J. Infrastruct. Policy Dev. **8**, 2569 (2023). https://doi.org/10.24294/JIPD.V8I1.2569
59. Maresova, P., Stemberkova, R., Fadeyi, O.: Models, processes, and roles of universities in technology transfer management: a systematic review. Adm. Sci. **9**, 67 (2019). https://doi.org/10.3390/ADMSCI9030067
60. Javaid, M., Haleem, A., Singh, R.P., Suman, R., Gonzalez, E.S.: Understanding the adoption of industry 4.0 technologies in improving environmental sustainability. Sustain. Oper. Comput. **3**, 203–17 (2022). https://doi.org/10.1016/J.SUSOC.2022.01.008
61. Kashezheva, M., Kedakhezheva, R., Morozov, V., Kolesnikov, R., Voropaev, V., Mezentsev, S., et al.: Harnessing nanocatalysts for sustainable chemical processes: innovations in green

chemistry. Asian J. Green Chem. **9**, 57–82 (2025). https://doi.org/10.48309/AJGC.2025.483172.1564
62. Mehmood, S., Zaman, K., Khan, S., Ali, Z., Khan Hur, R.: The role of green industrial transformation in mitigating carbon emissions: exploring the channels of technological innovation and environmental regulation. Energy Built Environ. **5**, 464–79 (2024). https://doi.org/10.1016/J.ENBENV.2023.03.001
63. Shayan, N.F., Mohabbati-Kalejahi, N., Alavi, S., Zahed, M.A.: Sustainable development goals (SDGs) as a framework for corporate social responsibility (CSR). Sustainability **14**, 1222 (2022). https://doi.org/10.3390/SU14031222
64. Zhao, X., Castka, P., Searcy, C.: ISO standards: a platform for achieving sustainable development goal 2. Sustainability **12**, 9332 (2020). https://doi.org/10.3390/SU12229332
65. Padilla-Rivera, A., Hannouf, M., Assefa, G., Gates, I.: Enhancing environmental, social, and governance, performance and reporting through integration of life cycle sustainability assessment framework. Sustain. Dev. (2024). https://doi.org/10.1002/SD.3265
66. Islam, M.T., Ali, A.: Sustainable green energy transition in Saudi Arabia: characterizing policy framework, interrelations and future research directions. Next Energy **5**, 100161 (2024). https://doi.org/10.1016/J.NXENER.2024.100161
67. Amoneit, M., Weckowska, D., Spahr, S., Wagner, O., Adeli, M., Mai, I., et al.: Green chemistry and responsible research and innovation: Moving beyond the 12 principles. J. Clean. Prod. **484**, 144011 (2024). https://doi.org/10.1016/J.JCLEPRO.2024.144011
68. Aal, E., Ali, R.S., Meng, J., Khan, M.E.I., Jiang, X.: Machine learning advancements in organic synthesis: a focused exploration of artificial intelligence applications in chemistry. Artif. Intell. Chem. **2**, 100049 (2024). https://doi.org/10.1016/J.AICHEM.2024.100049
69. G PC, V RB, DSVNM R, G R, SVGVA P.: AI and machine learning in biotechnology: a paradigm shift in biochemical innovation. Int. J. Plant Anim. Environ. Sci. **14**, 70–80 (n.d)
70. Campo Gay, I., Hvam, L., Haug, A., Huang, G.Q., Larsson, R.: A digital tool for life cycle assessment in construction projects. Dev. Built Environ. **20**, 100535 (2024). https://doi.org/10.1016/J.DIBE.2024.100535
71. Nawaz, T., Gu, L., Hu, Z., Fahad, S., Saud, S., Zhou, R.: Advancements in synthetic biology for enhancing cyanobacterial capabilities in sustainable plastic production: a green horizon perspective. Fuels **5**, 394–438 (2024). https://doi.org/10.3390/FUELS5030023
72. Gaur, S., Kaur, M., Kalra, R., Rene, E.R., Goel, M.: Application of microbial resources in biorefineries: current trend and future prospects. Heliyon **10**, e28615 (2024). https://doi.org/10.1016/J.HELIYON.2024.E28615
73. de Souza, F.M., Gupta, R.K.: Bacteria for bioplastics: progress, applications, and challenges. ACS Omega **9**, 8666–8686 (2024). https://doi.org/10.1021/ACSOMEGA.3C07372/ASSET/IMAGES/LARGE/AO3C07372_0009.JPEG
74. Saxena, S., Moharil, M.P., Jadhav, P.V., Ghodake, B., Deshmukh, R., Ingle, A.P.: Transforming waste into wealth: Leveraging nanotechnology for recycling agricultural byproducts into value-added products. Plant Nano Biology **11**, 100127 (2025). https://doi.org/10.1016/J.PLANA.2024.100127
75. Irfan, F., Tasnim, N., Razzak, S.A., Uddin, S.: Advances in capturing carbon and producing bioplastics through the microalgal approach towards a sustainable future: a review. J. Clean. Prod. **486**, 144576 (2025). https://doi.org/10.1016/J.JCLEPRO.2024.144576
76. Bell, E.L., Finnigan, W., France, S.P., Green, A.P., Hayes, M.A., Hepworth, L.J., et al.: Biocatalysis. Nat. Rev. Methods Prim. **1**, 1–21 (2021). https://doi.org/10.1038/s43586-021-00044-z
77. Sheldon, R.A.: Green chemistry and biocatalysis: engineering a sustainable future. Catal. Today. Today **431**, 114571 (2024). https://doi.org/10.1016/J.CATTOD.2024.114571
78. Xu, R., Fang, Y., Zhang, Z., Cao, Y., Yan, Y., Gan, L., et al.: Recent advances in biodegradable and biocompatible synthetic polymers used in skin wound healing. Materials **16**, 5459 (2023)
79. Filippi, M., Mekkattu, M., Katzschmann, R.K.: Sustainable biofabrication: from bioprinting to AI-driven predictive methods. Trends Biotechnol.Biotechnol. (2024). https://doi.org/10.1016/J.TIBTECH.2024.07.002

80. Rafeeq, H., Afsheen, N., Rafique, S., Arshad, A., Intisar, M., Hussain, A., et al.: Genetically engineered microorganisms for environmental remediation. Chemosphere **310**, 136751 (2023). https://doi.org/10.1016/J.CHEMOSPHERE.2022.136751
81. Molaiyan, P., Bhattacharyya, S., dos Reis, G.S., Sliz, R., Paolella, A., Lassi, U.: Towards greener batteries: sustainable components and materials for next-generation batteries. Green Chem. **26**, 7508–7531 (2024). https://doi.org/10.1039/D3GC05027K
82. Satyam, S., Patra, S.: Innovations and challenges in adsorption-based wastewater remediation: a comprehensive review. Heliyon **10**, e29573 (2024). https://doi.org/10.1016/J.HELIYON.2024.E29573

Chapter 3
Green Chemistry in Industrial Catalysis: Enhancements and Innovations

B. Namratha, Swathi Salian, and Santosh L. Gaonkar

Abstract Catalysis occupies a distinctive role in sustainable industrial practices that have been integrated into the chemical sector over the preceding five decades. Nonetheless, these catalysts continue to exhibit a deficiency in selectivity within the industrial context. Green catalysts possess the capability to facilitate a variety of chemical reactions, thereby representing a crucial alternative to environmentally detrimental chemical processes. In this regard, the various types and challenges associated with conventional catalytic methodologies will be examined. Innovative catalytic materials, including those derived from biomass and renewable sources, advancements in biocatalysis, and metal-free catalysts, will be elucidated in detail. The examination of energy-efficient catalytic processes, along with waste minimization and by-product management within catalysis, will be illustrated through classic examples. Noteworthy instances of green catalytic innovations pertaining to the pharmaceutical, petrochemical, and fine chemical sectors will be thoroughly explored. This chapter addresses the emerging materials and technologies pertinent to green catalysis.

Keywords Green catalysts · Green chemistry · Biocatalysis · Sustainable · Biomass · Renewable sources

B. Namratha · S. Salian
Department of Chemistry, The Yenepoya Institute of Arts, Science, Commerce and Management, Yenepoya (Deemed to be University), Mangaluru, Karnataka, India

S. L. Gaonkar (✉)
Department of Chemistry, Manipal Institute of Technology, Manipal Academy of Higher Education, Manipal, Karnataka, India
e-mail: sl.gaonkar@manipal.edu

P. Jain et al. (eds.), *Towards Green Chemical Processes: Strategies and Innovations*, Intelligent Systems Reference Library 277,
https://doi.org/10.1007/978-3-031-94090-3_3

3.1 Introduction to Green Chemistry in Catalysis

Green chemistry is a branch of chemical engineering and research that focusses on creating goods and procedures that use and produce less harmful ingredients. Its objective is to reduce waste and eliminate potential risks at the design stage by producing safer and more effective chemical entities while utilising the safest and most efficient synthesis processes. The field of sustainability is highlighted by green chemistry, which likewise focusses on the effects of chemical production and use on the environment and human health. Green Chemistry aims to enhance ecosystem growth and encourage sustainable development. Paul Anastus was a pioneer in the field of green chemistry, which developed prior to the 1990s from fundamental ideas like catalysis and atom economy. It was fuelled by growing worries about chemical pollution and the depletion of natural resources [1].

3.1.1 Principles of Green Chemistry

A collection of 12 fundamental principles that foresee the reduction, elimination, or deletion of hazardous byproducts throughout the synthetic route serve as the foundation for the idea of green chemistry (Fig. 3.1).

Preventing the production of waste: The primary tenet of green chemistry is thought to be waste prevention. Preventing trash production is significantly more advantageous than repair later on.

Atom Economy: Since it focuses the response flow towards zero-waste or minimal waste creation, the idea of the "atom economy" is inextricably tied to the waste prevention principle. It offers the best way to gauge a chemical reaction's efficiency in terms of using all of the raw components and guaranteeing that they are fully incorporated into the final product.

$$Atom\ Economy\ (\%) = \frac{Molecular\ weight\ of\ desired\ product}{Sum\ ofmolecular\ weights\ of\ all\ reactants} \times 100$$

This aids in maximising the utilisation of raw materials and minimising waste. The most environmentally friendly and sustainable approach is one with a high atom economy.

Less dangerous chemical synthesis: This idea relates to the intention of minimising the production of harmful compounds during the chemical synthesis process. It enforces the requirement to replace hazardous chemical entities with safer ones in order to produce materials that are less or completely harmful to the environment and human health.

Creating safer chemical designs: The foundation of creating a greener environment is designing safe chemicals with the goal of minimising the negative effects that chemical processes or products have on the environment and human health. The

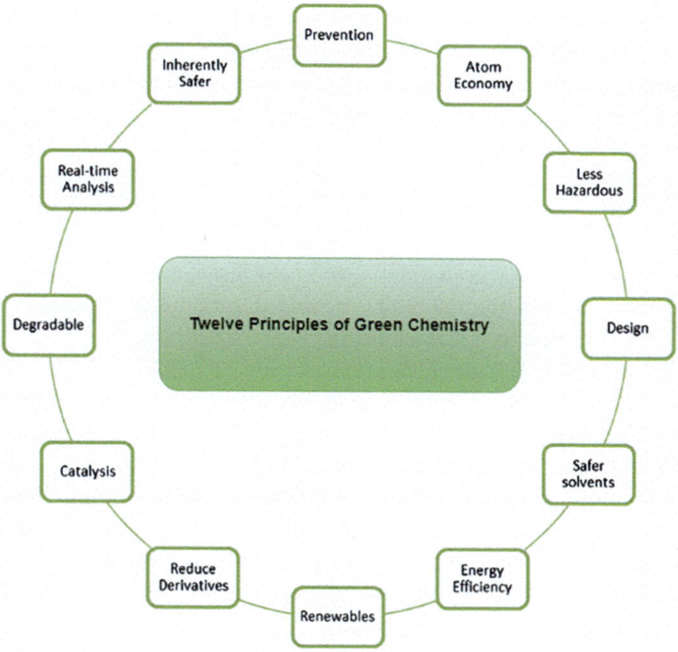

Fig. 3.1 Twelve principles for a greener world

product of a specific chemical process needs to be made in a way that maintains its effectiveness and aids in lowering its toxicity.

Safer solvents and auxiliaries: Since long-term reactions typically require large volumes of solvents, it is imperative to adapt safer solvents. Therefore, choosing a solvent is likewise a matter of debate. Solvents must be selected to reduce risk, with a preference for bio-based and renewable sources, solvents that are recyclable and recoverable, and solvents that don't bio-accumulate. Safer auxiliaries are being developed, such as solvent-free reactions, catalysts, and surfactants for process intensification. Hazardous organic solvents can be reduced with the aid of a phase transition catalyst.

Boost energy efficiency: Increasing energy efficiency is a simple way to save energy for greater uses. The need for energy is a hot topic in the expanding world of today. Therefore, energy efficiency will undoubtedly be the topic of conversation throughout this hour. The main goal is to design compounds that are active at room temperature and use catalysts to optimise the reaction pathway and significantly reduce energy requirements. implementing solvent-free processes, employing biocatalysts, raising potentials by electrical energy, etc.

Utilising renewable feedstocks: By reducing their negative effects on the environment, they help process key products and play a significant part in replacing fossil fuels.

Reduce chemical derivatives: By merely cutting them off with a protective substance, unnecessary derivatives can be minimised or decreased. Indeed, this can contribute to improving the atom economy and energy efficiency.

Using catalysts: Catalysts assist in guiding reactions in a more environmentally friendly direction. Catalysts are crucial in lowering energy needs, negative environmental effects, the use of hazardous materials, etc. They serve as the cornerstones of green chemistry's sustainable future. By encouraging the use of biodegradable catalysts, the catalysis principle reduces the amount of electricity used, eliminates the need for organochlorine compounds, and uses less water or waste water. Since bio-catalysis is a biomimetic technique entirely dependent on natural or modified enzymes, it is an example of green chemistry.

Design for degradation: Chemical goods should be made so that, as their properties expire, they decompose into compounds that are no longer harmful to the environment. The ability to transform into innocuous products is the aim of creating degradable chemical substances and goods.

Real-time analysis to prevent pollution: Analytical techniques must also advance to enable real-time process scanning, monitoring, and management prior to the creation of hazardous materials. Because research-focused analytical methodologies have a high potential for developing certain toxicants or unexpected scenarios, industries must make it a necessary process.

Pass on safer chemistry to avert mishaps: Reducing the usage of compounds that can have negative effects in chemical processes is the fundamental necessity. As an illustration, consider the growing usage of supercritical CO_2, which takes the role of organic solvents and is environmentally acceptable despite not being harmful or explosive. Several degrees of the least amount of personal protective equipment are used to accomplish safety, which is defined as the control of recognised dangers by attaining an acceptable level of risk [2].

3.1.2 Importance of Catalysis in Sustainable Industrial Processes

Chemistry's foundational field is catalysis. The Greek word katalysis, which indicates dissolution or breakdown, is where the word catalysis comes from. By adding a tiny amount of alleged catalyst that doesn't change during the reaction, catalysis speeds up the reaction. Creating sustainable catalysts that facilitate more environmentally friendly industrial processes has been a major area of focus. The use of renewable feedstocks, effective reaction conditions, and the reduction of waste creation are just a few of the ways that sustainable catalysts can support green chemistry. As businesses become more aware of the financial and ecological advantages of implementing sustainable practices, green chemistry's industrial applications have been growing. Because it makes greener and more energy-efficient industrial practices possible, catalysis is essential to the advancement of sustainable chemistry [3].

3.2 Overview of Conventional Catalysis and Limitations

3.2.1 Conventional Catalysis: Types and Environmental Impacts

Conventional catalysis, also known as organocatalysis, is a kind of chemical reaction in which the reaction is accelerated by the use of organic catalysts. A well-known natural phenomena, catalysis is the basis for chemical reactions that take place more quickly in both natural processes and a variety of real-world applications. It is also an effective way to guide and manage a chemical reaction. There are three different kinds of catalysts: enzymatic, heterogeneous, and homogeneous [4].

Without being consumed during the reaction, a catalyst quickens or enhances the rate of a reaction. They work by increasing the likelihood of a reaction by reducing the activation energy needed for it to occur. A faster response rate results from using less energy to start the reaction. A catalyst mixture can be an enzyme-based media, an acidic medium, or an alkaline medium. With the particular reactant, the catalyst frequently creates a few transient intermediate chemicals. These intermediates might be more easily transformed into products and have unique properties. The catalyst has the ability to regenerate once products are formed, making it usable in further reactions [5]. More advanced and modern catalysts are included in the broader phrase "conventional catalyst." Because of their recent development and origins, as well as their intricate compositions, they are widely employed. Metals or sulphides, halides, or oxides of metallic and semi-metallic elements, such as silicon, boron, or aluminium, make up the majority of solid catalysts [5].

3.2.1.1 Homogeneous Catalysis

Both the catalyst and the reactants are present in the same phase during homogeneous catalysis. An intermediate complex is produced when the catalyst interacts directly with the reactants. It frequently produces cleaner reactions due to its strong activity and selectivity. Acid-base catalysts, which are employed in numerous organic reactions including esterification and hydrolysis, are among the examples. A variety of processes, such as oxidation, polymerisation, and hydrogenation, can be catalysed by transition metal complexes [6].

3.2.1.2 Heterogeneous Catalysis

The catalyst and reactant are in distinct phases during heterogeneous catalysis. In most cases, the reactants are liquids or gases, while the catalyst is a solid. Reactant molecules absorb, react, and then desorb as products at the catalyst's surface, where the reaction takes place. The process is more ecologically friendly and efficient. Compared to homogeneous catalysts, this kind of catalyst may be less selective, and other outside variables may have an impact on its activity. The metal in the conventional heterogeneous catalyst is in the form of clusters of atoms or nanoparticles, which results in a low or minimal utilisation of the metal atoms. However, a single atom catalyst can achieve a high dispersion of metal atoms or particles on the support material's surface, which not only maximises the use of metal atoms but also reduces metal consumption, increasing the density of active sites and improving chemical reaction selectivity and activity [7].

Classic examples of heterogeneous catalysis include the Haber-Bosch process for ammonia production and catalytic converters found in automobiles. Another significant example is the employment of metal oxide catalysts in oxidation and reduction reactions, such as the contact process for the generation of sulphuric acid. A heterogeneous acid catalyst might be easily added to a packed bed continuous flow reactor to streamline the product's separation and purification process and reduce waste generation. More reactant molecules will be able to collide with the catalyst when the surface area is big, speeding up the reaction. In many different sectors, both homogeneous and heterogeneous catalysts are useful and play significant roles. The particular reactions, selectivity, intended product, and financial considerations are the primary determinants of the catalyst selection. The ability to be readily filtered out at the end of the process is one benefit of heterogeneous catalysts [8].

3.2.1.3 Environmental Impacts of Conventional Catalyst

Although the catalyst system's effects on the environment can differ, there are a few important characteristics that are more frequently linked to its use. Heavy metals like palladium, platinum, and nickel that are included in catalysts have the potential to leak into the environment. These metals have the potential to pollute water sources, endangering aquatic life as well as human health. Additionally, it pollutes the soil, which impacts plant development and may cause it to enter the food chain [9].

Particulate matter produced by specific catalytic processes has the potential to pollute the air and cause respiratory issues. Volatile organic molecules are released into the environment by certain catalysts. Certain catalysts have the potential to produce hazardous waste that needs to be disposed of carefully. To avoid contaminating the environment, catalysts that are no longer in use or inactive must be disposed of carefully. Long-term ecological effects and accumulation may result from certain catalysts' prolonged environmental persistence. It can generate various chemical waste products which include solvents and other byproducts [10]. High temperatures and pressures are necessary for some catalytic reactions, which need a large amount of energy. They raise the chemical process's carbon footprint. Fossil fuels, which contribute to greenhouse gas emissions and climate change, are frequently the source of this energy.

3.2.2 Challenges of Traditional Catalytic Methods

Traditional catalysts are materials that were employed to speed up chemical reactions in the past. By reducing the activation energy needed for the reaction to proceed, they can be metals, metal oxides, or typically inorganic compounds that offer a surface or area for the reactants to interact. It may be obtained by a straightforward chemical process or from natural sources. While not all conventional catalysts are traditional, all traditional catalysts are conventional. By accelerating reactions without being eaten, it has transformed a variety of industries, including the chemical, food, automotive, and petroleum refining sectors, among many others [11].

Even though traditional catalytic methods are essential for many processes, they have a number of disadvantages. Among its many challenges is the deactivation of the catalyst through sintering, cocking, or poisoning. When impurities from the feedstock connect to the active sites, they can reduce the catalyst's activity. The catalyst particles may clump together as a result of high sintering temperatures, reducing their surface area and catalytic activity. When carbon is deposited on the surface, it blocks the active sites and hinders mass movement. Selectivity issues include product distribution and adverse reactions [12].

Catalysts commonly initiate a variety of reactions that frequently result in undesirable byproducts. Controlling product distribution can also be challenging, especially for complex processes. Heat transfer and diffusional resistance are barriers to mass transfer. Slow diffusion of reactants and products from the active sites can limit the overall reaction rate. Controlling heat generation inside the catalyst is crucial to preventing deactivation and enhancing performance. Many catalytic reactions require high temperatures and pressures, which consume a lot of energy. Deactivated catalyst regeneration can be time-consuming and energy-intensive [12]. Because they include valuable metals, the majority of catalysts, especially those utilised in precision chemical synthesis, are costly. Catalyst recovery and recycling can be challenging, particularly for complex systems. It affects the environment as well. Because catalyst waste may be toxic to the environment, it must be disposed of appropriately. The energy-intensive nature of catalytic processes may contribute to greenhouse gas emissions. Catalysts are necessary for many industrial processes, but if they are not handled properly, they can potentially have negative environmental effects. The mining and exploitation of these metals can have detrimental impacts on the environment, resulting in greenhouse gas emissions, habitat destruction, and water contamination. Additionally, catalysts have the ability to release harmful metals into the environment, contaminating soil and water sources and endangering human health and ecosystems [13].

3.3 Innovative Catalytic Materials

Future applications of different catalytic processes depend on the creation and synthesis of new and affordable catalytic materials. Advances in metal-free catalysts and biocatalysts, heterogeneous versus homogeneous catalysts in green applications, and biomass-derived and renewable catalysts are examples of unique catalytic materials.

3.3.1 Biomass Derived and Renewable Catalysts

The organic mass produced directly by photosynthesis is referred to as "biomass." Known as biomass, photosynthesis stores solar energy in vegetative products like grass, plants, trees, and so on. This biomass produces unconventional energy and is renewable. These days, biomass energy can be used directly for burning and indirectly for the production of biofuels instead of fuels, making it a renewable energy source. Since solid catalysts derived from biomass are non-toxic, non-corrosive, and prevent effluent production, they offer an environmentally beneficial substitute. Additionally, it is mostly made from biomass, which is thought to be an inexpensive and easily accessible commodity [14].

Because of a range of agricultural techniques, biomass is a readily available material that is plentiful in nature. Due to its affordability and accessibility, biomass has recently drawn interest as a potential source for the manufacturing of biodiesel. Biomass is mostly composed of carbon compounds that contain cellulose, hemicellulose, and lignin. Because of these characteristics, researchers employ a range of methods to exploit biomass of high-value added materials. The structure of biomass char is asymmetrical, and its pores are enormous. Pyrolysis, direct combustion, hydro gasification, gasification, anaerobic digestion, liquefaction, trans-esterification, esterification, and alcoholic fermentation are the primary methods for producing energy from biomass. The sulfonated catalysts are made from biomass using a number of techniques [15].

The biomass has high levels of carbon and oxygen but low levels of calcium, potassium, magnesium, and salt. Sulfonated carbon catalysts can generally be made by two processes: pyrolysis and hydrothermal carbonisation (HTC). Heterogeneous and homogeneous catalysts are the two categories of catalysts generated from biomass [16]. They are as follows:

- Homogeneous catalysts are substances with specific catalytic activity that are based on lignin, cellulose, or functionalised sugars.
- Heterogeneous catalysts: To increase catalytic activity, carbon-based catalysts are activated carbons derived from biomass that have been doped with metals or heteroatoms like nitrogen, phosphorus, or sulphur. The process of pyrolysing biomass produces biochar catalysts, which are then often functionalised to enhance their catalytic properties. To improve recyclability and reduce metal

waste, metal oxide nanoparticles are created via biogenic processes or bio templates [18].

3.3.2 Homogeneous Versus Heterogeneous Catalysts in Green Applications

Although heterogeneous and homogeneous catalysts differ substantially in their properties, modes of operation, and suitability for certain processes, both are necessary for environmentally acceptable applications. Nowadays, a lot of biodiesel is produced industrially using homogeneous catalysts. A homogeneous base catalyst, like potassium hydroxide (KOH) or sodium hydroxide (NaOH), provides a number of benefits, such as rapid reaction times, moderate operating conditions, high catalytic activity, and widely accessible and reasonably priced raw materials. Water and free fatty acids (FFA) have a major effect on the homogeneous base catalyst process. For example, the saponification and neutralisation processes used in the manufacturing of soap. The production process's intricate extraction and purifying procedures raise operating expenses. Additionally, a considerable amount of effluent is produced. This catalyst proves to be environmentally fatal in this way [17].

Naturally very acidic and reactive are the homogeneous acid catalysts. Product separation and purification in a homogeneous operation necessitate several steps. This raises operating expenses and produces a lot of waste water. Homogeneous catalyst recovery and recycling are difficult, expensive, impractical, and require extra processing. Research on the use of solid or heterogeneous catalysts in the production of biodiesel has significantly increased in recent years. Heterogeneous catalysts and reactants exist in multiple states or phases. When these catalysts combine with their reactants, they frequently create active sites [17].

Heterogeneously catalysed methanolysis is a very complicated process since it takes place in a three-phase system that includes a solid heterogeneous catalyst and two immiscible liquid phases, like methanol and oil. Apart from methanolysis, other methods include the neutralisation of FFAs with the help of a catalyst and the saponification of glycerides and methyl esters [19]. Since the catalysts are not broken down or destroyed in the reaction mixture, it is easier to remove them from the final product. Additionally, less catalyst is required overall, which lowers costs, because the recovered catalyst may be used again in the process. Among its many benefits are the heterogeneous-based operation's longer cost endurance, ease of separation, and noncorrosive nature. Heterogeneous catalysts include alkaline earth metals such as MgO and SrO. CaO is most commonly used as a catalyst for transesterification. There are however certain disadvantages, such as the fact that heterogeneous catalysis requires greater temperatures and oil/alcohol ratios than homogeneous catalysis [19].

Aluminium oxides, such as Al_2O_3, supported by MgO and CaO, are employed as heterogeneous base catalysts. Several naturally available calcium sources, including eggshell, mollusc shell, and bones, are frequently employed as the starting materials for the production of catalysts because they can reduce wastes and create catalysts with a high degree of cost competitiveness. Catalysts for basic-catalyzed processes include metal oxide, mixed oxide, and hydrotalcite. On the other hand, various forms of carbon, zeolites, ion-exchange resin, and transition metal oxides act as catalysts for acidic activity [20].

3.3.3 Advances in Metal-Free Catalysis and Biocatalysts

One significant effort to reduce or even eliminate the usage of catalysts based on precious metals has been the development of non-precious metal catalysts for bifunctional or multifunctional catalysis. Thus far, electrocatalysis has showed promise for transition metal sulphides, selenides, nitrides, carbides, phosphides, borides, oxides, hydro oxides, single atoms, and their alloys or complexes. Among these metals are Ni, Co, and Fe. Since the discovery of the first carbon-based metal-free electrocatalyst (VA-NCNT), research on carbon-based metal-free catalysts (C-MFC) has expanded internationally [21].

Compared to traditional metal-based electrocatalysts, C-MFCs offer a number of additional benefits in addition to cost savings. C-MFCs have superior stability, activity, and selectivity compared to conventional metal-based electrocatalysts. They are free from metal poisoning and dissolution and can be widely altered due to their broad surface chemistries. C-MFCs can have several catalytic roles by adding structural defects or co-doping to generate a spectrum of co-existing active sites, whereas metal-based catalysts find this difficult, if not impossible. This important benefit makes it possible to create novel C-MFCs that can catalyse multiple electrochemical reactions at once [21].

Because it enables the production of products that cannot be produced chemically, the use of substitute raw materials, reduced operating costs, low fixed-cost infrastructure, and enhanced eco-efficiency, biocatalysis is especially alluring to industry. The conference schedule includes a special lecture session on enzyme cascades, which are particularly interesting since they do not require intermediary purification and separation, which reduces waste, manufacturing costs, and operation time. The development of primary taste determinants in hopped beer and the synthesis of opioids in yeast employing 21–23 plant enzymes are recent examples of cascades that have been effective. The topic of biocatalysis is advancing due to the emergence of biological processes and continuous advancements in enzyme engineering. Furthermore, it is anticipated that more recent structural methods will shed light on enzyme catalysis [22].

3.4 Energy-Efficient Catalytic Processes

The Swedish chemist Berzelius initially proposed the idea of catalysis in 1835. Later in 1894, Ostwald developed this concept further, characterising catalysis as the process by which a chemical reaction is accelerated by a foreign material. In essence, catalysis is the process by which a catalyst lowers the energy needed for a chemical reaction to occur, hence accelerating the reaction. The catalyst does not change during the process; instead, it adheres to the substrate. The term "catalysis at ambient conditions" describes a variety of chemical reactions that are aided by catalysts and take place at room temperature and pressure. Ambient circumstances simplify the process and reduce energy consumption. Numerous biological activities are catalysed by enzymes, which can speed things up by 20 times compared to their uncatalyzed counterparts. Catalysis is more significant and one of the main drivers of social progress in our society due to environmental concerns [23].

A chemical reaction is catalysed by photocatalysis, a process in which light activates the catalyst. Electronic excitation is a characteristic of photochemical reactions. On the other hand, electrocatalysis speeds up the chemical reaction by using electrical energy. In green chemistry, photocatalysis and electrocatalysis are crucial for ecologically friendly chemical processes and sustainable development. One of the most significant areas of green chemistry is photocatalysis and electrocatalysis, which offer atom-economical, selective, and energy-efficient answers to numerous significant industrial issues. A chemical process that uses less energy is referred to as a low energy catalytical pathway [24].

3.4.1 *Catalysis at Ambient Conditions*

Chemical processes that are aided by catalysts that take place at room temperature and pressure are referred to as ambient catalysis. Ambient conditions simplify the procedure and reduce energy use. The Arrhenius equation can be used to describe activation energy, which is mostly how catalysis works.

$$k = A.e^{(-Ea/RT)} \tag{3.1}$$

where Ea is the activation energy, A is the high temperature limit of k, also known as the pre-exponential factor, R is the gas constant, T is the absolute temperature, and k is the rate constant of an activated process of any type of reaction. A straight line with (-Ea)/RT as the slope and ln (A/unit) as the intercept is produced when the Eqn. (3.1) is plotted in its logarithmic form as a function of 1/T [23].

3.4.1.1 Enzymatic Catalysis (Biocatalysis)

The chemical, pharmaceutical, and organic synthesis industries, among others, now heavily rely on biocatalysis. Numerous chemical processes require enzymes, such as the production of acrylamide by nitrile hydratases and the addition of enzymes to detergents. For biocatalysis, enzymes that are involved in the production of specific metabolites, or natural products, are very helpful as beginning points. Methyltransferases, Diels-Alderases, halogenases, uridine diphosphate-dependent glycosyltransferases, laccases, and pyridoxal phosphate-dependent enzymes are examples of natural product biosynthetic enzymes that are employed in chemical synthesis [25].

3.4.1.2 Homogenous and Heterogenous Catalysis

A process known as heterogeneous catalysis occurs when the catalyst and reactants are in distinct stages. A functional substance that continuously forms active sites with its reactants under reaction circumstances is known as a heterogeneous catalyst [26].

Oxides catalyse oxidation events, while sulphide catalysts catalyse desulphurization reactions. Dehydration is catalysed by insulator oxides, and acid/base solids interact with carbocation/carbanionic intermediates in processes. Oil refinery, energy and transportation, bulk chemicals, polymers and detergents, textiles, fine chemicals, catalyst development, and catalytic processes are the main industrial applications of heterogeneous catalysis. Homogeneous catalysis occurs when the catalyst and reagent are in the same phase. While homogeneous catalysts dissolve in the reaction medium, making all of their catalytic sites available for reaction, heterogeneous catalysts are insoluble in the reaction medium during the reaction, allowing gaseous or liquid reagent reactions to occur at the surface. For instance, a chiral diphosphine ligand and a rhodium(I) complex catalyse the asymmetric hydrogenation of prochiral olefins [27].

3.4.1.3 Organocatalysis

Lewis bases, Lewis acids, Brønsted bases, and Brønsted acids are the four categories of organo-catalysts. By adding or subtracting protons or electrons from a substrate or transition state, these catalysts start their catalytic cycles. Lewis acid catalysts like carbonyl compounds are infrequently employed in organo-catalysis, but Lewis base catalysts like amines and carbenes predominate. The application of chiral organic Brønsted acid catalysts is a very recent idea [28].

3.4.2 *Photocatalysis and Electrocatalysis in Green Chemistry*

Since photocatalysis uses photons to activate the catalyst and runs at ambient temperature without the requirement for thermal energy, it is very interesting to green and environmental chemists. Simple oxides like TiO_2, doped oxides, sulphides, etc., are examples of photocatalysts. Photocatalysis is polyphasic, consisting of solid catalysts, liquid or gaseous reactants, and electromagnetic photons from light or UV rays. Applications for photocatalysis are numerous and include organic synthesis, the recovery of contaminated wastewater, and water hydrolysis for the production of hydrogen as fuel. Electronic excitation is a characteristic of photochemical reactions. Chemical molecules that are challenging to generate using other traditional methods of organic synthesis have their chemical reactivity significantly altered by electronic stimulation. Photooxygenation is one of the best instances of a combination of green and photoorganic chemistry [29]. Combining photocatalysis with enzyme catalysis can simplify and enhance it. By offering atom-economical, selective, and energy-efficient solutions to numerous industrially significant problems, photocatalysis continues to be one of the most significant areas of green chemistry [30].

Electrical energy is used in electrocatalysis to speed up chemical reactions. Green chemistry is greatly advanced by electrocatalysts, which aid in sustainable and effective chemical reactions. The Swedish chemist Berzelius initially proposed the idea of catalysis in 1835. Later, in 1894, Ostwald developed this concept further, characterising catalysis as the process by which a chemical reaction is accelerated by a foreign material. It is essential for energy conversion and storage technologies such as fuel cells, batteries, and electrolysers. Moreover, electrocatalysis can reduce waste and reliance on fossil fuels by promoting the environmentally friendly production of chemicals and breaking down pollutants such as organic compounds and heavy metals [31].

3.4.3 *Low Energy Catalytic Pathways*

Chemical reactions that need less energy and are aided by catalysts are referred to as low-energy catalytic pathways. Energy-efficient and sustainable processes depend on these pathways (Fig. 3.2).

- Enzymatic Catalysis: This process accelerates chemical reactions by using enzymes as catalysts. Classic examples for such catalysts include methyltransferases, Diels-Alderases, halogenases, glycosyltransferases [32].
- Organocatalysis: This technique uses tiny organic molecules as catalysts to increase the rate and effectiveness of chemical reactions. A chiral phosphoric acid catalyst can be used to asymmetrically hydrogenate prochiral olefins, a common use of organocatalysis that results in extremely efficient and selective reactions [27].

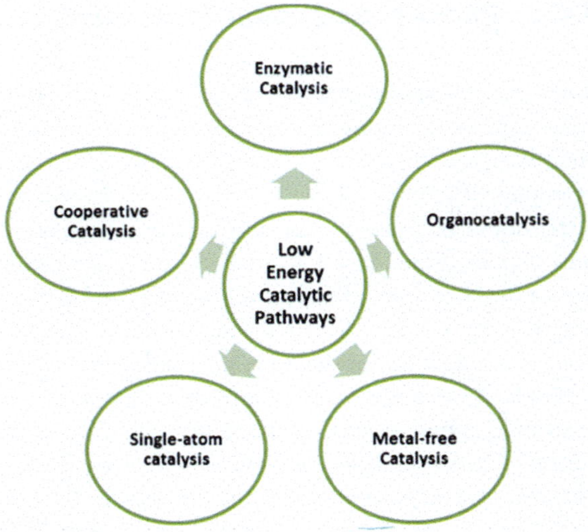

Fig. 3.2 Visualising low energy catalytic pathways for enhanced reactivity

- Metal-free Catalysis: A class of catalytic processes known as "metal-free catalysis" uses non-metallic materials, such as carbon-based materials, to promote chemical reactions and convert materials. The hydrogenation of carbon dioxide to formic acid, which may be effectively catalysed by a graphene-based material, is a noteworthy example of metal-free catalysis and demonstrates the potential of non-metallic catalysts in sustainable chemistry [33].
- Single-atom Catalysis: This innovative method uses individual metal atoms as catalysts to achieve previously unheard-of levels of efficiency and selectivity. The hydrogenation of carbon dioxide to methanol, which may be accomplished using a copper-based single-atom catalyst, is a powerful illustration of the enormous potential of this novel technique [31].
- Cooperative Catalysis: This synergistic method uses the combined strength of several catalysts to speed up chemical reactions, frequently resulting in increased productivity, efficiency, and selectivity [31].

3.5 Waste Reduction and Byproduct Management in Catalysis

The key components of catalysis that will result in resource efficiency and environmental sustainability are waste reduction and by-product management. Additionally, catalysis has numerous industrial, environmental, and economic benefits. Because catalysis helps to improve the efficiency of chemical processes by lowering the creation of unwanted by-products that are detrimental to the environment, it promotes environmental stability by minimising pollution [34]. Waste reduction is linked to green chemistry principles, which seek to lessen the impact of chemical manufacturing on the environment. Economic advantages include reduced production costs, increased product yields, and cost savings. Industries can avoid fines and enhance their market standing by adhering to waste and emission standards. High selectivity is emphasised in catalyst design advancements to avoid waste, and reusable catalysts lower material requirements [34].

3.5.1 Minimising Toxic By-products

The environment and human health are seriously threatened by toxicants produced in several industrial sectors. The degree of hazard depends on the kind and length of exposure. Heavy metals like lead, mercury, and organic compounds can contaminate water, upsetting aquatic ecosystems and leading to bioaccumulation and biodiversity loss. Greenhouse gases like methane and nitrous oxide contribute to climate change, while toxic waste degrades soil by reducing fertility and damaging essential microorganisms [35]. Inhaling contaminants like carbon monoxide can cause cardiovascular and respiratory issues in people. Phthalates impair reproductive health, and toxic byproducts like benzene and dioxins can cause cancer. Chronic sickness will result with prolonged exposure. Hazardous by-products should be minimised in order to safeguard human health and the environment. Create effective and selective catalysts to cut down on unwanted reactions and combine many steps into a single process to save energy and waste [35].

3.6 Role of Catalysis in CO_2 Utilization and Carbon Capture

Because carbon emissions have detrimental effects and are intimately linked to global warming, they are a major problem for the environment. The main source of carbon emissions brought on by human activity is carbon dioxide. A treatment approach that involves capturing, storing (either permanently or temporarily), or using CO_2 from different point sources, including industry, is one of the objectives for stabilising CO_2 levels. Three basic steps make up CO_2 treatment strategies: catch, concentrate, and use or store for a short or extended period [36].

3.6.1 Catalytic Pathway for CO_2 Conversion

Alcohols, acids, hydrocarbons, polymers, carbon monoxide, carbon nanotubes, liquid fuels, fertiliser, polymers, and secondary chemicals are all products of the chemical conversion of carbon dioxide. Since CO_2 is thermodynamically stable and requires an energy-intensive procedure to convert, catalysts are needed to overcome kinetic barriers [36].

3.6.1.1 CO_2 Conversion by Thermal Catalysis

The most inventive catalytic process for turning CO_2 into chemicals and fuels is the thermochemical pathway. This process uses a high temperature and a catalyst to break through the strong stability of CO_2 and convert it into products like methanol, syngas, and hydrocarbons.

$$CO_2 + H_2 \rightarrow CO + H_2O \tag{3.2}$$

The RWGS (Reverse Water Gas Shift) reaction, which is the process that turns CO_2 into CO, is represented by Eqn. (3.2). Using a catalyst and a high temperature—typically above 500 °C—CO_2 and H_2 are transformed into CO and H_2 in the reaction described above. A common tactic to boost conversion is the use of significant extra CO_2 or H_2 or the continuous removal of products. Usually, copper, iron, nickel, or ceria are used as catalysts.

Carbon monoxide (CO), which is thus created, is used as a crucial feedstock in the synthesis of numerous chemicals and fuels that we use on a daily basis. H_2 is created when water is electrolysed by electricity generated by solar cells or wind turbines. Methane, which can be used as fuel, is then created by the reaction of H_2 and CO_2. Ru, Rh, Ni, and Pd are the catalysts used in the reaction, and their temperatures range from 200 to 300 ºC. The resulting methanol can be utilised as a chemical feedstock for polymers, solvents, and other products, as well as fuel [37].

3.6.1.2 Carbon Dioxide Conversion by Electrochemical Catalysis

The reduction of CO_2 is accomplished using electrocatalysis, in which the reaction is powered by electricity. Reducing atmospheric CO_2 and storing energy can generally be achieved through the electrocatalytic conversion of CO_2 to products like methane, ethylene, or formic acid. employing water as the proton donor and an external electric field as the energy source for the carbon atom's nucleophilic assault. Cu, Ag, Au, and other catalysts are employed. Water serves as the proton donor in electrocatalytic conversion rather than H_2. This includes recyclable catalyst and electrolyte, normal temperature and pressure, and adjustable reaction conditions and rate. Compared to thermal catalysis, the reaction is more cost-effective because it operates with basic equipment [38].

3.6.1.3 Photocatalytic Conversion of CO_2

When the heterogeneous photocatalysts are exposed to sunlight, electron–hole pairs are produced, simulating natural photosynthesis. TiO_2, ZnO, and perovskites are semiconductor materials that have been employed as photocatalysts. The redox reaction that the photogenerated electrons cause in CO_2 produces CO, formic acid, hydrocarbons, and alcohol. It happens in three stages: charge separation and transfer, catalytic reduction of CO_2, oxidation of H_2O, and absorption of sunlight by the photocatalyst or catalyst and photosensitiser [39].

3.6.2 Green Catalysts for Carbon Capture and Storage (CCS)

- Alkali and Alkaline Earth Metal Catalysts: They are made from naturally occurring metals that are plentiful in the periodic table's s- block. For instance, potassium, sodium, calcium, magnesium. They are employed in mineralisation reactions and CO_2 absorption. They also make procedures like air capture less expensive [40].
- Carbon-Based Catalysts: Examples of carbon-based catalysts utilised in a range of reactions and industries include graphene, graphene oxide, carbon nanotubes, and carbon nanofibers. To absorb CO_2, solid adsorbents such as charcoal, graphene oxide, or activated carbon (AC) are employed [41].
- Catalyst based on enzymes: One enzyme that acts as a catalyst in the capture and storage of carbon is carbonic anhydrase (CA). It functions in mild weather [42].
- Amine-functionalized sorbents: They are solid supports functionalised with amines that have a lower heat capacity, are tuneable, and chemically absorb CO_2. They may also be reversed, making them environmentally benign [43].
- Metal–Organic Frameworks (MOFs): These are the crystalline materials composed of organic ligands and metal clusters that combine to create a porous structure. MOFs based on aluminium, cerium, chromium, and other elements are a few examples [44].

3.7 Industrial Applications of Green Catalytic Innovations in Pharmaceutical, Petrochemical and Fine Chemical Industries

When used properly, green catalysts are a sustainable development tool that aids the petrochemical, fine chemical and pharmaceutical industries in meeting their environmental objectives. Green chemistry, fundamentally involving the conceptualization of chemical products and processes that significantly reduce or completely eliminate the utilization and generation of hazardous substances, empowers industries to develop superior products through environmentally sustainable methodologies [45].

The majority of sectors typically use one of two green catalytic processes. For example, biocatalysis is one of the primary green chemistry technologies used by the pharmaceutical, petrochemical, and fine chemicals industries to produce compounds with improved yields. When it comes to implementing green catalytic processes and designing synthetic routes to integrate them, several of these industries are typically deemed inadequate. Therefore, a number of studies and inventions are underway to enhance synthetic routes in order to safeguard the environment and promote economic growth [46].

Chemocatalysis has advanced to the point where it is now widely employed in production. It encompasses both organocatalysis and several types of metal catalysis. The primary goal of the pharmaceutical industry is to lower disease rates in order to raise patient standards of living. Typically, pharmaceutical companies generate more waste per kilogramme of product than other businesses like bulk, petrochemical, fine chemical, and polymer companies [47].

Two measures of the amount of waste produced by chemical processes were primarily adopted by the chemical industries. The PMI (unit of raw materials utilised per unit) and the E-factor (unit of waste generated per unit of product) are the two. To get a desired product and to safeguard the environment, both of these values should be lower. For a sustainable development, this is the aim of any industry. Green catalytic approaches can be used to accomplish this. China and India are at the forefront of advancing green chemistry in the manufacturing of petrochemical products, fine chemicals, and pharmaceuticals [48].

3.7.1 Pharmaceutical Industry

The pharmaceutical sector has implemented green catalytic innovations to improve the sustainability, environmental friendliness, and cost-effectiveness of drug development and production processes. These advances primarily involve lowering waste or byproducts, using no toxic solvents or reactants, and using less energy [49].

- Biocatalysis: The pharmaceutical sector relies heavily on enzymes. When synthesising active pharmaceutical ingredients (APM), enzymatic techniques can boost sustainability; it's crucial to employ proven tools to measure their influence. Water is frequently used as the reaction medium in enzyme or cell-based synthetic processes, which are usually conducted at room temperature and in mild circumstances. The great selectivity of enzymes eliminates the requirement for protection/deprotection procedures and functional group activation, which are often necessary in conventional chemical synthesis. Enzymes are already scalable since they eliminate the requirement to produce a salt with several chiral chemicals, boosting the yield from 50 to 100% [50].
- Flow chemistry: A sustainable production technique with great potential for scaling up APIs used in continuous manufacturing processes is flow chemistry. Compared to other techniques, flow chemistry enables better reactions as well

as faster rates of mass and heat transfer, resulting in safe and cost-effective operations. Flow chemistry allows for an even greater degree of reactor design adjustment. Each of these elements helps to reduce waste production and increase productivity [51].

- Parallel screening: API molecules are typically novel intermediates, and documented processes never function as intended. Therefore, it could be necessary to alter or modify current circumstances in order to make them more feasible. When properly designed and used, parallel reaction screening saves a significant amount of time and offers timely actions throughout the research process. Numerous reactions can be carried out concurrently through screening and automation in processes like weighing, quenching, and analysis. Many green catalytic developments are currently being used in the pharmaceutical industry to eliminate waste products and contaminants [52].

3.7.2 Fine Chemistry Industry

The contract manufacturing organisations (CMOs) that produce APIs for major pharmaceutical companies are commonly referred to as the fine chemical industry. Numerous significant industrial sectors rely on the valuable goods produced by the fine and speciality chemical industry. Primarily, the food, beverage, pharmaceutical, flavour, and fragrance industries have more demand worldwide and want eco-friendly products. Thus, catalytic advances in fine chemicals and green production are popular. Successful immobilisation and utilisation of catalysts under conditions of continuous, increased flow will enable the production of multifunctional goods that are necessary for both developed and emerging economies. Metal nanoparticles, organic molecules, enzymes, single atoms, and molecular metal complexes are examples of immobilised catalytic species. There are several financial and environmental benefits if these catalytic materials exhibit consistent catalytic activity [53].

It is impossible to overstate the significance of heterogeneous catalysis to the fine chemical industry. Solid catalysts that are stable and selective are perfect for continuous processes that can be readily scaled up to create the required amounts of fine chemicals or pharmaceuticals. In the synthesis of chiral and nonchiral compounds with new innovations pertinent to industries, homogeneous organometallic catalysts are frequently used in fine chemicals for processes ranging from asymmetric hydrogenation to carbon-carbon and carbon-nitrogen cross-coupling reactions. The thermally sensitive catalyst needs to be removed from the product once the reaction is complete. In the fine chemical industries, the sole heterogeneously catalysed reaction that uses heterogenized metal nanoparticles is catalytic hydrogenation over supported metal nanoparticles. These are extensively used in research and development as well as in the manufacturing of vitamins, nutraceuticals, and organic processes [53].

3.7.3 Petrochemical Industry

The petroleum sector, which is known for its complexity and diversity, regularly deals with issues including gas leaks, pipe corrosion, and energy loss during reactions brought on by subpar catalysts. All of which make handling necessary products more challenging. The distinctive characteristics of nanomaterials set them apart from conventional materials from a scientific standpoint. Changes in pore volume and surface area at the nanoscale modify their basic characteristics, giving catalysts more active sites and increasing their effectiveness in petrochemical processes [54].

Enhancing the cracking and reforming processes of hydrocarbons is one of the main functions of catalysts in molecular metamorphosis. Large hydrocarbon molecules are broken up into smaller, more valuable ones by a process called cracking. Zeolite-based catalysts are used in catalytic cracking, where molecules are selectively broken down to produce high-demand petrochemical products like olefins and petrol [55].

3.8 Future Perspectives

Enhancements in the composition and efficacy of catalytic materials are imperative for addressing various contemporary challenges, including the advancement of green fuel technologies and the elimination of ecologically detrimental chemical and pharmaceutical procedures. Porous materials that consist of organic linkers interconnected by metal ions serve as framework scaffolds for heterogeneous catalysis predicated on organic or inorganic constituents. An array of methodologies may be employed to functionalize the former or to generate coordinatively unsaturated metal sites. In the realm of C-H bond activation, a category of materials known as meta-organic frameworks (MOFs) is currently drawing significant scholarly attention [56]. Single-atom catalysts (SACs) possess the capacity to fabricate atom-efficient, precisely defined, and strategically designed catalysts with active sites tailored for reaction selectivity, thereby enhancing the efficiency and environmental sustainability of numerous chemical transformations prevalent in contemporary society. To fully realize the potential of SACs, it is essential to comprehend their atomic-scale architecture and chemistry, alongside the fundamental physicochemical behaviors inherent to single-atom tailored catalytic systems [57].

The development of novel magnetic materials to facilitate Ortho-Para conversion for hydrogen gas and liquid, aimed at long-term energy storage, also garners considerable scholarly interest. To appreciate the catalyst's compositional and morphological evolution, one must grasp the atomic-level magnetic interactions that occur between the hydrogen molecule and its magnetically engineered open surface. As the global community endeavors to attain a more sustainable future, the domain of green catalysis has emerged as a potent remedy [58]. Among the most promising materials in the field of catalysts—integral to fostering environmentally benign and sustainable chemical reactions—is zirconia. Zirconia, chemically recognized as zirconium dioxide, serves as an exemplary catalyst for green chemistry attributable to its distinctive amalgamation of properties. Its intrinsic stability and resistance to chemical degradation empower it to endure rigorous reaction conditions without undergoing disintegration. The substantial specific surface area of zirconia, which offers a plethora of active sites for molecular interactions, further facilitates effective catalytic reactions. Due to these attributes, zirconia catalysts demonstrate resilience against fouling from contaminants frequently present in impure feedstocks, thereby ensuring their sustained efficacy [59].

The longevity and recyclability of catalysts present additional challenges, particularly in industrial contexts where prolonged stability is paramount. The discipline of green catalysis is anticipated to expand significantly as artificial intelligence (AI) and machine learning (ML) technologies are integrated into the design and optimization of catalysts. These advanced technologies assist in expediting the discovery of novel catalysts while simultaneously reducing the temporal and financial expenditures associated with experimental investigations through the estimation of ideal configurations and mechanistic routes [60].

3.9 Conclusion

Green catalysis represents a pivotal advancement in the domain of sustainable chemistry, providing innovative solutions to pressing environmental and energy-related challenges. The development of environmentally benign catalytic methodologies has attained heightened importance as global industries endeavor to mitigate their ecological footprint. Green catalysis, which entails the formulation and utilization of catalysts that facilitate chemical reactions, is congruent with the principles of green chemistry, which prioritize the elimination of hazardous substances and the innovation of safer, more sustainable industrial practices. Recently, green catalysis has transformed numerous sectors, including materials science, energy production, and pharmaceuticals, by presenting safer and more efficient alternatives to conventional chemical methodologies. Catalysts play an essential role in expediting reactions while diminishing the necessity for superfluous raw materials and toxic solvents. Advancements within this sphere have ushered in new avenues for sustainable manufacturing techniques, such as the utilization of renewable feedstocks and bio-based catalysts. Given their capacity to reduce pollution and waste generation, enhance atom economy to maximize reactant conversion into the desired product, and facilitate the recovery and reuse of products or emissions from catalyzed reactions, green catalysts are deemed environmentally advantageous.

References

1. Linthorst, J.A.: An overview: origins and development of green chemistry. Found. Chem. **12**, 55–68 (2009)
2. Anita, I., Ana, D., Anita, M.B., Stanislava, T.: Review of 12 principles of green chemistry in practice. Int. J. Sustain. Green Energy **6**, 39–48 (2017)
3. Peralta, Y.M., Molina, R., Moreno, S.: Rice HUSK silica: a review from conventional uses to new catalysts for advanced oxidation processes. J. Environ. Manage. **370**, 122735 (2024)
4. Melnyk, N., Iribarren, I., Mates-Torres, E., Trujillo, C.: Theoretical perspectives in organocatalysis. Chemistry **28**, e202201570 (2022)
5. Wang, P., Ma, Y., Shi, Y., Duan, F., Wang, M.: Application of metal-based catalysts for semi-hydrogenation of alkynol: a review. Materials (Basel) **16**, 7409 (2023)
6. García-Álvarez, J.: Special issue: advances in homogeneous catalysis. Molecules **25**, 1493 (2020)
7. Feng, H., Zhang, Y., Liu, J., Liu, D.: Towards heterogeneous catalysis: a review on recent advances of depositing nanocatalysts in continuous-flow microreactors. Molecules **27**, 8052 (2022)
8. Zaera, F.: Designing sites in heterogeneous catalysis: are we reaching selectivities competitive with those of homogeneous catalysts? Chem. Rev. **122**, 8594–8757 (2022)
9. Centi, G., Ciambelli, P., Perathoner, S., Russo, P.: Environmental catalysis: trends and outlook. Catal. Today **75**, 3–15 (2002)
10. Ferenc, E.K., Milenko, J., Goran, C.B.: Economic and ecological aspects of biodiesel production over homogeneous and heterogeneous catalysts. Fuel Process. Technol. **91**, 1316–1320 (2010)
11. Mulvihill, M.J., Beach, E.S., Zimmerman, J.B., Anastas, P.T.: Green chemistry and green engineering: a framework for sustainable technology development. Annu. Rev. Environ. Resour. **36**, 271–293 (2011)
12. Yentekakis, I.V., Fan, D.: Grand challenges for catalytic remediation in environmental and energy applications toward a cleaner and sustainable future. Front. Environ. Chem. **1** (2020)
13. Daniel, J.H.: An examination of biorefining processes, catalysts and challenges. Catal. Today **145**, 138–151 (2009)
14. Belluati, M., Silvia, T., Emanuela, C.G., Giancarlo, C.: Maela Manzoli: biomass-derived carbon-based catalysts for lignocellulosic biomass and waste valorisation: a circular approach. Green Chem. **26**, 8642 (2024)
15. Wang, Y., Zhang, M., Shen, X., Wang, H., Wang, H., Xia, K., Yin, Z., Zhang, Y.: Biomass-derived carbon materials: controllable preparation and versatile applications. Small **17**, e2008079 (2021)
16. Sharifah, H.Y.S.A., Nur, H.M.H., Azman, A., Roslan, U., Hafizan, J., Helena, K., Azizah, E.: A review of biomass-derived heterogeneous catalyst for a sustainable biodiesel production. Renew. Sustain. Energy Rev. **70**, 1040–1051 (2017)
17. Rothenberg, G.: Catalysis: concepts and green applications. Wiley, US (2017)
18. Li, C., Yan, L. (eds.): Bridging heterogeneous and homogeneous catalysis: concepts, strategies, and applications. Wiley, US (2014)
19. Wang, K., Jonathan H., Nan A., Adelina V.: Homogeneous versus heterogeneous catalysts for acceptorless dehydrogenation of biomass-derived glycerol and ethanol towards circular chemistry. Green Chem. **26**, 3546–3564 (2024)
20. Kumar, A., Mohammed, A.H.S.S., Siham, Y.A.: Applications of heterogeneous catalysts in green chemistry. Catalysts **14**, 699 (2024)
21. Cai, M., Lei, X., Jinjia, G., Xiaobin, Y., Xuexia, H., Peng, H.: Recent advances in metal-free electrocatalysts for the hydrogen evolution reaction. J. Mater. Chem. A **12**, 592–612 (2024)
22. Sheldon, R.A., Brady, D., Bode, M.L.: The Hitchhiker's guide to biocatalysis: recent advances in the use of enzymes in organic synthesis. Chem. Sci. **11**, 2587–2605 (2020)

23. Schultz, B.J., Kim, S.Y., Lau, W., Sattely, E.S.: Total biosynthesis for milligram-scale production of etoposide intermediates in a plant chassis. J. Am. Chem. Soc. **141**, 19231–19235 (2019)
24. Fechete, I., Ye, W., Jacques, C.V.: The past, present, and future of heterogeneous catalysis. Catal. Today **189**, 2–27 (2012)
25. Marsh, C.O., Nicholas, R.L., Li-Chen, H., Matthew, J.B., Sbusisiwe, Z.M., Laurence, M., Sebastian, P., Philip, W.D., James, E.M.S., Paul, C., Christine, L.W., Paul, R.R.: A natural diels-alder biocatalyst enables efficient [4 + 2] cycloaddition under harsh reaction conditions. Chem. Cat. Chem. **11**, 5027–5031 (2019)
26. Fisher, B.F., Snodgrass, H.M., Jones, K.A., Andorfer, M.C., Lewis, J.C.: Site- selective C-H halogenation using flavin-dependent halogenases identified via family-wide activity profiling. ACS Cent. Sci. **5**, 1844–1856 (2019)
27. Anastas, P.T.: Introduction: green chemistry. Chem. Rev. **107**, 2167–2168 (2007)
28. List, B.: Introduction: organocatalysis. Chem. Rev. **107**, 5413–5415 (2007)
29. Schlögl, R.: Heterogeneous catalysis. Angew. Chem. Int. Ed. **54**, 3465–3520 (2015)
30. Michelin, C., Norbert, H.: Photocatalysis applied to organic synthesis—a green chemistry approach. Curr. Opin. Green Sustainable Chem. **10**, 1–6 (2018)
31. Das, N., Nupur, O., Sanjeeb, K.M.: Wastewater treatment using plant-derived bioflocculants: green chemistry approach for safe environment. Water Sci. Technol. **83**, 1797–1812 (2021)
32. Schramm, V.L.: Introduction: principles of enzymatic catalysis. Chem. Rev. **107**, 1301–1302 (2007)
33. Jamesh, M., Arumugam, A., Dhakshinamoorthy, S., Karunanidhi, G.P., Vetrivel, S.: Revathi: fabrication of earth-abundant electrocatalysts based on green-chemistry approaches to achieve efficient alkaline water splitting—a review. Sustain. **14**, 16359 (2022)
34. Marafi, M., Stanislaus, A.: Spent catalyst waste management: a review: Part I—developments in hydroprocessing catalyst waste reduction and use. Resour. Conserv. Recycl. **52**, 859–873 (2008)
35. Pandian, A.M.K., Rajamehala, M., Singh, M.V.P., Sarojini, G., Rajamohan, N.: Potential risks and approaches to reduce the toxicity of disinfection by-product—a review. Sci. Total. Environ. **822**, 153323 (2022)
36. Liotta, L.F., Hongjing, W: CO_2 capture, utilization and storage: catalysts design. Catalysts **14**, 80 (2024)
37. Yongjun, J., Kangzhou, W., Yuan, W., Zhihao, L., Xinhua, G., Jianli, Z., Qingxiang, M., Subing, F., Tian-Sheng, Z., Min, Y.: Recent advances in thermocatalytic hydrogenation of carbon dioxide to light olefins and liquid fuels via modified Fischer-Tropsch pathway, J. CO_2 Util. **67**, 102321 (2023)
38. Sajna, M.S., Sifani, Z., Anton, P., Peter, K., Ali, A., Udeogu, O., Miao, W., Hyunwoong, P., Dong, S.H.: Electrochemical system design for CO_2 conversion: a comprehensive review. J. Environ. Chem. Eng. **11**, 110467 (2023)
39. Zhiyan, F., Qi, Y., Zhan, L., Fei, C., Fubing, Y., Ting, X., Yu, Z., Dongbo, W., Jing, L., Xiaoming, L., Guangming, Z.: Photocatalytic conversion of carbon dioxide: from products to design the catalysts, J. CO_2 Util. **34** 63–73 (2019)
40. Tsiotsias, A.I., Nikolaos, D.C., Ioannis, V.Y., Maria, A.G.: The role of alkali and alkaline earth metals in the CO_2 methanation reaction and the combined capture and methanation of CO_2. Catalysts **10**, 812 (2020)
41. Ali, Z., Hammouda, S., Jie, S., Xiaolei, W., Xia, L., Zhi, C.: Carbon-based materials for CO_2 capture: their production, modification and performance. J. Environ. Chem. Eng. **11**, 109741 (2023)
42. Pragya, S., Rohit, S.: Utilization of bio-inspired catalyst for CO_2 reduction into green fuels: recent advancement and future perspectives. J. CO_2 Util. **53**, 101748 (2021)
43. Anish, M.V., Georgios, N.K.: CO_2 capture adsorbents functionalized by amine—bearing polymers: a review. Int. J. Greenh. Gas Control **96**, 103005 (2020)
44. Zhousheng, M., Saira, A., Mohammad, T., Anuj, K., Ghulam, Y., Wei, Z.: Metal-organic frameworks-based advanced catalysts for anthropogenic CO_2 conversion toward sustainable future. Fuel Process. Technol. **244**, 107705 (2023)

45. Messire, G., Emma, C., Sabine, B.: Green catalysts and/or green solvents for sustainable multi-component reactions. Catalysts **14**, 593 (2024)
46. Yi, D., Thomas, B., Christoffel, P.S.B., Shuke, W., Mark, D., Matthias, H., Uwe, T.B.: Recent trends in biocatalysis. Chem. Soc. Rev. **50**, 8003–8049 (2021)
47. Gröger, H., Fabrice, G., Bruce, H.L.: Where chemocatalysis meets biocatalysis: in water. Chem. Rev. **123**, 5262–5296 (2022)
48. Andraos, J.: Relationships between step and cumulative PMI and E-factors: implications on estimating material efficiency with respect to charting synthesis optimization strategies. Green Process. Synth. **8**, 324–336 (2019)
49. Hayler, J.D., David, K.L., Eric, M.S.: A pharmaceutical industry perspective on sustainable metal catalysis. Organometallics **38**, 36–46 (2018)
50. Rosenthal, K., Lütz, S.: Recent developments and challenges of biocatalytic processes in the pharmaceutical industry. Curr. Opin. Green Sustain. Chem. **11**, 58–64 (2018)
51. Bogdan, A.R., Amanda, W.D.: Emerging trends in flow chemistry and applications to the pharmaceutical industry. J. Med. Chem. **62**, 6422–6468 (2019)
52. Murugan, N.A., Artur, P., Davide, G., Emanuele, V., Gianluca, P., Stefano, M.: A review on parallel virtual screening softwares for high-performance computers. Pharmaceuticals **15**, 63 (2022)
53. Pollak, P.: Fine chemicals: the industry and the business. Wiley, US (2011)
54. Hassani, H., Emmanuel, S.S., Ahmed, M.A.K.: The role of innovation and technology in sustaining the petroleum and petrochemical industry. Technol. Forecast. Soc. Change. **119**, 1–17 (2017)
55. Sterba, M.J., Vladimir, H.: Catalytic reforming. Ind. Eng. Chem. Prod. Res. Dev. **15**, 2–17 (1976)
56. Holst, J.R., Abbie, T., Andrew, I.C.: Porous organic molecules. Nat. Chem. **2**, 915–920 (2010)
57. Cheng, N., Lei, Z., Kieran, D., Xueliang, S.: Single-atom catalysts: from design to application. Electrochem. Energy Rev. **2**, 539–573 (2019)
58. Yang, L., Xinbao, L., Ying, C., Xiaoling, Z., Kai, S.: Ortho to para hydrogen conversion over bimetallic iron and cobalt catalysts. Sci. Rep. **14**, 20925 (2024)
59. Tharp, W.F., Lekaa, K.A.K.: Green synthesis, characterization, antimicrobial and anticancer studies of zirconium oxide nanoparticles using thyme plant extract. Mor. J. Chem. **12**, 643–656 (2024)
60. Benavides-Hernández, J., Franck, D.: From characterization to discovery: artificial intelligence, machine learning and high-throughput experiments for heterogeneous catalyst design. ACS Catal. **14**, 11749–11779 (2024)

Chapter 4
The Sustainable Microbial Processes: An Emerging Trends for Green Chemical Approach

Anil Kumar Jangir and Ritu Jangid

Abstract Green chemistry protects our environment while making biological processes more sustainable including bioremediation, biofuel production and agriculture systems. This chapter looks at green chemistry's key part in microbe systems by lowering toxic substance usage and waste production while making their interactions more successful. It includes research on how green chemistry helps make sustainable microbial processes. Green chemistry's principles help microbes successfully replace hazardous chemicals because they process solar-powered materials that support environmentally friendly biology. Sustainable microbial processes work through varied natural processes that create minimal damage and produce less hazardous waste to meet green chemistry standards. By creating more efficient bio-based materials and reducing resource usage and waste this sustainable microbial approach supports the fundamental principles of the circular economy. A healthy microbiome functions better under green chemistry principles helping scientists make safe biodegradable materials and eco-friendly antibacterials which help protect our environment. Green chemistry practices support farmers by recommending natural pest control methods while urging reduced use of environmentally harmful fertilizers that harm soil organisms. This chapter demonstrates that eco-friendly chemical methods can preserve natural microbial processes that maintain environmental wellness. In creative applications, sustainable microbial techniques create better ecosystem health while encouraging their natural activities.

Keywords Sustainable microbial applications · Green chemistry · Biomaterials · Microbial defense systems · Natural pesticides

---

A. K. Jangir
Department of Chemistry, Shri Jagdishprasad Jhabarmal Tibrewala University, Vidyanagri, Jhunjhunu, Rajasthan, India

R. Jangid (✉)
Department of Science, Biyani Girls College, Jaipur, Rajasthan, India
e-mail: ritu.jangid010@gmail.com

P. Jain et al. (eds.), *Towards Green Chemical Processes: Strategies and Innovations*, Intelligent Systems Reference Library 277,
https://doi.org/10.1007/978-3-031-94090-3_4

4.1 Introduction

Researchers today use sustainable chemistry or "green chemistry" to develop products and processes that eliminate dangerous substances from the production cycle [1]. This scientific area advances rapidly because society demands safer and greener chemical methods. In response to environmental issues that come from making synthetic materials and processes, scientists created "green chemistry" which focuses on protecting the environment.

In 1991 P.T. Anastas created green chemistry for EPA USA to produce safer less toxic and effective chemical processes. Scientists develop better ways to manufacture chemicals with minimum danger to people and ecology. Green chemistry describes chemistry operations that support sustainable planet health. Our present efforts toward science and industry need to incorporate effective solutions for environmental risks. The development of sustainable chemicals and reduced environmental impact through chemical processing demands a new approach identified as green chemistry [3]. This discipline focuses on twelve main concepts that teach us how to decrease waste creation and use energy efficiently while lowering toxic effects and adopting renewable materials. The guidelines now guide companies toward more innovative eco-friendly industrial production. Complementing green chemistry with microbial systems creates new possibilities. Organisms naturally perform biocatalytic actions and create better eco-friendly solutions that replace traditional energy-hungry chemical methods frequently using dangerous materials. Through microbial systems, industries bring together economic and environmental benefits that drive sustainable progress [4].

This chapter studies the transformative potential of green chemistry working with microorganisms to make industrial and farming tasks more eco-friendly. The chapter shows how combining sustainable microbial processes with green technology will create an eco-friendly pathway to a future that supports both people and nature.

4.1.1 The Use of the Principles of Green Chemistry

Green chemistry as sustainable chemistry presents a groundbreaking way to design chemical products and systems by minimizing or eliminating hazardous materials. This system provides structured approaches to better protect workers and resources while making operations more environmentally friendly. It uses the twelve main concepts identified by Anastas and Warner [2]. These standards help develop materials that decompose safely while using sustainable renewable resources plus they promote energy saving and pollution management. They form the basis for building more sustainable and environmentally friendly chemical production methods [5]. Biochemical engineering fits naturally with green chemistry principles because microbes act as eco-friendly catalysts. Organisms meet sustainable chemistry needs

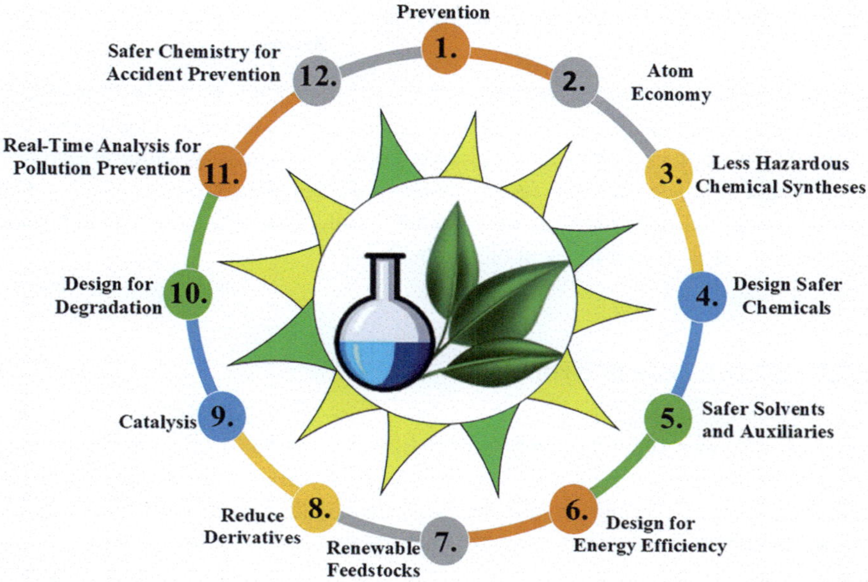

Fig. 4.1 A graphical depiction of the twelve principles of green chemistry

through their effective conversion of renewable resources while handling gentle reaction processes and producing minimal waste. Incorporating microbiological methods with green chemistry principles improves both environmental sustainability and industrial efficiency for improved sustainable development. This section analyses how applying twelve green chemistry rules with microbial systems helps transform chemical industry practices. Employing a microorganism's unique features enables scientists and industrial specialists to develop new eco-friendly production practices. As seen in Fig. 4.1 this principle also emphasizes the importance of interdisciplinary cooperation in creating safer, non-toxic materials.

4.2 The Role of Microbial Processes in Green Chemistry

Microbial processes represent a cornerstone of green chemistry due to their inherent sustainability, specificity, and efficiency. Microorganisms act as nature's chemists and catalyze complex biochemical reactions under mild conditions, creating environmentally friendly alternatives to traditional rather than often hazardous chemical methods.

Microbial systems naturally align with the principles of green chemistry in the below-mentioned domains particularly:

4.2.1 Bio-based Chemical Production

Microbial fermentation processes are widely used to produce biofuels (bioethanol, biobutanol), organic acids (such as lactic acid, and citric acid), and other bio-*based* chemicals. These processes utilize renewable feedstocks such as agricultural residues which reduce reliance on fossil fuels. For example, *Saccharomyces cerevisiae* and *Zymomonas mobilis* are employed in bioethanol production, while *Corynebacterium glutamicum* is used for amino acid biosynthesis [6].

4.2.2 Biocatalysis in Synthesis

Microbial enzymes offer an eco-friendly alternative to hazardous chemical catalysts. For instance, lipases are used for biodiesel production via transesterification of vegetable oils. Laccases and peroxidases facilitate oxidative reactions in the synthesis of fine chemicals, dyes, and polymers and amylases and cellulases are integral to bioethanol production, breaking down starches and cellulose into fermentable sugars [7].

4.2.3 Environmental Remediation

Microbes play a vital role in degrading pollutants, detoxifying heavy metals, and breaking down plastics. For example, *Pseudomonas putida* is known for degrading hydrocarbons in oil spills, while *Ideonella sakaiensis* breaks down polyethylene terephthalate (PET) plastics into biodegradable monomers [8].

4.2.4 Energy Conservation and Process Efficiency

Microbial processes operate under ambient temperatures and pressures, drastically reducing energy input compared to traditional chemical reactions. This supports the green chemistry principle of reducing energy consumption while maintaining or improving process efficiency.

4.3 Benefits of Microbial Processes in Green Chemistry

4.3.1 Sustainability

Microbial systems provide a sustainable alternative by leveraging renewable feedstocks such as agricultural residues, lignocellulosic biomass, and organic waste to synthesize valuable products. Unlike traditional petrochemical processes, which rely on depleting non-renewable resources, microbial processes contribute to a circular economy.

- *Production of Bioethanol*: Sugars from renewable crops such as corn, sugarcane, or lignocellulosic biomass can be effectively converted into bioethanol, a clean and sustainable energy source, through microbial fermentation, especially with *Saccharomyces cerevisiae*. As a result, greenhouse gas emissions are decreased and reliance on fossil fuels is lessened [9].
- *Bioplastics Synthesis*: Using renewable substrates, microorganisms like *Cupriavidus necator* create biodegradable polymers like PHAs, providing a long-term solution to the world's plastic pollution problem [10]. Industries can lessen environmental damage and their carbon footprint by implementing these microbiological processes.

4.3.2 Reduced Hazardous Reagents

By replacing conventional chemical catalysts which frequently use hazardous chemicals or heavy metals microbial biocatalysts improve safety and cut down on waste.

- *Production of Biodiesel*: Under mild conditions, a class of enzymes called lipases catalyzes the conversion of vegetable oils to biodiesel, obviating the need for dangerous chemicals like sodium methoxide, which is extremely reactive and caustic [11].
- *Synthesis of Pharmaceuticals*: Transaminases and other microbial enzymes are used to produce chiral pharmacological intermediates, which have historically required heavy-metal catalysts. This improves reaction specificity and yield while reducing harmful byproducts [12]. In addition to lowering chemical waste and improving workplace safety, the application of microbial systems is consistent with green chemistry concepts.

4.3.3 Energy Efficiency

The energy footprint of industrial operations is greatly reduced by microbial processes, which naturally function in environments with low energy requirements, such as neutral pH, atmospheric pressure, and ambient temperatures.

- *Production of Biofuel*: A crucial stage in the production of bioethanol is the breakdown of complex polysaccharides into fermentable sugars by enzymes such as cellulases and amylases. When compared to chemical hydrolysis techniques, these enzymatic processes require a lot less energy [7].
- *Biogas Production*: Under anaerobic conditions, methane-producing archaea, such as *Methanobacterium,* turn organic waste into biogas. This procedure is a low-energy substitute for conventional techniques since it functions effectively without the need for high pressures or temperatures [13]. These procedures support the objective of green chemistry, which is to reduce energy consumption to reduce expenses and the impact on the environment.

4.3.4 Resource Optimization

Microbial systems work extraordinarily well with different materials while consuming raw materials more efficiently resulting in minimal leftovers.

- *Acetone-butanol-ethanol (ABE) Fermentation*: The organism *Clostridium acetobutylicum* transforms farming waste into acetone butanol and ethanol through one fermentation cycle while using raw materials effectively [14].
- *Enzymatic Degradation in Industry*: Cellulases produced by the fungus *Trichoderma reesei* decompose lignocellulosic waste in the textile and paper industries. This reduces the need for chemical agents while simultaneously recycling trash into products with added value [15]. These methods guide industrial businesses towards sustainability through intelligent resource management.

4.4 Application of Green Chemistry Principles to Microbial Systems

The primary goal of green chemistry remains to create methods and products that waste less hazardous materials while driving sustainability. Natural microbial systems contain biological methods that deliver modern eco-friendly solutions for many applications. In the following section, we look at how green chemistry principles work when applied to microbial systems and what they offer both our planet and our financial system.

4.4.1 *Bioremediation: Microbes as Natural Detoxifiers*

Green chemistry supports bioremediation since it aims to prevent waste production through biological catalysts that match the way microorganisms convert harmful materials into harmless molecules. Bacterial and fungal microorganisms transform pesticide chemicals plus metal pollutants and oil products into harmless substances like CO_2, water, and biomass through their natural activity. For instance, some fungal species can break down persistent organic pollutants, *Pseudomonas putida* is known to break down petroleum hydrocarbons. Microbes function as renewable, self-replicating catalysts that require little external input for destruction, in contrast to chemical therapies that depend on dangerous reagents. This procedure offers a sustainable substitute for energy-intensive or chemical-*based* remediation techniques, recovers polluted ecosystems, and stops the buildup of hazardous waste [16].

4.4.2 *Biofuel Production: Renewable Energy from Microbial Systems*

The utilization of renewable feedstocks, energy efficiency, and atom economy are all demonstrated by the synthesis of biofuel from microbes. Microbes such as algae, bacteria, and yeast convert renewable biomass like agricultural residues, algae, and food waste into biofuels, reducing dependence on fossil fuels. *Example*: Algae produce lipid-*based* biodiesel, while yeast like *Saccharomyces cerevisiae* ferment sugars into ethanol. Microbial systems operate at ambient temperatures and pressures, minimizing energy consumption compared to traditional fuel production processes. By converting feedstocks into fuels with minimal waste, microbial processes maximize resource utilization and reduce byproducts. Biofuels from microbial systems offer reduced greenhouse gas emissions, enhanced energy security, and compatibility with existing energy infrastructure [17].

4.4.3 *Biodegradable Polymers: Sustainable Alternatives to Plastics*

Microbial synthesis of biodegradable polymers, such as PHAs which align with green chemistry's design-for-degradation principle offering eco-friendly alternatives to conventional plastics. Certain bacteria, such as *Cupriavidus necator*, produce PHAs by converting sugars or lipids into bioplastics. These polymers naturally degrade in the environment into water and CO_2. Unlike petroleum-*based* plastics, microbial polymers decompose without releasing toxic residues, reducing environmental pollution and landfill accumulation. Biodegradable polymers minimize long-term ecological impacts, support circular economy goals, and reduce reliance on finite fossil resources [18].

4.4.4 *Agriculture: Enhancing Soil Health with Microbial Inputs*

Microbial processes contribute to green chemistry in agriculture by producing biofertilizers and biopesticides that reduce chemical-intensive practices and enhance sustainability. Microorganisms like *Rhizobium* fix atmospheric nitrogen, while phosphate-solubilizing bacteria convert insoluble phosphates into forms plants can absorb, reducing the need for synthetic fertilizers [19]. Bacteria like *Bacillus thuringiensis* produce bio-insecticides that target specific pests without harming non-target organisms, including beneficial insects and humans. By reducing chemical inputs, microbial solutions promote soil microbial diversity, improve nutrient cycling, and prevent soil degradation. These methods improve agricultural sustainability, lessen their negative effects on the environment, and produce healthier crops. Furthermore, Fig. 4.2 offers an overview of the publications on sustainable microbial processes using green chemistry principles over the past decades.

4.5 Sustainable Microbial Processes (SMP) Framework

The SMP approach uses green chemistry principles to make sustainable solutions in different market areas, particularly bioremediation, bioenergy, farming, and industries. Green chemistry principles work effectively with microbes through SMP. Through SMP technology green chemistry methods are employed in multiple industries like manufacturing, pollution treatment, fuel production, and farming. SMP replaces traditional chemical systems by processing sustainable materials while promoting sustainable practices.

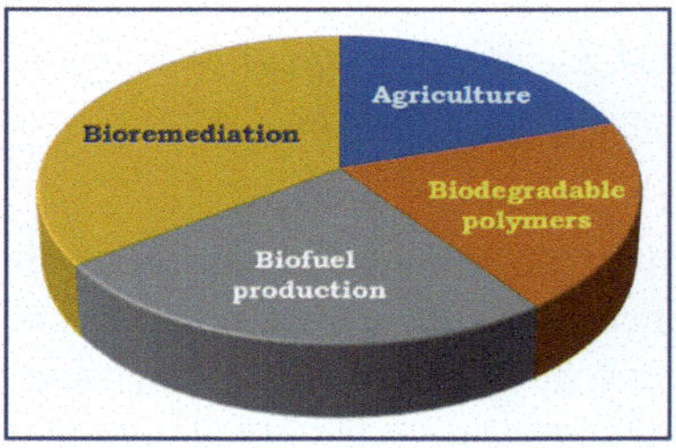

Fig. 4.2 The use of sustainable microbial processes in various eco-friendly biotransformation products

4.5.1 Bioremediation

A major part of SMP technology helps clean polluted environments through microbial action. Microorganisms help break down pollution while bringing back environmental health. Specific microorganisms can neutralize heavy metal pollution in soil along with cleaning oil spills [20]. The green chemistry principles help both protect the environment and reduce any release of new pollutants when used in these procedures.

4.5.2 Biofuel Production

The SMP framework serves the energy industry best by helping with biofuel production. Microorganisms transform biomass into ethanol biodiesel and biogas through their natural processing methods. Biofuels have less climate impact and refresh better than fossil fuels. The production process of biofuel through microbial fermentation displays core aspects of sustainable green chemistry operations. Microorganism species including *Rhodococcus opacus* and microalgae provide lipids for biodiesel production through a transesterification process. Also, *Saccharomyces cerevisiae* operates effectively as a major yeast species for diverse sugar fermentation into ethanol derived from biomass [21].

4.5.3 Agricultural Applications

Through SMP technology biofertilizers and biopesticides become easier to use in farming. These microbial-based products deliver eco-friendly alternatives to synthetic chemicals that damage soil microbiota and lower soil quality. Biopesticides target pests effectively while remaining harmless to living animals but biofertilizers bring more nutrients into the soil for plants to use. Our microbial system helps create stronger and healthier natural ecosystems above and below farming surfaces. Microbes that use phosphorus struggle can update phosphorus better than other microbes while *Rhizobium* bacteria fix nitrogen for legumes, and certain *Bacillus thuringiensis* produce proteins that naturally kill specific pests instead of chemical pesticides [19].

4.5.4 Industrial Manufacturing

Many industries now depend on microbe-based production methods to make bio-based products enzymes and chemicals. Companies use microbes in fermentation to make eco-friendly solvents and biodegradable plastic materials. These accomplishments lower our need for petrochemicals while helping us use resources better and reduce waste. Biotechnology helps create safer biosurfactants in microbial fermentation because they generate less environmental damage than synthetic surfactants. The bacterium *Cupriavidus necator* produces PHAs as an eco-friendly alternative to petroleum polymers [22]. Also, Table 4.1 depicts various sustainable microbial processes and their comparison with the conventional methods.

Table 4.1 Sustainable process stands apart from regular microbial operations

Aspect	Sustainable microbial processes	Conventional processes
Raw materials	Use of renewable resources like agricultural waste, CO_2, and bio-*based* materials	Dependence on non-renewable resources like petroleum and coal
Energy consumption	Low energy requirements due to ambient operational conditions (e.g., room temperature, normal pressure)	High energy input is required for high-temperature and high-pressure operations
Waste generation	Minimal waste generation with high product specificity and biodegradability	Large quantities of hazardous and non-biodegradable waste
Environmental impact	Reduced carbon footprint and eco-friendly emissions	Significant carbon emissions and potential environmental degradation
Process efficiency	High efficiency through tailored microbial strains and optimized fermentation processes	Moderate to low efficiency, often requiring extensive purification steps
Economic viability	Cost-effective in the long term with sustainable resource utilization	High initial productivity but unsustainable resource depletion raises long-term costs
Scalability	Gradual improvements in scalability with advancements in biotechnology	Well-established scalability but at the cost of environmental harm
Innovation potential	Opens avenues for bio-*based* products like biofuels, bioplastics, and green chemicals	Limited scope for innovation, primarily focused on chemical derivatives
Health and safety	Lower risk due to reduced use of toxic chemicals and safer by-products	High risks associated with toxic intermediates and hazardous waste
Regulatory compliance	Aligns with global sustainability goals and regulations like the Paris Agreement	Increasing regulatory challenges due to environmental and health concerns
Circular economy compatibility	Supports circular economy principles by recycling waste into value-added products	Linear economy model, resulting in resource depletion and waste accumulation

4.6 Benefits of Reducing Hazardous Chemical Use and Waste Generation

Integrating green chemistry into microbiological systems provides benefits across multiple areas including regulations industry profitability and environmental safety. A thorough examination of these benefits is provided below.

4.6.1 Environmental Protection

The health of the ecosystem is significantly impacted by the combination of microbial processes and green chemistry. In addition to lowering pollution and promoting biodiversity, sustainable microbiological techniques avoid the discharge of hazardous byproducts. These methods support the preservation of a healthy ecosystem by substituting microbial solutions for chemical-intensive procedures. For example, in agriculture, microbial biopesticides and biofertilizers promote soil health and keep soil microbial communities from being disturbed. Microbial systems help to reduce pollution in soil, water, and air by getting rid of harmful chemicals and byproducts. In sectors including waste treatment, tanning, and textile dyeing, microbial procedures take the place of dangerous chemicals, lowering the release of toxic materials. Microbial enzymes are used in wastewater treatment to break down organic contaminants. Bioremediation processes clean up oil spills, heavy metal-contaminated soils, and pesticide-polluted water bodies, restoring natural ecosystems [23].

4.6.2 Health and Safety

Replacing hazardous chemicals with microbial alternatives significantly reduces risks to human health and safety. Workers in industries adopting microbial processes face reduced exposure to toxic substances, such as carcinogenic solvents or heavy metals. Public health outcomes are improved when hazardous agrochemical use is decreased because there is less chance of contamination in food and water supplies [24].

4.6.3 Resource Efficiency

Microbial systems optimize the use of raw materials, ensuring minimal waste generation and maximum resource utilization. Microbes convert renewable feedstocks like lignocellulosic biomass into valuable products with minimal loss. Many microbial processes operate under mild conditions, conserving energy and water compared to traditional industrial processes [25].

4.6.4 Cost Savings

Green chemistry-*based* microbial processes are often more cost-effective due to lower energy requirements and reduced reliance on expensive raw materials. Ambient temperature and pressure conditions in microbial systems significantly lower operational costs. Industries reduce waste disposal and treatment costs by producing fewer hazardous wastes [26].

4.6.5 Adherence to Regulations

Industries that use green microbiological methods are better able to comply with strict environmental standards. Microbial substitutes comply with laws designed to lower waste and harmful emissions, such as the EPA guidelines in the US and the Registration, Evaluation, Authorisation, and Restriction of Chemicals (REACH) rule of European Regulations. Industries benefit from compliance by avoiding the penalties and harm to their brand that come with non-compliance [27].

4.6.6 Sustainability

By using renewable feedstocks and generating biodegradable resources, microbial systems support long-term sustainability. Microbes reduce reliance on non-renewable fossil fuels by using sustainable resources such as food waste and agricultural residues. Products like PHAs degrade naturally, aligning with sustainability goals and reducing environmental burden [28].

4.6.7 Innovation and Growth

The intersection of green chemistry and microbial science drives innovation, creating new eco-friendly technologies and expanding market opportunities. Advances in synthetic biology and bioengineering are enabling the development of custom-designed microbes for specific industrial applications. Consumer demand for sustainable and green products is driving growth in industries utilizing microbial systems, such as biodegradable packaging, green energy, and eco-friendly agriculture [29].

4.6.8 Microbial Biocatalysts for Sustainable Chemical Reactions

Microbial biocatalysts, comprising enzymes or whole microbial cells, facilitate highly specific biochemical reactions under environmentally friendly conditions. Their use enhances atom economy, reduces waste, and limits energy consumption, making them indispensable for sustainable chemical manufacturing. Microbial biocatalysts are remarkably selective, targeting specific chemical bonds or functional groups while avoiding unwanted side reactions. This specificity ensures minimal waste generation and higher product purity. Lipases derived from microbes like *Candida antarctica* catalyze the transesterification of triglycerides in oils into fatty acid methyl esters (biodiesel). Unlike traditional chemical catalysts, microbial lipases operate without harsh reagents or excessive byproducts, making biodiesel production cleaner and safer. Microbial enzymes such as reductases and transaminases facilitate the synthesis of optically pure pharmaceutical intermediates. These processes eliminate the need for additional purification steps, thereby conserving resources and reducing waste. Unlike conventional chemical processes that require extreme temperatures, pressures, or pH levels, microbial biocatalysts operate efficiently under ambient or near-ambient conditions. As a result, less energy and dangerous chemicals are needed. Energy-intensive synthetic methods are replaced by *S. cerevisiae*, which manufactures ethanol in anaerobic circumstances at moderate temperatures [30]. Greenhouse gas emissions and resource use are reduced by this microbial process. Under mild conditions, microbial cellulases convert lignocellulosic biomass into fermentable sugars, allowing agricultural leftovers to be converted into biofuel. This method saves energy and avoids the need for harsh chemical treatments. Microbial biocatalysts are used in a wide range of businesses, promoting sustainability in every field, by eliminating dangerous chemical processes, enzymes like penicillin acylase allow for the environmentally friendly manufacture of medicines like amoxicillin. Reliance on fossil fuels is decreased by microbes like *Clostridium acetobutylicum*, which ferment renewable feedstocks into bioethanol and biobutanol. In cheese manufacturing, brewing, and taste development, microbial lipases and proteases are often utilized in place of chemical additions [31].

4.6.9 Using Microbiological Alternatives in Place of Toxic Reagents

Microbial systems offer safer and more environmentally friendly synthesis pathways for industrial applications by serving as environmentally friendly alternatives to dangerous chemicals. They lower operational risks and environmental contamination by taking the place of traditional synthetic processes that depend on hazardous reagents. Scientists used to make citric acid through harmful environmental processes with sulfuric acid and lime as the main ingredients. Under controlled mild conditions, *Aspergillus niger* microbes eliminate harmful substances by making citric acid. The food, medicine, and beauty sectors heavily rely on microbial production to make citric acid in environmentally friendly operations that save resources. *Rhizobium* bacteria save high-energy Haber–Bosch processing by converting atmospheric nitrogen directly into ammonia. The biological production method for ammonia uses less energy and produces no harmful CO_2 emissions while reducing the environmental impacts of this process by 95%. Microorganisms help us avoid dangerous chemical processes by making complex organic chemicals that nature already produces. Microbes of the Streptomyces species family use their metabolic processes to produce polyketides such as erythromycin and anticancer medications. Microorganisms produce these substances easily because chemical creation remains challenging [32]. The biological production approach decreases environmental impacts from drug manufacturing and replaces risky chemicals entirely. *Bacillus subtilis* microorganisms can produce vitamin B2 better and cleaner than conventional chemical approaches. By using microorganisms instead of dangerous chemicals this method reduces harm to both the environment and our reliance on toxic substances. Biodegradable polymers develop through microbial processes creating eco-friendly replacements for regular plastics.

Bacteria like *Ralstonia eutropha* turn renewable agricultural waste into PHAs which serve as biodegradable plastic materials. Plastic products made from PHAs break down into water and carbon dioxide faster than fossil fuel-based plastics and generate less pollution. Bacteria make PLA through the fermentation of sugars taken from sustainable materials. Unlike petroleum-*based* plastics, PLA is biodegradable and can be used in various applications, including packaging, textiles, and medical implants. Its production reduces dependence on petrochemicals, offering a more sustainable alternative. Microbial processes inherently minimize waste generation by replacing stoichiometric reagents with catalytic processes. Microbial proteases used in detergents replace phosphate-*based* additives, which can contribute to eutrophication in aquatic ecosystems. Protease-*based* detergents help reduce harmful environmental impacts while being effective in stain removal [33]. Amylases, produced by microbes, are used to remove starch from fabrics during processing. This process eliminates the need for harsh chemicals, reducing both chemical and water waste in textile industries. Microbial systems maximize the use of renewable feedstocks, minimize energy consumption, and support resource efficiency. Microbes efficiently process renewable raw materials, transforming them into valuable products. Microbial cellulases break down lignocellulosic biomass into fermentable sugars, enabling biofuel production from agricultural residues [34]. This process supports sustainable energy production while reducing waste and reliance on fossil fuels. Microorganisms can convert food waste into bioethanol or biogas, addressing both energy production and waste management challenges. This method not only recycles food waste but also reduces the environmental footprint of landfill disposal. Microbial processes support material recycling and upcycling, crucial for a circular economy. *Ideonella sakaiensis*, a bacterium, can metabolize PET plastics into reusable monomers. This ability enables plastic recycling and contributes to reducing plastic waste.

Methanogenic archaea convert CO_2 into methane, providing a renewable energy source while mitigating greenhouse gas emissions. This process supports carbon capture and utilization, reducing the impact of climate change. Microbial processes exhibit high atom efficiency, converting nearly all substrate atoms into the desired product with minimal byproducts. Yeasts like *Saccharomyces cerevisiae* efficiently convert sugars into ethanol, with minimal waste generation. This process is widely used in the biofuel industry, providing a sustainable alternative to fossil fuels. *Corynebacterium glutamicum* produces amino acids such as glutamate and lysine with near-total substrate utilization. These amino acids are critical in the food and pharmaceutical industries, providing high-efficiency production with minimal waste. Microbial processes often require less water and energy compared to traditional chemical methods. Microbial fuel cells (MFCs) provide a twofold advantage for water and energy sustainability by producing electricity and purifying wastewater. This procedure concurrently lowers wastewater pollution and energy use. Compared to conventional biofuel crops, microalgae may generate biofuels from saline or wastewater, using less freshwater and land. This approach promotes environmentally friendly, sustainable bioenergy production [35].

4.6.10 Role of Microbes in Degrading Pollutants and Heavy Metals

Microorganisms can metabolize a variety of pollutants, including hydrocarbons, heavy metals, and synthetic chemicals, into less harmful substances. Microbes such as *Pseudomonas* and *Alcanivorax* degrade petroleum hydrocarbons, essential in cleaning oil spills. These bacteria utilize hydrocarbons as a carbon source, converting them into water and CO_2. Fungi such as *Phanerochaete chrysosporium* produce lignin-degrading enzymes to break down persistent organic pollutants like polychlorinated biphenyls (PCBs) and dioxins. Bacteria such as *Shewanella* and *Geobacter* reduce toxic heavy metals like uranium and chromium into less soluble and less toxic forms. Algae and fungi trap metals such as mercury and cadmium through bioaccumulation, preventing their spread in the ecosystem [36].

4.6.11 Role of Microbes in Soil, Water, and Air Purification

Rhizosphere bacteria like *Rhizobium* break down agricultural chemicals, including pesticides and herbicides, improving soil quality. Phytoremediation enhanced with rhizobacteria accelerates the degradation of contaminants by using plants and their associated microbial communities. Anaerobic and aerobic bacteria such as *Nitrosomonas* and *Methylococcus* remove organic pollutants, nitrates, and methane from wastewater. Microbial consortia in bioreactors target effluents from industries for pollutant removal. Biofilters containing microbial biofilms degrade VOCs like benzene and toluene. Methanotrophic bacteria oxidize methane, reducing its emission as a potent greenhouse gas [37].

4.6.12 Role of Microbes in Biofuel Production

Microbial processes in biofuel production contribute to renewable energy development, reducing dependency on fossil fuels while lowering greenhouse gas emissions. *Saccharomyces cerevisiae* ferments sugars from crops like sugarcane and corn to produce bioethanol. Lignocellulosic ethanol production involves cellulolytic microbes such as *Trichoderma* and *Clostridium*, breaking down cellulose into fermentable sugars. Microbial lipases catalyze the transesterification of oils, producing biodiesel efficiently without the need for chemical catalysts. Algae like *Chlorella* synthesize high lipid content, offering a sustainable biodiesel source. Methanogenic archaea such as *Methanosarcina* break down organic waste in anaerobic digesters to produce methane-rich biogas. Co-digestion of agricultural waste and municipal sludge maximizes biogas yield [38].

4.6.13 Role of Microbes in Agriculture

Microbial applications in agriculture enhance productivity and sustainability by reducing chemical inputs and improving soil health. *Rhizobium* fixes atmospheric nitrogen in symbiosis with legumes, reducing the need for synthetic fertilizers. Bacteria like *Pseudomonas* convert insoluble phosphorus into plant-available forms. Mycorrhizal fungi enhance plant nutrient uptake, particularly phosphorus, zinc, and water. *Bacillus thuringiensis* (Bt) produces toxins specific to insect pests, offering an alternative to chemical pesticides [39]. Fungal biopesticides like *Trichoderma* suppress soil pathogens through antifungal metabolites.

4.6.14 Role of Microbes in Industrial Applications

Microbes revolutionize industrial processes by offering eco-friendly alternatives for producing chemicals, enzymes, and materials. Microbial fermentation produces organic acids like citric acid and lactic acid, widely used in food, pharmaceuticals, and polymers. Butanol production by *Clostridium acetobutylicum* serves as a green alternative to petrochemical-derived solvents. Industrial enzymes, including amylases, cellulases, and proteases, are produced by microbes such as *Aspergillus* and *Bacillus*. These enzymes are employed in detergent, textile, and food industries to

Table 4.2 The diverse and essential roles microbes play in various environmental and industrial processes

Category	Role of microbes	Examples
Degrading pollutants and heavy metals	Microbes can break down or convert pollutants and heavy metals into less toxic forms	*Pseudomonas* species for hydrocarbon degradation *Bacillus* species for heavy metal detoxification (e.g., lead, mercury)
Soil, water, and air purification	Microbial processes help in the detoxification and purification of contaminated environments	*Rhizobium* species in soil nitrogen fixation *Deinococcus radiodurans* for radiation and heavy metal resistance in contaminated water
Biofuel production	Microbes are used to produce biofuels by fermenting organic matter	*Saccharomyces cerevisiae* for ethanol production *Clostridium acetobutylicum* for butanol production
Agriculture	Microbes enhance soil health, promote plant growth, and protect crops from pathogens	*Azospirillum* for nitrogen fixation and promoting plant growth *Trichoderma* species for controlling soil-borne plant diseases
Industrial applications	Microbes are used in the production of enzymes, chemicals, and bioproducts in various industries	*Aspergillus* species for enzyme production in food processing *Escherichia coli* for the production of recombinant proteins

replace harsh chemicals. Microbial synthesis of PHAs provides biodegradable alternatives to conventional plastics [40]. Algal cellulose is utilized for bio-composites, offering sustainable replacements for synthetic fibers. The role of microbes in various applications is summarised with examples in Table 4.2.

4.7 Advantages of Green Chemistry in Sustainable Microbial Processes

Green chemistry emphasizes designing products and processes that reduce or eliminate hazardous substances, aligning closely with microbial processes. By leveraging the innate capabilities of microorganisms, sustainable microbial processes reduce harmful byproducts, conserve resources, and foster eco-friendly innovations across industries. Below, each advantage is elaborated on in detail, exploring the transformative potential of green chemistry in sustainable microbial processes.

4.7.1 Reduction of Harmful Byproducts

Microbial processes utilize enzymes with remarkable specificity, enabling reactions to proceed with high efficiency and minimal side reactions. This precision minimizes the generation of chemical waste, a stark contrast to traditional chemical methods, which often yield significant amounts of undesirable byproducts. Lipases are used in esterification reactions to produce esters like ethyl acetate. Unlike chemical catalysts, lipases operate under mild conditions and avoid generating toxic byproducts. The resulting high-purity products require minimal purification, further reducing waste. Microbial synthesis eliminates the use of toxic and corrosive reagents, reducing the risks associated with chemical synthesis. This also minimizes the need for extensive handling, storage, and disposal of hazardous substances. The microbial production of citric acid by *Aspergillus niger* avoids sulfuric acid used in traditional chemical processes. This substitution results in cleaner production with fewer environmental hazards. Microorganisms can break down toxic compounds into benign substances, offering a sustainable solution to pollution. These natural processes ensure that harmful substances do not accumulate in ecosystems. In wastewater treatment, bacteria like *Pseudomonas* degrade phenolic compounds into CO_2 and water, avoiding the production of secondary pollutants often associated with chemical treatments [41].

4.7.2 Lower Environmental Impact and Resource Consumption

Microbial processes operate at ambient or moderate temperatures and pressures, requiring significantly less energy compared to conventional chemical methods. This efficiency reduces the carbon footprint of industrial operations. The enzymatic hydrolysis of cellulose into glucose for bioethanol production occurs at 50–60 °C whereas, chemical hydrolysis requires temperatures exceeding 150 °C and harsh acidic or basic conditions. Microbial processes rely on renewable feedstocks such as biomass, agricultural residues, and organic waste. This approach conserves non-renewable resources like petroleum and reduces dependence on fossil fuels. Algal biomass, rich in lipids and carbohydrates, is converted into biofuels and bioplastics, offering a sustainable alternative to petrochemical-derived products [42]. Microbial processes produce fewer greenhouse gas emissions and toxic pollutants, supporting global efforts to combat climate change and reduce air pollution. Bioethanol production via microbial fermentation emits CO_2 absorbed during crop growth, resulting in a near-neutral carbon cycle. This contrasts with fossil fuel combustion, which introduces previously sequestered carbon into the atmosphere. Efficient water use and recycling are inherent features of microbial processes, making them suitable for regions facing water scarcity. In anaerobic digestion systems, water is recycled within the system, reducing overall consumption. The treated water can also be reused in agriculture or industrial processes [43].

4.7.3 Contribution to Circular Economy Goals

Microbial processes transform waste into valuable products, promoting resource recovery and reducing the burden on landfills. This aligns with the principles of a circular economy, where waste is treated as a resource. Food waste is converted into biogas through anaerobic digestion. The resulting biogas provides renewable energy, while the digestate serves as an organic fertilizer [44]. The bacterium *Ideonella sakaiensis* breaks down PET plastic into its monomers, which can be reused to manufacture new plastic, closing the loop on plastic recycling. Microbial processes support closed-loop systems, where materials are continually reused with minimal waste. This ensures maximum resource efficiency and aligns with sustainability goals. During bioethanol production, fermentation byproducts, such as distillers' grains, are repurposed as animal feed or soil conditioners, ensuring that no part of the process goes to waste. Microorganisms synthesize biodegradable polymers like PHAs, which serve as eco-friendly alternatives to traditional plastics. These materials degrade naturally, reducing long-term environmental impacts. *Cupriavidus necator* produces PHAs from renewable feedstocks. These bioplastics are compostable, breaking down in soil or water without leaving harmful residues. By utilizing agricultural residues, industrial byproducts, and other non-traditional feedstocks, microbial processes conserve virgin resources and reduce environmental degradation associated with resource extraction. Lignocellulosic biomass, such as corn stover or sugarcane bagasse, is converted into bioethanol or bioplastics, offering a sustainable use for agricultural byproducts that might otherwise go to waste [45].

4.7.4 Broader Societal and Environmental Benefits

The reduced use of toxic reagents and solvents in microbial processes minimizes occupational hazards and environmental risks. Biopesticides produced by microbes, such as *Bacillus thuringiensis*, target pests without harming humans, non-target species, or beneficial insects, offering a safer alternative to synthetic pesticides. The adoption of microbial technologies promotes the growth of bio-*based* industries, creating jobs and supporting sustainable economic development. The production of biofuels and bioplastics supports rural economies by creating demand for agricultural residues and waste materials, diversifying income sources for farmers [46].

4.8 Innovative Applications and Emerging Trends in Sustainable Microbial Processes

Microbial processes have emerged as powerful tools for addressing pressing global challenges in environmental sustainability, public health, and resource management. By leveraging microorganisms' natural abilities and integrating advanced technologies, researchers are creating eco-friendly solutions and pioneering trends that redefine traditional practices. This chapter delves deeply into the mechanics, uses, and importance of two important topics: biodegradable polymers and environmentally friendly antimicrobial agents.

4.8.1 Eco-friendly Antimicrobial Agents

Although they have made a substantial contribution to the fight against microbial threats, traditional antimicrobial agents such as chemical disinfectants and antibiotics have come at the expense of environmental damage, antimicrobial resistance, and public health issues. An inventive method for creating environmentally friendly substitutes that lessen these difficulties is provided by sustainable microbial procedures.

4.8.1.1 Microbial Biopesticides

Microbial biopesticides target particular pests by using naturally occurring microbes like *Bacillus thuringiensis* (Bt.). These biopesticides solve the drawbacks of synthetic chemical pesticides, which can damage beneficial creatures, contaminate soil and water, and build up the food chain. They are selective, biodegradable, and environmentally safe. The alkaline stomach environment of target pests activates the insecticidal proteins (Cry and Cyt poisons) produced by *Bacillus thuringiensis* (Bt.). These proteins bind to receptors in the pest's gut cells, causing cell lysis and death. The process is species-specific, ensuring non-target organisms remain unharmed. Its uses are in agricultural pest control, reducing reliance on synthetic pesticides, and Integrated pest management programs for crops like cotton, maize, and rice. Bt-*based* formulations are extensively used to control *Helicoverpa armigera* (cotton bollworm), reducing crop losses while maintaining ecological balance [47].

4.8.1.2 Natural Antimicrobial Peptides (AMPs)

AMPs are small, naturally occurring molecules produced by microorganisms, plants, and animals. They exhibit broad-spectrum activity against bacteria, fungi, and viruses, making them promising candidates for replacing synthetic antimicrobial agents. AMPs interact with microbial membranes, disrupting their integrity by forming pores or causing membrane depolarization. Unlike traditional antibiotics, they target fundamental structures in microbes, reducing the likelihood of resistance development. Effectively used in wound dressings and coatings for medical implants to prevent infections, also used as food preservation like nisin, derived from *Lactococcus lactis*, inhibits spoilage bacteria in dairy products and is also used as a natural preservative in formulations to prevent microbial contamination. *Lactobacillus*-derived AMPs are used in the dairy industry to maintain product safety and inhibit pathogens like *Listeria monocytogenes* without compromising product quality [48].

4.8.1.3 Microbial Production of Organic Acids

Microorganisms such as *Lactobacillus*, *Acetobacter*, and *Aspergillus* spp. produce organic acids with antimicrobial properties, which can replace synthetic chemical preservatives and disinfectants. These organic acids are biodegradable, less toxic, and environmentally friendly. Organic acids disrupt microbial metabolism by reducing intracellular pH or interfering with enzymatic activities. They can inhibit microbial growth in food, personal care products, and cleaning agents. Used in organic acids like lactic acid and acetic acid act as natural preservatives, extending shelf life. Organic acids are used in hand sanitizers and surface disinfectants. Organic acids enhance product safety while being gentle on the skin and hair. Lactic acid produced by *Lactobacillus spp.* is widely used as a preservative in food products like pickles, cheese, and yoghurt, providing antimicrobial benefits while maintaining natural flavours [49].

4.8.2 Biodegradable Polymers

Plastic pollution has become a global crisis, with traditional petroleum-*based* plastics accumulating in landfills and oceans, causing long-term environmental damage. Biodegradable polymers produced by microorganisms offer sustainable alternatives that degrade naturally, reducing ecological harm.

4.8.2.1 Polyhydroxyalkanoates (PHAs)

PHAs are a family of biodegradable polyesters synthesized by microorganisms like *Cupriavidus necator* under nutrient-limited conditions with excess carbon sources. These polymers mimic the properties of conventional plastics while being compostable. Microorganisms metabolize renewable feedstocks, such as plant oils, agricultural residues, or industrial waste, to accumulate PHAs as intracellular granules. PHAs are extracted and processed into films, fibers, or molded products. Used in food containers, bags, and films to reduce reliance on single-use plastics. Applied in sutures, drug delivery systems, and implants due to biocompatibility. Used in mulch films and seed coatings to prevent soil contamination. PHAs are used in making compostable packaging materials that degrade in soil or water within months, offering an eco-friendly alternative to polyethylene [50].

4.8.2.2 Polylactic Acid (PLA)

PLA is a biodegradable polymer produced through microbial fermentation of sugars derived from renewable resources like corn starch, sugarcane, or cassava. It is one of the most widely used biodegradable plastics. Microbial fermentation of sugars produces lactic acid, which is polymerized into PLA through a condensation reaction or ring-opening polymerization. PLA can be molded into various shapes or extruded into films and fibers. PLA is used in food packaging, disposable cutlery, and cups, replacing petroleum-*based* plastics. PLA is used in surgical implants, drug delivery systems, and tissue engineering scaffolds due to its biocompatibility and biodegradability. PLA fibers are used in clothing, carpets, and upholstery as sustainable alternatives to synthetic fibers. PLA-*based* packaging for fresh produce ensures product freshness while offering a compostable alternative to conventional plastic wraps [51].

4.9 Advances in Bioengineering and Synthetic Biology for Sustainable Practices

Bioengineering and synthetic biology are at the forefront of innovations aimed at optimizing microbial processes and developing sustainable solutions for various industries. These fields merge biology, chemistry, and engineering to create tailored microorganisms with enhanced capabilities for specific applications. Both eco-friendly antimicrobial agents and biodegradable polymers reduce the environmental footprint by minimizing toxic residues and addressing plastic pollution. Natural antimicrobial agents reduce the risk of antimicrobial resistance and exposure to toxic chemicals. The growing demand for sustainable alternatives creates new markets and job opportunities in bio-*based* industries. These innovations support the United Nations Sustainable Development Goals (SDGs), particularly those related to responsible production, climate action, and life below water [52].

4.9.1 Microbial Engineering for Enhanced Production

Bioengineering techniques allow for the modification of microbial genomes to enhance their natural metabolic pathways, making them more efficient for producing high-value chemicals, biofuels, and pharmaceuticals. These advancements contribute to sustainable practices by reducing resource consumption and increasing the yields of desired products. Genetic modifications enable the rerouting of microbial metabolic pathways to enhance the production of biofuels, pharmaceuticals, and biochemicals. The metabolic engineering of *Escherichia coli* has enabled the efficient production of biofuels like butanol and biodiesel from renewable feedstocks [53]. Through synthetic biology, microorganisms are engineered to produce complex chemicals, such as antibiotics, vitamins, and industrial enzymes, more sustainably compared to traditional chemical synthesis. *Saccharomyces cerevisiae* has been genetically engineered to produce high-value bio-*based* chemicals like 1,3-propanediol and succinic acid, reducing reliance on petroleum-*based* processes. Bioengineering can also optimize microbial strains to enhance their ability to degrade environmental pollutants, such as heavy metals, plastics, and industrial waste. Genetically engineered *Pseudomonas* strains have been developed to detoxify heavy metals like mercury and cadmium in contaminated water bodies [54].

4.9.2 Synthetic Biology for Design of New Biological Systems

Synthetic biology goes beyond modifying existing organisms and aims to create entirely new biological systems and pathways. These systems can be designed to perform specific tasks such as pollutant degradation, carbon capture, or sustainable biofuel production. Scientists can design artificial metabolic pathways within microorganisms that allow them to synthesize non-natural products or degrade pollutants that would otherwise be difficult to address. The engineering of *Escherichia coli* to produce bio-*based* nylon precursors represents a major step in sustainable chemistry, as it eliminates the need for petroleum-*based* feedstocks. Synthetic biology techniques are being explored to enable microorganisms to capture and utilize CO_2 as a feedstock for the production of biofuels, chemicals, and other high-value products. The engineering of *Cyanobacteria* to efficiently fix atmospheric CO_2 and convert it into biofuels or biodegradable plastics is a promising approach to reducing carbon emissions [55].

4.9.3 Integration of Artificial Intelligence and Machine Learning in Optimizing Microbial Processes

Artificial intelligence (AI) and machine learning (ML) are transforming microbial process optimization by providing tools for data-driven insights, predictive modelling, and automation. The integration of AI/ ML into microbial engineering and biotechnology enhances the scalability, efficiency, and sustainability of microbial processes. ML algorithms can analyze vast datasets from genetic, metabolic, and environmental sources to predict the optimal genetic modifications or environmental conditions for enhanced microbial performance. This allows for rapid strain development and process optimization. AI models can predict how genetic modifications will affect microbial metabolism and product yields. ML tools have been used to predict the best metabolic engineering strategies for increasing biofuel production in *Saccharomyces cerevisiae* [56]. AI-powered automation platforms can rapidly screen thousands of microbial strains or genetic variants to identify the most promising candidates for industrial applications. High-throughput robotic systems combined with AI allow for the rapid identification of engineered strains that produce high levels of desired bio-*based* chemicals. AI and ML are used to optimize bioreactor conditions in real-time, improving microbial growth, and product yields, and reducing resource consumption. Sensors, combined with AI algorithms, can adjust parameters such as temperature, pH, oxygen levels, and nutrient availability to optimize microbial processes. AI algorithms analyze real-time data from bioreactors to optimize conditions for microbial fermentation processes. AI-driven systems are used to optimize ethanol production by adjusting fermentation parameters in real-time, leading to improved yields and energy efficiency. AI can predict equipment failures before they occur, minimizing downtime and ensuring optimal operation. Machine learning algorithms are used to monitor microbial growth in bioreactors and predict when maintenance or nutrient adjustments are needed [57].

4.10 Role of Microbial Processes in Promoting Biodiversity

Microbial biodiversity is fundamental to ecosystem functioning, contributing to nutrient cycling, decomposition, symbiosis, and the regulation of harmful species. Microorganisms such as bacteria, fungi, and viruses maintain ecological balance and promote biodiversity through their interactions with plants, animals, and other organisms.

4.10.1 Microbial Interactions and Nutrient Cycling

Microbial processes are essential for nutrient cycling, including the nitrogen, carbon, sulfur, and phosphorus cycles, which support ecosystem productivity and biodiversity. In the nitrogen cycle, microorganisms perform critical roles such as nitrogen fixation, where atmospheric nitrogen is converted into forms usable by plants, and denitrification, which returns nitrogen to the atmosphere. These processes ensure nitrogen availability, sustaining plant growth and indirectly supporting higher trophic levels. In the carbon cycle, microbes decompose organic matter and recycle carbon, converting plant and animal biomass into CO_2 for use in photosynthesis. Microbial decomposition prevents the accumulation of organic material and maintains soil fertility. Microbial roles in phosphorus and sulfur cycles involve processes like mineralization and immobilization, ensuring the availability of these nutrients for plants and maintaining ecosystem health [58].

4.10.2 Soil Microbial Communities and Plant Symbiosis

Soil microorganisms form symbiotic relationships with plants, enhancing nutrient uptake and health. Mycorrhizal fungi associate with plant roots, improving phosphorus and water absorption while helping plants resist environmental stresses. These fungi support the growth of diverse plant communities, particularly in nutrient-poor soils. Nitrogen-fixing bacteria such as *Rhizobium* form nodules on legume roots, converting atmospheric nitrogen into bioavailable forms. This enriches soil fertility and supports plant diversity, ultimately benefiting the entire ecosystem [59].

4.10.3 Microbial Degradation and Bioremediation

Microbes degrade pollutants like petroleum hydrocarbons, heavy metals, and pesticides, mitigating environmental contamination. Bioremediation restores polluted ecosystems, enabling the re-establishment of plant and animal life and promoting long-term biodiversity conservation [60].

4.10.4 Microbial Disease Regulation and Pest Control

Microorganisms regulate pathogens and pests, preventing disease outbreaks and maintaining ecosystem balance. *Bacillus thuringiensis* (Bt), a microbial biological pesticide, targets pests without harming non-target species, preserving biodiversity and promoting healthier ecosystems. These microbial processes are indispensable for sustaining biodiversity, maintaining ecosystem functions, and fostering resilience in natural environments.

4.11 Innovations Facilitated by Green Chemistry

Green chemistry has significantly advanced microbial science leading to the development of environmentally sustainable technologies and products. The development of environmentally friendly antibacterial agents is one of the major innovations. These substances are made to minimize negative impacts on ecosystems while addressing the growing danger of antibiotic resistance. Making antimicrobial compounds through green chemistry improves their environmental impact by creating them from renewable natural materials that easily break down in the environment. Microbial and plant natural antibacterial materials appear effective at limiting microbial drug resistance. Research has shown that plant polyphenols from natural sources alongside synthetic peptides made by microbes work effectively as antibiotics and pose a lower environmental threat [61].

The production of biodegradable polymers through microbial fermentation demonstrates remarkable results in green chemistry. PLA and PHA biodegradable polymers serve as environmentally friendly replacements for conventional plastic materials. These biodegradable bioplastics from microbial production help protect our environment by reducing plastic waste accumulation. Fermentation processes and feedstock selection now help create biodegradable polymers both cheaper and at larger production scales.

Through green chemistry scientific advances synthetic biology now lets us build precise microbial systems for different needs. The combination of synthetic biology techniques and green chemistry principles produces microbial systems that use resources more effectively while minimizing waste and generating precisely targeted outputs. Scientists use modified microorganisms to turn renewable materials into important chemicals, pharmaceuticals, and fuel products. Microbes now support more business areas including healthcare and agriculture through their ability to manage metabolic routes. Synthetic biology lets scientists create effective biological antimicrobials using sustainable chemistry practices that support circular economic principles. Recent manipulations of living organisms have produced natural materials that automatically break down into the environment without long-term environmental impact. Scientists have created biopolymers and bacteriocins using this process [62].

4.12 Addressing Scalability and Cost Challenges Towards Circular Economy Using Microbial Processes

Biotechnology's main challenge lies in taking microbe processes from lab research to industrial applications. While these processes have shown remarkable potential at small scales, translating their efficiency to large-scale production involves overcoming issues related to bioreactor design, substrate cost, and downstream processing. Exact control over environmental variables is beneficial for laboratory-scale microbial operations. However, issues including ineffective oxygen transmission, difficulties with heat dissipation, and shear stresses are introduced when scaling up to industrial bioreactors, and they can have a detrimental effect on the growth and productivity of microorganisms. Optimizing microbial strains and bioreactor designs is necessary to maintain homogeneity in large-scale systems. Limitations in oxygen transport are a serious problem since aerobic microbial metabolism depends on oxygen. The oxygen requirements of dense cultures are frequently not met by industrial bioreactors, which results in lower yields. Similarly, when microbial activity produces a lot of heat that is more difficult to control in bigger volumes, heat dissipation becomes an issue. Another problem is shear stress, which can harm cells and jeopardize survival, particularly for delicate microbes [63]. Solutions to these challenges include designing advanced bioreactor systems with enhanced mixing, aeration, and temperature control capabilities. For instance, bioreactors with improved impeller designs or airlift systems can enhance oxygen and heat distribution. Additionally, microbial strain engineering offers potential by creating robust strains capable of tolerating shear forces and thriving in suboptimal conditions. Despite their sustainability advantages, microbial processes often face high production costs, driven by the expense of substrates, energy consumption, and specialized equipment. The economic feasibility of microbial bioproduction is further limited when expensive pure substrates, such as glucose, are required. Energy-intensive operations, such as maintaining optimal fermentation conditions, further escalate costs [64]. Using waste materials as substrates offers a promising solution. Agricultural residues, lignocellulosic biomass, food industry by-products, and even CO_2 are being explored as low-cost, renewable feedstocks. These materials not only reduce substrate costs but also contribute to waste vaporization and circular economy objectives. Optimizing microbial metabolic pathways through systems biology and genetic engineering can also enhance substrate utilization efficiency and increase product yields, offsetting production costs [65]. Downstream processing, including the separation and purification of bio-*based* products from microbial cultures, is often the most expensive step in microbial processes. The complexity of microbial fermentation broths, which contain a mix of cells, metabolites, and by-products, poses significant purification challenges. The toxicity of certain products to the producing microorganisms further complicates production and recovery. Integrated bioreactor systems, which combine fermentation and product recovery in a continuous process, have emerged as a solution. The new systems help companies produce their products through fewer sub-steps while spending less money. New membrane filtration methods alongside extraction solvents and adsorptive chromatography give manufacturers better ways to purify products while using less energy and money [66].

SMP developments permit better control over waste reduction plus material reuse benefits. SMP links industrial production to environmental protection through the natural abilities of microorganisms in metabolism. This approach uses microbe systems to convert waste into valuable products while creating a sustainable recycling loop in resource usage. Microorganisms lead in converting industrial and agricultural waste to high-value chemicals such as biofuels and bioplastics. Microbial enzymes can turn agriculture waste including corn stover and sugarcane bagasse into fermentable sugars through their natural conversion process. Engineered microbes convert organic waste into butanol and ethanol biofuels to decrease fossil fuel consumption. Fermentation methods let microorganisms produce degradable plastics. Through bioengineering *Cupriavidus necator* and *Ralstonia eutropha* microorganisms transform organic waste into PHAs. These bioplastics use plant-based materials and break down naturally making them both environment-friendly and petroleum-free replacements for regular plastics. The circular economy's focus on managing waste and pollution before it appears best demonstrates this solution [67].

Microbial processes help industries and factories both recycle materials and reuse natural elements. Organic waste processes with several microbes make digestate with plant nutrients plus useful gas energy. Digestate from anaerobic digestion serves as a sustainable plant fertilizer and reduces our need for manufactured fertilizers while biogas supplies clean renewable energy. Agricultural resource cycles when combined with microbe activities form a sustainable waste treatment cycle [68].

A part of the circular economy focuses on extracting valuable materials from waste streams and streams. Metals and minerals can be extracted from industrial effluents and electronic trash using microbial methods. To recover metals like copper and gold from electronic waste, for example, bioleaching processes employing bacteria like *Acidithiobacillus ferrooxidans* eliminate the need for mining operations that harm the environment.

Furthermore, the principles of the circular economy are well aligned with the employment of microbial systems for carbon capture and conversion. CO_2 is being captured by engineered microbes, including modified cyanobacteria, and transformed into biofuels or bio-*based* compounds. This reinforces the circular economy's focus on keeping materials in use by addressing greenhouse gas emissions and repurposing carbon as a resource [69].

4.13 Future Prospects and Possible Solutions in Sustainable Microbial Processes

The potential for using SMP to address global issues in chemical manufacturing, waste management, and energy is enormous. Notwithstanding the encouraging developments, several obstacles prevent their broad use and industrial scalability. These obstacles are found in the fields of technology, economics, regulations, and society. SMP's chances for the future depend on creating plans to get over these obstacles and incorporate scientific and technological developments. The high cost of production associated with industrial biotechnology is a major obstacle to scaling microbial technologies. When compared to traditional chemical synthesis, microbial-*based* production is less competitive due to factors including substrate costs, microbial strain development, and process optimization. For example, when high-purity substrates are needed for fermentation operations, feedstock costs frequently make up a significant number of overall expenditures [70]. The industry faces major difficulties due to the weak performance and lower effectiveness of microbial organisms When microorganisms thrive their efficiency decreases because they face contamination poor conditions and toxic products. Despite recent synthetic biology breakthroughs scientists find it hard and expensive to develop superior microbial strains for industrial processes. Regulatory requirements make it challenging to turn microbial production into profitable market businesses [71]. Several countries control how GMOs function through strict rules that demand total safety research before use. These operations make production more costly and slow down product development. Additionally, there is a lack of uniformity in international regulations, creating uncertainties for companies operating in multiple markets. Environmental and Social Acceptance is another critical factor. Public perception of GMO-*based* technologies and microbial processes is often influenced by misconceptions about biosafety and environmental impact. Gaining societal acceptance for the adoption of microbial technologies, particularly in food production and waste management, requires transparent communication and education [72].

The future of sustainable microbial processes lies in addressing these challenges through technological and policy innovations. Advancements in Systems and Synthetic Biology offer solutions to many technical challenges. By integrating omics technologies and AI-driven tools, researchers can design microbial strains with enhanced metabolic capabilities, resilience to industrial stressors, and the ability to utilize low-cost, renewable feedstocks such as agricultural residues or industrial byproducts. These approaches will make microbial processes more cost-effective and efficient. The development of Circular Economy Frameworks can further enhance the sustainability of microbial processes. For example, integrating SMP into biorefineries allows for the simultaneous production of biofuels, bioplastics, and value-added chemicals from waste streams. This holistic approach not only reduces costs but also supports waste minimization and resource recovery [73]. Innovative Policy and Regulatory Mechanisms are essential for fostering the adoption of microbial technologies. Governments and international organizations should streamline the regulatory approval process for GMOs while ensuring safety standards. Incentivizing companies through subsidies, tax benefits, and public–private partnerships can accelerate the commercialization of microbial innovations. Education and Public Engagement are crucial for gaining societal acceptance. Creating awareness about the environmental and economic benefits of microbial processes, along with transparent communication about biosafety measures, can alleviate public concerns and foster trust. Integration of AI and Machine Learning in microbial process development is expected to

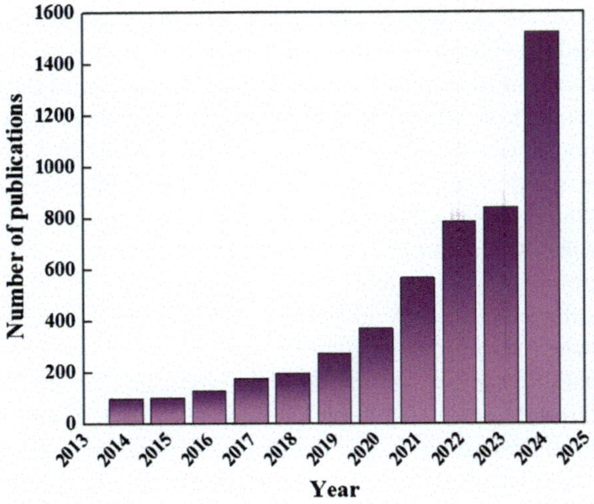

Fig. 4.3 The number of publications illustrating the increasing trend in research output, highlighting the growing significance of sustainable microbial processes in green chemistry from 2014 to 2024. (*Source* Elsevier and Web of Science journals)

revolutionize strain optimization, bioreactor monitoring, and process control. These technologies can significantly reduce development timelines and operational costs, making microbial processes more accessible to industries [74]. The steady rise in publications over the past decade, as illustrated by the bar graph below in Fig. 4.3 underscores the growing importance and potential of sustainable microbial processes in green chemistry pointing to a promising future for further research and innovative solutions in this field.

4.14 Conclusion and Long-Term Benefits

The commercialization of microbial processes hinges on addressing regulatory and environmental concerns. While microbial systems offer sustainability advantages, ensuring their safety for humans and ecosystems and meeting regulatory standards crucial for their widespread adoption. To guarantee their safety, microbial-*based* products used in food, medicine, or the environment must pass rigorous testing and approval procedures and their uses are enforced by regulatory organizations. While microbiological methods pose known risks, they still outperform traditional chemical processing methods when it comes to environmental impact. Our environment could get changed when genetically modified microorganisms enter nature plus genetic material moves between creatures. The disposal of waste materials from microbiological work has a significant environmental impact. When working to safeguard the environment we need to put strict safety practices in place to track and hold hazardous materials. Scientists recently developed techniques to make engineered microbes survive naturally only in specific conditions. Actively applying circular economy techniques makes microbial processes more environmentally friendly. Producing microbial functions helps keep ecosystems strong and healthy over many years. Through their effects on nutrient flow and soil quality while cleaning up unhealthy areas microorganisms keep ecosystems healthy under all conditions. Bacteria, fungi, and protozoa work on earth materials to make rich humus that feed plants while giving back needed nutrients to the soil. Soil microorganisms improve earth conditions making it more sustainable for farming by maintaining quality soil and decreasing artificial fertilizer use. Microorganisms maintain stability under environmental disruptions like droughts floods and pollution. Microorganisms conduct two main carbon functions in the environment by breaking down organic materials into soil and sediment while making long-lasting organic bonds. Through soil microbial processes carbon remains buried in the ground which eases climate impacts by lowering CO_2 amounts in the atmosphere. MFCs produce electrical power by microorganisms during their energy conversion process and simultaneously clean organic waste. Scientists now reengineer microbes to take in CO_2 from the atmosphere and use it to produce bioplastics and fuel sources. The technologies provide carbon–neutral and carbon-negative methods that help industries lower their carbon emissions and fight climate change. New technology based on AI and machine learning helps companies optimize their microbial operations by improving speed in developing microbial strains and monitoring processes in real-time along with predicting environmental impacts. Successful research shows that microbial technology can help us build more sustainable ecosystems while protecting natural life.

References

1. Matlack, A.: Introduction to Green Chemistry, 2nd ed. CRC Press (2010)
2. Anastas, P.T., Tracy, C.W.: Green chemistry: an overview. ACS Publications. 1–17 (1996)
3. Gupta, M.: Principles and issues of green chemistry. Int. J. Phys. Math. Sci. **3**(2), 19–22 (2014)
4. Clark, J.H.: Green chemistry: challenges and opportunities. Green Chem. **1**(1), 1–8 (1999)
5. Namieśnik, J., Wardencki, W.: Solventless sample preparation techniques in environmental analysis. J. High Resol. Chromatogr. **23**(4), 297–303 (2000)
6. Ikeda, M., Nakagawa, S.: The corynebacterium glutamicum genome: features and impacts on biotechnological processes. Appl. Microbiol. Biotechnol. **62**(2–3), 99–109 (2003)
7. Lynd, L.R., Weimer, P.J., Van Zyl, W.H., Pretorius, I.S.: Microbial cellulose utilization: fundamentals and biotechnology. MMBR. **66**(3), 506–577 (2002)
8. Wackett, L.P.: Pseudomonas putida—a versatile biocatalyst. Nat. Biotechnol. **21**(2), 136–138 (2003)
9. Hahn-Hägerdal, B., Galbe, M., Gorwa-Grauslund, M.F., Lidén, G., Zacchi, G.: Bio-ethanol–the fuel of tomorrow from the residues of today. Trends Biotechnol. **24**(12), 549–556 (2006)
10. Koller, M., Maršálek, L., de Sousa Dias, M.M., Braunegg, G.: Producing microbial polyhydroxyalkanoate (PHA) biopolyesters in a sustainable manner. N. Biotechnol. **37**, 24–38 (2017)
11. Wu, H., Zong, M.H., Lou, W.Y.: Efficient production of biodiesel using lipase immobilized on magnetic nanoparticles in a solvent-free system. Bioresour. Technol. **161**, 162–165 (2014)
12. Martínez, C.A., Hu, S., Dumond, Y., Tao, J., Kelleher, P., Tully, T.P.: Biocatalytic preparation of atorvastatin intermediates. Appl. Environ. Microbiol. **74**(6), 1917–1928 (2008)
13. Karthikeyan, O.P., Visvanathan, C.: Bio-energy recovery from high-solid organic substrates by dry anaerobic bio-digestion: a review. Rev. Environ. Sci. Biotechnol. **12**(3), 257–284 (2013)

14. Lee, S.Y., Jang, S.H.: Microbial production of butanol and other higher alcohols: a review. Biotechnol. Bioeng. **101**(2), 209–228 (2011)
15. Bischof, R.H., Ramoni, J., Seiboth, B.: Cellulases and beyond: the first 70 years of the enzyme producer *Trichoderma reesei*. Microb. Cell Factories **15**(1), 106 (2016).
16. Ramos, J.L.: *Pseudomonas putida*: a versatile biocatalyst for industrial and environmental applications. Nat. Rev. Microbiol. **9**(6), 439–450 (2011)
17. Demain, A.L.: Biosolutions to the energy problem. J. Ind. Microbiol. Biotechnol. **36**(3), 319–332 (2009)
18. Anderson, A.J., Dawes, E.A.: Occurrence, metabolism, metabolic role, and industrial uses of bacterial polyhydroxyalkanoates. Microbiol. Rev. **54**(4), 450–472 (1990)
19. Vessey, J.K.: Plant growth promoting rhizobacteria as biofertilizers. Plant Soil **255**(2), 571–586 (2003)
20. Das, N., Chandran, P.: Microbial degradation of petroleum hydrocarbon contaminants: an overview. Biotechnol. Res. Int. **2011**(1), 941810 (2011)
21. Hahn-Hägerdal, B., Karhumaa, K., Larsson, C.U., Gorwa-Grauslund, M., Görgens, J., Van Zyl, W.H.: Role of cultivation media in the development of yeast strains for large scale industrial use. Microb. Cell Factories **4**, 1–16 (2005)
22. Gumel, A.M., Annuar, M.S.M., Chisti, Y.: Recent advances in the production, recovery and applications of polyhydroxyalkanoates. J. Polym. Environ. **21**, 580–605 (2013)
23. Singh, S., Gupta, V.K.: Biodegradation and bioremediation of pollutants: perspectives strategies and applications. IJPBS **10**(1), 53 (2016)
24. Marteel-Parrish, A.E., Abraham, M.A.: Green chemistry and engineering: a pathway to sustainability. JWS (2013)
25. Liao, Y., Zhao, M.: Green chemistry and engineering: a sustainable approach to industrial chemistry. Wiley (2015)
26. Singh, B., Sundar, V.: Waste management and environmental impact assessment. Springer (2017)
27. Havelaar, A.H., Vazquez, K.M., Topalcengiz, Z., Muñoz-Carpena, R., Danyluk, M.D.: Evaluating the US food safety modernization act produce safety rule standard for microbial quality of agricultural water for growing produce. J. Food Prot. **80**(11), 1832–1841 (2017)
28. Mathuriya, A.S., Yakhmi, J.V., Martínez, L.M.T., Kharissova, O.V., Kharisov, B.I.: Polyhydroxyalkanoates: biodegradable plastics and their applications. Handb. Ecomater. **4**, 2873–2900 (2017)
29. Arunachalam, P.: Polymer-based nanocomposites for energy and environmental applications. In: Polymer-based Nanocomposites for Energy and Environmental Applications. Woodhead Publishing, pp 185–203 (2018)
30. Matsushika, A., Inoue, H., Kodaki, T., Sawayama, S.: Ethanol production from xylose in engineered Saccharomyces cerevisiae strains: current state and perspectives. Appl. Microbiol. Biotechno. **84**, 37–53 (2009)
31. Zhang, H., Yang, P., Wang, Z., Li, M., Zhang, J., Liu, D., Ying, H.: Clostridium acetobutylicum biofilm: advances in understanding the basis. Front. Bioeng. Biotechnol. **9**, 658568 (2021)
32. Risdian, C., Mozef, T., Wink, J.: Biosynthesis of polyketides in Streptomyces. Microorganisms **7**(5), 124 (2019)
33. Gupta, R., Beg, Q.K., Lorenz, P.: Bacterial alkaline proteases: molecular approaches and industrial applications. Appl. Microbiol. Biotechnol. **59**(1), 15–32 (2002)
34. Himmel, M.E., Ding, S.Y., Johnson, D.K., Adney, W.S., Nimlos, M.R., Brady, J.W., Foust, T.D.: Biomass recalcitrance: engineering plants and enzymes for biofuels production. Science **315**(5813), 804–807 (2007)
35. Chisti, Y.: Biodiesel from microalgae. Biotechnol. Adv. **25**(3), 294–306 (2007)
36. Gadd, G.M.: Biosorption: critical review of scientific rationale, environmental importance, and significance for pollution treatment. J. Chem. Technol. Biotechnol. **84**(1), 13–28 (2009)
37. Murrell, J.C., Jetten, M.S.: The microbial methane cycle. Environ. Microbiol. **1**(5), 279–284 (2009)

38. Karthikeyan, O.P., Visvanathan, C.: Bio-energy recovery from high-solid organic substrates by dry anaerobic bio-conversion processes: a review. Rev. Environ. Sci. Biotechnol. **12**(3), 257–284 (2013)
39. Bravo, A., Gill, S.S., Soberón, M.: Mode of action of bacillus thuringiensis cry and cyt toxins and their potential for insect control. Toxicon **49**(4), 423–435 (2007)
40. Chen, G.Q.: A microbial polyhydroxyalkanoates (PHA) based bio- and materials industry. Chem. Soc. Rev. **38**(8), 2434–2446 (2009)
41. Haritash, A.K., Kaushik, C.P.: Biodegradation aspects of polycyclic aromatic hydrocarbons (PAHs): a review. J. Hazard. Mater. **169**(1–3), 1–15 (2009)
42. Brennan, L., Owende, P.: Biofuels from microalgae—a review of technologies for production, processing, and extractions of biofuels and co-products. Renew. Sust. Energ. Rev. **14**(2), 557–577 (2010)
43. Holm-Nielsen, J.B., Al Seadi, T., Oleskowicz-Popiel, P.: The future of anaerobic digestion and biogas utilization. Bioresour. Technol. **100**(22), 5478–5484 (2009)
44. Appels, L., Baeyens, J., Degrève, J., Dewil, R.: Principles and potential of the anaerobic digestion of waste-activated sludge. Prog. Energy Combust. Sci. **34**(6), 755–781 (2008)
45. Duarte, A., Uribe, J.C., Sarache, W., Calderón, A.: Economic, environmental, and social assessment of bioethanol production using multiple coffee crop residues. Energy **216**, 119170 (2021)
46. Naik, S.N., Goud, V.V., Rout, P.K., Dalai, A.K.: Production of first and second generation biofuels: a comprehensive review. Renew. Sust. Energ. Rev. **14**(2), 578–597 (2010)
47. Bravo, A., Likitvivatanavong, S., Gill, S.S., Soberón, M.: Bacillus thuringiensis: a story of a successful bioinsecticide. Insect Biochem. Mol. Biol. **41**(7), 423–431 (2011)
48. Klaenhammer, T.R.: Bacteriocins of lactic acid bacteria. Biochimie **70**(3), 337–349 (1988)
49. Datta, R., Henry, M.: Lactic acid: Recent advances in products, processes and technologies—a review. J. Chem. Technol. Biotechnol. **81**(7), 1119–1129 (2006)
50. Virov, P.: Polyhydroxyalkanoates: biodegradable polymers and plastics from renewable resources. Mater. Technol. **47**, 5–12 (2013)
51. Geyer, R., Jambeck, J.R., Law, K.L.: Production, use, and fate of all plastics ever made. Sci. Adv. **3**(7), e1700782 (2017)
52. Atsumi, S., Hanai, T., Liao, J.C.: Non-fermentative pathways for synthesis of branched-chain higher alcohols as biofuels. Nature **451**(7174), 86–89 (2008)
53. de Jong, B., Siewers, V., Nielsen, J.: Systems biology of yeast: enabling technology for development of cell factories for production of advanced biofuels. Curr. Opin. Biotechnol. **23**(4), 624–630 (2012)
54. Kind, S., Neubauer, S., Becker, J., Yamamoto, M., Völkert, M., von Abendroth, G., Wittmann, C.: From zero to hero–production of bio-based nylon from renewable resources using engineered corynebacterium glutamicum. Metab. Eng. **25**, 113–123 (2014)
55. Cheng, Y., Bi, X., Xu, Y., Liu, Y., Li, J., Du, G., Liu, L.: Machine learning for metabolic pathway optimization: a review. Comput. Struct. Biotechnol. J. **21**, 2381–2393 (2023)
56. Carbonell, P., Currin, A., Jervis, A.J., Rattray, N.J., Swainston, N., Yan, C., Breitling, R.: Bioinformatics for the synthetic biology of natural products: integrating across the design–build–test cycle. Nat. Prod. Rep. **33**(8), 925–932 (2016)
57. Khan, R.A.: Natural products chemistry: the emerging trends and prospective goals. Saudi Pharm J. **26**(5), 739–753 (2018)
58. Mendes, R., Garbeva, P., Raaijmakers, J.M.: The rhizosphere microbiome: significance of plant beneficial, plant pathogenic, and human pathogenic microorganisms. FEMS Microbiol. Rev. **37**(5), 634–663 (2013)
59. Oldroyd, G.E., Dixon, R.: Biotechnological solutions to the nitrogen problem. Curr. Opin. Biotechnol. **26**, 19–24 (2014)
60. Das, N., Chandran, P.: Microbial degradation of petroleum hydrocarbons: an overview. Biotechnol. Res. Int. **2011**, 941810 (2011)
61. Papageorgiou, G.Z.: Thinking green: sustainable polymers from renewable resources. Polymers **10**(9), 952 (2018)

62. Limayem, A., Ricke, S.C.: Lignocellulosic biomass for bioethanol production: current perspectives, potential issues and future prospects. Prog. Energy Combust. Sci. **38**(4), 449–467 (2012)
63. Singh, V., Buehler, K.: Engineering microbial robustness for industrial bioprocesses. Curr. Opin. Biotechnol. **50**, 77–86 (2018)
64. Singhvi, M.S., Gokhale, D.V.: Lignocellulosic biomass: hurdles and challenges in its valorization. Appl. Microbiol. Biotechnol. **103**(23), 9305–9320 (2019)
65. Calvo-Flores, F.G., Martin-Martinez, F.J.: Biorefineries: achievements and challenges for a bio-based economy. Front. Chem. **10**, 973417 (2022)
66. Wiek, A., Zemp, S., Siegrist, M., Walter, A.I.: Sustainable governance of emerging technologies—critical constellations in the agent network of nanotechnology. Technol. Soc. **29**(4), 388–406 (2007)
67. Czekała, W.: Biogas as a sustainable and renewable energy source. Clean Fuels Mobil. 201–214 (2022)
68. Brierley, C.L.: Biohydrometallurgical prospects. Hydrometallurgy **165**, 85–95 (2017)
69. Shih, L., Van Winkle, A.: Improving cost-effectiveness in microbial biofuel production. Renew. Sust. Energ. Rev. **76**, 631–641 (2017)
70. Nielsen, J., Keasling, J.D.: Engineering cellular metabolism. Cell **164**(6), 1185–1197 (2016)
71. Newell, P.: Globalization and the governance of biotechnology. GEP. **3**(2), 56–71 (2003)
72. Qumsani, A.T.: The contribution of microorganisms to sustainable development: towards a green future through synthetic biology and systems biology. JUQUAS 1–17 (2024)
73. Bashor, C.J., Collins, J.J.: Understanding biological regulation through synthetic biology. Annu. Rev. Biophys. **47**(1), 399–423 (2018)
74. Schaepe, S., Kuprijanov, A., Sieblist, C., Jenzsch, M., Simutis, R., Lübbert, A.: Current advances in tools improving bioreactor performance. Curr Biotechnol. **3**(2), 133–144 (2014)

Chapter 5
Biocatalysis in Industrial Green Chemistry

Gaurav Uniyal, Chetan Kumar, Diya Vishwakarma, and Ritu Painuli

Abstract Based on the principles of green chemistry and sustainable development, biocatalysis have emerged as green and sustainable technology. Utilization of biomolecules as a catalyst assists inimitable and specific synthesis as the catalysts follow all the principle of "green chemistry." The green chemistry principle can increase efficacy and reduce environmental burden during chemical synthesis. Although the field of biocatalysts has made significant progress but its full potential is yet to appreciate. This is partly due to some significant difficulties in understanding the biochemistry of enzymes. Therefore, the present chapter delves into better understanding the principle a method of green chemistry and will highlight via examples, the ways by which the green chemistry an reduce the harmful impact of chemical process and its related technology on the environment. In addition, the application of biocatalysts in various fields like Biofuel production, Fine Chemicals, Pharmaceutical industry, Food and Beverage Processing Industry has also explored, thereby indicating the sustainability of biocatalytic reactions. Moreover, its broader application will further be encouraged in the future by the emerging biobased economy.

Keywords Green chemistry · Sustainability · Biocatalysts · Biofuel · Pharmaceutical industry · Food and beverage processing industry

---

G. Uniyal · D. Vishwakarma · R. Painuli (✉)
Department of Chemistry, School of Applied and Life Sciences, Uttaranchal University, Dehradun, Uttarakhand, India
e-mail: Ritsjune8.h@gmail.com

G. Uniyal
e-mail: gauravuniyal750@gmail.com

D. Vishwakarma
e-mail: vdiya0104@gmail.com

C. Kumar
School of Pharmaceutical and Populations Health Informatics, Faculty of Pharmacy, DIT University, Dehradun, Uttarakhand, India
e-mail: kumarbelwal@gmail.com

P. Jain et al. (eds.), *Towards Green Chemical Processes: Strategies and Innovations*, Intelligent Systems Reference Library 277,
https://doi.org/10.1007/978-3-031-94090-3_5

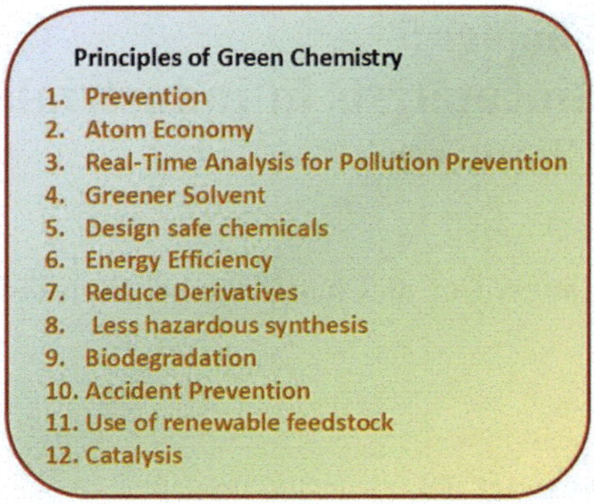

Fig. 5.1 12 fundamentals of green/clean/sustainable chemistry

5.1 Introduction

5.1.1 Green Chemistry

According to the current scenario, the entire world shares a common goal: green sustainability. The perception of green sustainability varies according to the domain. For the last few decades, the chemistry discipline has chosen green chemistry over traditional chemical approaches to achieving green sustainability [1]. According to Anastas and Warner [2], within the field of chemical science, this eco-friendly concept is based on the 12 principles that either directly or indirectly contribute to achieving goals of technological, economic, and ecological sustainability [2]. The term "green reaction" is introduced by sustainable chemistry, a new multidisciplinary approach to chemistry. The use of a catalyst or biocatalyst (which encourages the reaction to produce a selective product that increases the atom economy of the reaction by eliminating or reducing the formation of byproducts), greener solvents (such as water, supercritical carbon dioxide, ionic liquid, and ethanol), biomass as an eco-friendly reactant, and renewable energy sources (such as sunlight, ultrasound, and microwave) to provide activation energy are all factors that make that reaction green or sustainable [3, 4]. The ultimate goal of eco-friendly, clean, greener, or sustainable chemistry is to provide an ideal chemical reaction. An ideal reaction includes features such as 100% atom economy (A measure on how in any chemical reaction the starting material converts into the desired products) or null E-factor value, single step, ease, safety, and eco-acceptability (Fig. 5.1).

5.1.2 Biocatalysis as a Green Chemistry Tool

In the fine chemicals and pharmaceutical industries prior to the last two decades, inorganic reagents such as metals, metal hydrides, and oxidizing agents were used in the synthesis of complex molecules. Because of the complexity of molecules, their synthesis requires multiple steps, each of which produces byproducts. These multi-steps produce more waste (high E-factor) with a low atom economy [1]. The green chemistry approach offers a solution to this problem, which is catalysis. Berzelius, a Swedish chemist, first used the term "catalysis" in 1836. The term "catalyst" refers to a chemical species (molecules, metal, or ion) whose involvement allows the reaction to choose the path that requires the least amount of activation energy to overcome the energy barrier without being consumed or altered. This mechanism primarily accelerates up the chemical conversion. According to Ostwald (1895), catalysts can't alter the direction of equilibrium [5]. Catalysis is a cyclic chemical conversion process involving the adsorption of reactants on the catalyst's surfaces which facilitates the formation of an intermediate reactant-catalyst (highly unstable), followed by a transformation into the adsorbed product on the catalyst's surface. Finally, there is product desorption and catalyst regeneration (Fig. 5.2) [6–8].

The term "biocatalysis" refers to all catalyzed reactions that use natural substances (derived from biological sources) to accelerate the reaction rate. Biocatalytic methodologies are more environmentally friendly, and cost-effective, and do not require additional investment because they operate under mild conditions [9]. Table 5.1 shows the advantages of biocatalysis with green and sustainable chemistry. The pharmaceutical, fine chemical, agrochemical, cosmetic, flavor, and fragrance industries have all embraced green chemistry or green approaches over the past 20 years in place of the conventional synthesis method using the catalyst, particularly a biocatalyst. Biocatalysts assist in minimizing industrial pollution and the depletion of our ecosystem brought on by the careless disposal of effluent [1].

In this chapter, the function of biocatalysts in sustainable industrial processes—including those involving food, fine chemicals, pharmaceuticals, and biofuel is

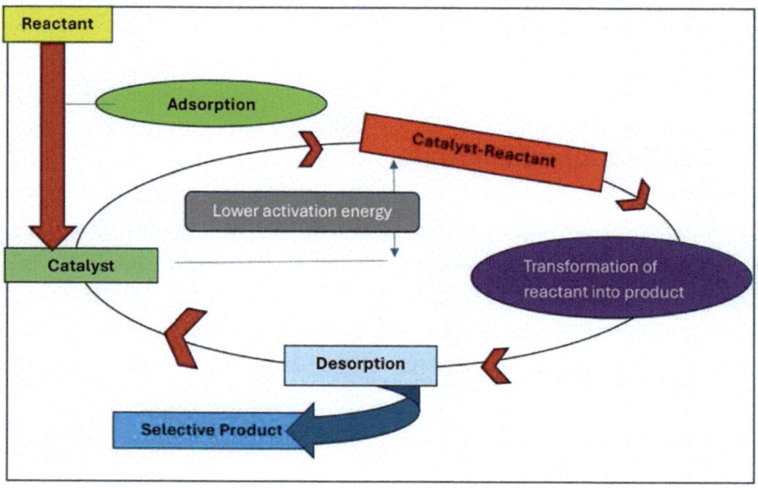

Fig. 5.2 Catalytic cycle mechanism

Table 5.1 Alignment of advantages of biocatalysis with green and sustainable chemistry

Green chemistry approaches fundamental	Biocatalysis's merits
Waste prevention	Efficiently reduced waste generation
Atom economy	The selective nature of the reaction reduces byproduct formation, making the step more cost-effective
Greener solvent	Usually performed in water (less volatile, non-toxic, and universal solvent)
Energy efficiency	Operate at low temperature or mild conditions
Catalysis	Enzymes are generally used as a catalyst
Real-time analysis	Applicability to biocatalytic process
Use of renewable feedstocks	Because of their renewable resource (living biological system)
Reduce derivatives	Avoids protection/deprotection steps
Accident prevention	Due to the mild and safe operating conditions, such as low temperature and pressure, inherently safer process

explored. It also explains its limitations, difficulties, and methods for resolving these issues as well.

5.2 Fundamental Principles of Biocatalysis

The word "enzyme" is derived from Latin and means yeast. This class of catalysts consists of protein-based biological macromolecules that are produced by living things [10].

5.2.1 Classification of Enzymes

Although there are many ways to group biocatalysts into different categories, since we are concentrating on their significance in chemical transformation, Therefore, the International Union of Biochemistry and Molecular Biology (IUBMB) enzymes division is considered. According to this, enzymes are divided into seven major classes based on the type of catalyzed reaction, as indicated as indicated in Table 5.2 [11].

Table 5.2 Enzyme classification according to the nature of the catalyzed reaction [10, 11]

Enzyme's class	Nature of catalyzed reaction	Examples
Oxidoreductase (dehydrogenases, oxidases, oxygenase and peroxidases)	Redox reactions (involve the transfer of hydrogen, oxygen, and electrons between the molecules)	Catalase (H_2O_2 Oxidoreductase), lactoperoxidase, lipoxygenase, laccases
Transferase	Intertransfer of atoms or groups (expects H, O and water) between two molecules	Glucotransferase (used in the modification of starch into cyclodextrins), transaminase
Hydrolases	Hydrolysis	Lactase, lipase, proteases, trypsin, ficin, chymosin
Lyases	Catalyze the additional of groups to double bonds and its vice versa	Aldolases, decarboxylases, and dehydratases
Isomerases—(racemases, epimerases, cis–trans isomerases, tautomerases, mutases and cyclo-isomerases)	Isomerization (rearrangement of the atom or group of atoms from one position to another position within a molecule)	Topoisomerase, Glucose isomerase,
Ligases (Synthetases)	Catalyze the synthesis of covalent bond formation (C–C, C-O, C-S and C-N) between two molecules via condensation reaction with simultaneous breakdown of ATP into ADP	Glutathione synthase, Aminoacyl t RNA synthetase
Translocases	Catalyze the movement of the ions or molecules across membranes or their separation within membranes	Ascorbate, ferrireductase, Ubiquinone reductase

5.2.2 *Model of Enzyme Action and Mechanism of Enzyme Action*

The term "substrate" is now synonymous with "reactant" in enzyme-catalyzed reactions, and the "active site" of the enzyme (catalyst) is the specific three-dimensional area of the catalyst where substrate binding phenomena take place and product formation is facilitated [10]. The enzyme action model explains how enzymes catalyzes reactions by interacting with their substrates (Table 5.2).

Fischer [12] assumed rigidity in enzymes (in the sense that the enzyme's active site cannot change shape during binding with a substrate) and compared the enzyme's active site to a lock and the substrate to a key because a specific key unlocks only that specific lock whose shape was identical to the key's shape. Because of this

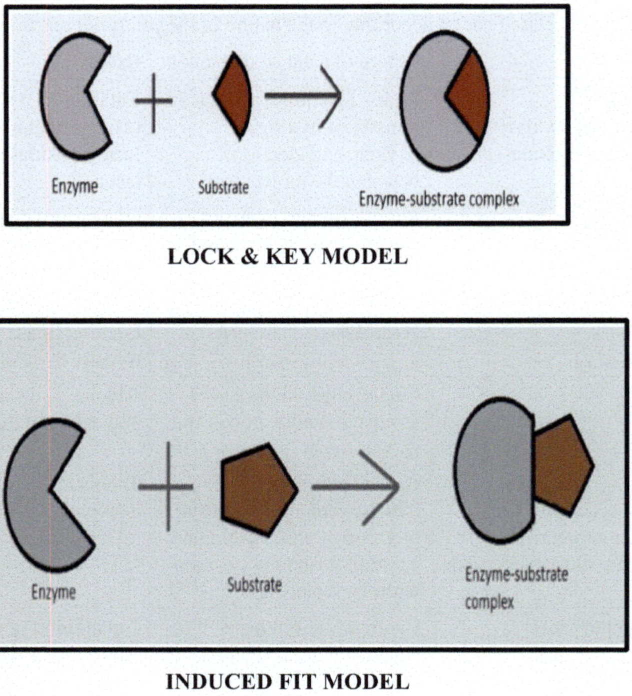

Fig. 5.3 Model of enzymatic action

comparison, Fischer's hypothesis is known as **the lock and key model** [10, 12] (Fig. 5.3).

Koshland's induced fit model, published in 1958, assumed flexibility in an enzyme's active site rather than rigidity. The flexibility of the active site allows for configurational change as the substrate approaches the enzyme. This flexibility refers to the broad substrate specificity [10, 13, 14] (Fig. 5.4 shows mechanism of enzyme action).

5.3 Applications in Different Industries

5.3.1 Biofuel Production

One of the potential solutions to the world's biggest problems is biofuels or biodiesel. These include air pollution, greenhouse gas emissions, sustainable development, and energy security. The combustion characteristics of biodiesel and fossil diesel are comparable; still, biofuels are non-toxic and biodegradable [15].

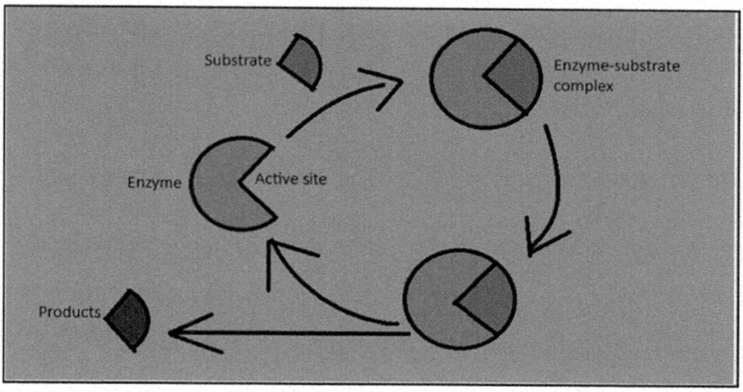

Fig. 5.4 Mechanism of enzymatic catalysis

Biodiesel is a product of transesterification or alcoholysis. Transesterification (Fig. 5.5) or alcoholysis terminology refers to that reaction of oil or fatty acid (sources are vegetable oils, recycled waste vegetable oils, animal fats and fish oil) with an excess amount of alcohol (methanol, ethanol, propanol, butanol, and amyl alcohol) in the presence of a catalyst (acid, base or biocatalyst) to form fatty acid alkyl esters. Methyl or ethyl ester of fatty acids used as biodiesel. Lipase is an enzyme that catalyzes transesterification rather than an inorganic catalyst or a heterogeneous enzyme because catalyzed alcoholysis has many advantages over chemical catalysis transesterification [15]. Enzymatic catalysis is more selective and specific than chemical catalysis, resulting in higher yields of high-quality biodiesel that lack the need for purification. Biocatalysis is a biocompatible, environmentally friendly, and energy-efficient methodology [16, 17]. However, this enzymatic transesterification transformation addresses a few issues, including lower lipase activity than other types of catalysts (such as acid, base, or heterogeneous catalysts), higher cost of biodiesel's product compared to other catalysis approaches as well as time-consuming way. Table 5.3 shows a comparative study for the transesterification methods for biodiesel's production.

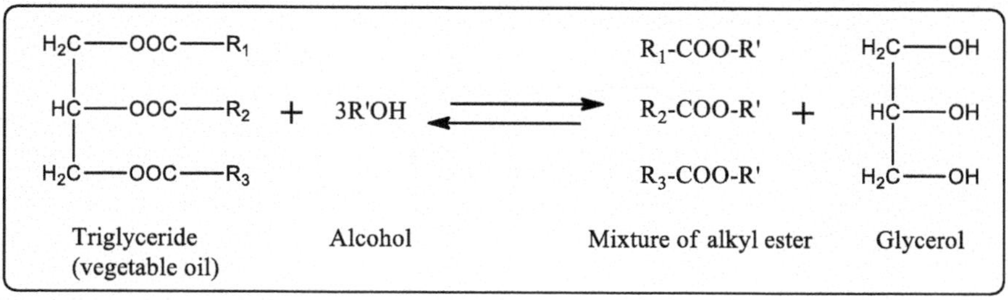

Fig. 5.5 Transesterification methods for biodiesel's production: a comparative study

Table 5.3 Transesterification methods for biodiesel's production: a comparative study [18–20]

Variable	Alkali catalysis	Acid catalysis	Lipase catalysis
Reaction temperature (°C)	60–70	55–80	30–40
Free fatty acid in raw materials	Saponified products	Esters	Methyl esters
Water in raw materials	Interference with reaction	Interference with reaction	No influence
Yields of biodiesel	Normal	Normal	Higher
Recovery of glycerol	Difficult	Difficult	Easy
Purification of biodiesel	Repeated washing	Repeated washing	None
Production cost of catalyst	Inexpensive	Inexpensive	Relatively inexpensive
Reaction time	Short	Short (9 h)	Long (36 h)

5.3.2 *Fine Chemicals*

One of the most notable examples of biocatalysis' role in the fine chemicals manufacturing industry is the production of acrylamide. Acrylamide is synthesized using conventional methodology, which involved the high-temperature (70–120 °C) hydration of acrylonitrile in the presence of an inorganic catalyst (Raney copper). That traditional method produced acrylamide with a low yield due to the generation of byproducts (which served as waste), and there was a significant need for purification due to acrylamide's tendency to polymerize. However, nitrile hydratases (Rhodococcus Rhodochrous J1) catalyzed the synthesis of acrylamide from acrylonitrile with 100% yield, and eliminating the purification steps helps to resolve all issues with the traditional method [21–25].

Nylon 6,6 is one of the most versatile polymers in the textile industry. Nylon 6,6 is a polyamide made up of two monomers: adipic acid and hexamethylene diamine. So, Adipic acid is currently produced at a massive scale of 2.2 million metric tons per year. The traditional method involves converting cyclohexane (derived from benzene) into cyclohexanol and cyclohexanone, which is then oxidized in nitric acid to produce adipic acid (Fig. 5.6). This method increases the concentration of nitrous oxide in the atmosphere by 10% globally. In the biocatalytic synthesis of adipic acid involves the conversion of glucose into cis, cis-muconic acid using engineered E. coli. WN1/pWN2.248 enzyme, followed by hydrogenation of cis, cis-muconic acid to produce adipic acid with a 97% yield while generating no nitrous oxide (primary air pollutant) (Fig. 5.7) [25–28].

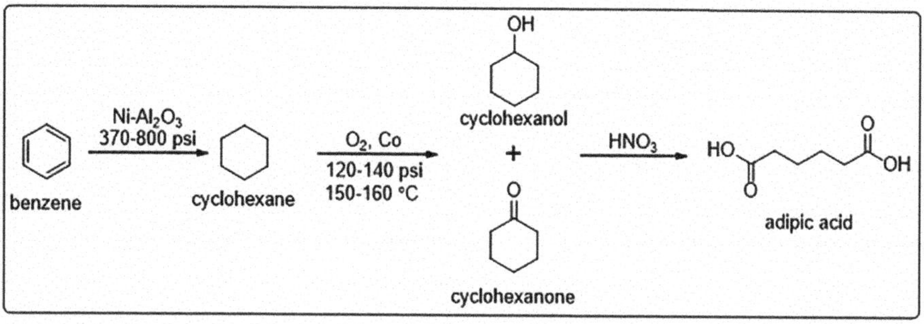

Fig. 5.6 Synthesis of adipic acid without biocatalysis

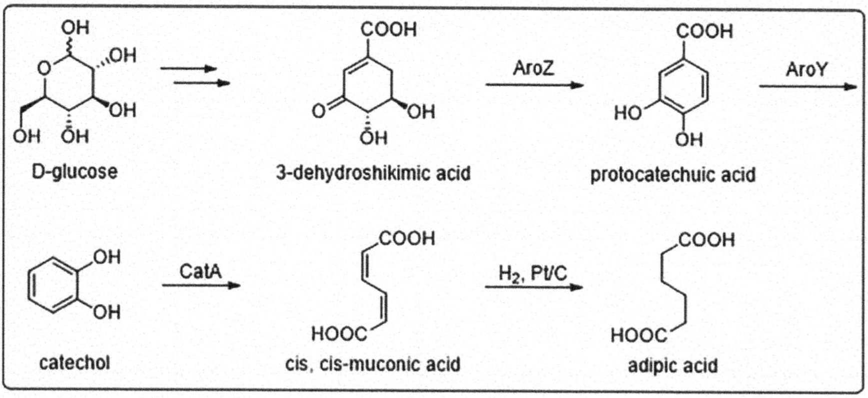

Fig. 5.7 Bio-catalyzed synthesis of adipic acid at industrial scale

5.3.3 Pharmaceutical Industry

Talampanol (a synthetic drug used for the treatment of epilepsy, brain tumor, and brain related disorders) is made using a conventional method that uses a lot of organic solvents and chromium oxide (an oxidizing agent) as well as 5-allyl-1,3-benzodioxole as a chemical precursor (Fig. 5.8). Talampanol was produced using this conventional method at a low yield of 16% per cycle. From an industrial standpoint, this approach was not commercially viable due to the low yield production. Biocatalytic production method is more environmentally friendly because it produces talampanol with a 51% increase in yield using *Zygosaccharomyces rouxii* while lowering the need for transition metal oxidants, solvents, and the overall reaction's E factor (reduces about 340,000 L of solvents and 3,000 kg of chromium waste per 1,000 kg of the drug) (Fig. 5.9) [25, 29–33].

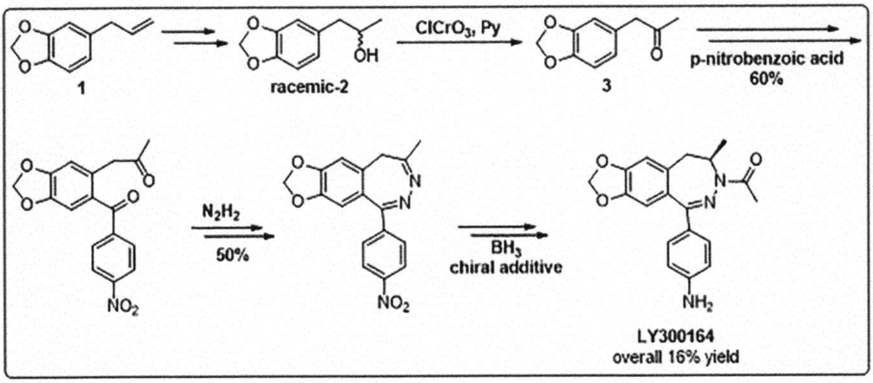

Fig. 5.8 Synthesis of talampanol without the usage of any enzymes

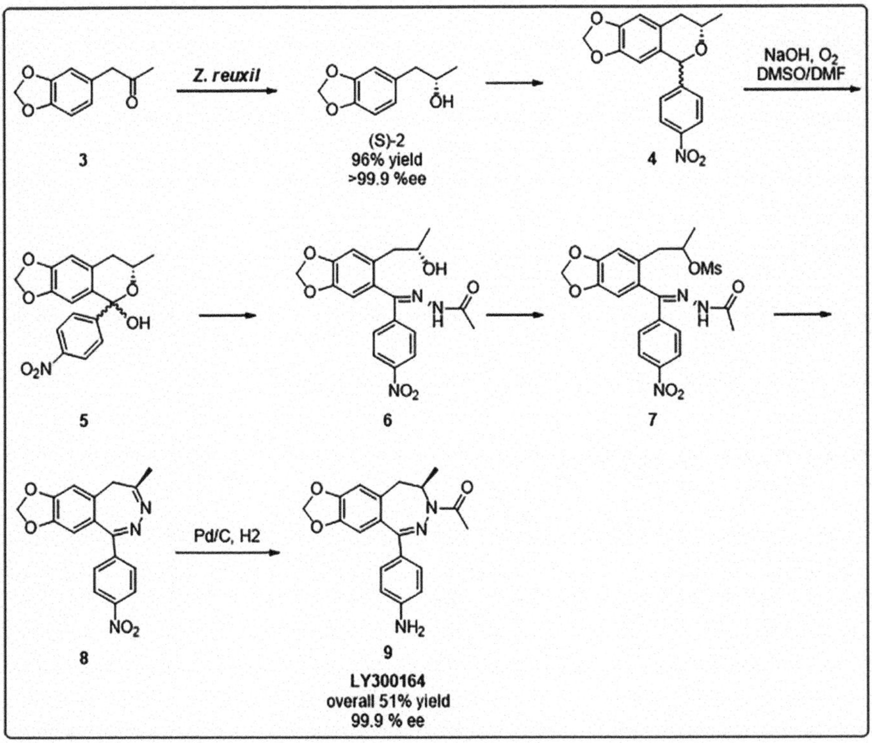

Fig. 5.9 Biocatalysised route of talampanol synthesis

5.3.4 Food and Beverage Processing Industry

Lignocellulose is known as the wood and non-wood plant biomass. Lignocellulolytic enzymes are usually employed for the hydrolysis of lignocellulose. It comprises of cellulose, lignin, pectin and hemicellulose. These enzymes are used for breaking the lignocellulosic substrates into soluble sugar for their application in the fruit and vegetable industry for adding flavor and texture to food products like oil, juice etc. [34].

For the extraction of vegetable oils usually chemical process is employed using hexane as a solvent. However, it can also be produced using pectinolytic enzymes, for disrupting the cell wall and increasing the phenolic compounds (thereby increasing the oil quality) [34, 35]. For preventing the acidity, randicity, and odor of olive oil (considered as best oil as it contains higher amount of monosaturated fats) [11, 36, 37].

Enzymes mixture of cellulase and hemicellulose are also used for stabilizing and improving nutritional value after extraction. Juices play major role in food and vegetable industry, owing to their nutritional values. The utilization of enzymes in these industries is a superior method as compare to traditional thermal and mechanical processing, as they increase juice extraction, and produce clear and pure juice [11, 38]. If fruits do not need to be peeled, extraction with pectinases enables the direct pulp separation by methods like centrifugation, sedimentation etc. A crucial step in the juice extraction process is peeling. The process of peeling involves using a knife, peeler, or other tool to remove the fruit's or other plant material's outer layer, or peel. In place of peeling tools, industries now use pectinases, which are enzymes. For example, apple juice extraction uses rhamnogalacturonans, which is derived from *Aspergillus aculeatus*, to peel apples [11, 39]. Pectinases, cellulases, and pectinases are used in the juice processing industry to improve organoleptic properties (such as appearance, aroma, taste, and mixture) [39].

For the first time in the dairy industry, a mixture of enzymes was used to produce cheese. Rennet was a mixture used as a coagulant that acted on milk proteins. Rennet is a mixture of chymosin and pepsin [38, 40]. Lactase, also known as β-galactosidase, is an enzyme that breaks down lactose into glucose and galactose. Its use makes dairy products compatible with lactose intolerant people (a condition in which a person is unable to digest lactose, sugar in milk, and other dairy products) [11, 40–42]. Traditionally, lactic yeast was the only source of this enzyme, but due to advances in technological and scientific research, it is now produced by genetically modified microorganisms (GMOs). A recent study used a GMO to produce β-D-fructofuranosidase for yogurt production. Table 5.4 shows application of different enzymes in food industry. This addition encouraged the growth of probiotics, which increased the nutritional value of the yogurt. Furthermore, the enzyme improved the gel formation process by shortening the time required for gelation, as well as changing the texture of the yogurt—specifically, decreasing elasticity and increasing viscosity [11, 42, 43].

Table 5.4 Applications of different enzymes in the production of food industry

S. no	Systemic name (other names)	Application in food industry	References
1	β-D-glucose: O_2 1-oxidoreductase (notalin; glucose oxidase)	In the meantime, being used as a food preservative and to remove oxygen from food packing, baking, the making of dry egg powder, and wine production	[44]
2	Thiol: O_2 oxidoreductase (sulphydryl oxidase)	Use in the bread and dairy industries; enhanced flavor	[45]
3	Donor: H_2 O_2 oxidoreductase (lactoperoxidase)	Food deterioration can be avoided by cold sterilizing milk	[46]
4	Linoate: O_2 13-oxidoreductase (lipoxygenase)	Bread whitening and dough strengthening	[47]
5	o-Diphenol: O_2oxidoreductase (laccases)	The manufacturing and processing of beverages, such as beer, wine, and fruit juice; the baking sector; and flavor enhancer	[48]
6	D-hexose: oxygen1-oxidoreductase (hexose oxidase)	It is used as an oxygen scavenger in the manufacturing of food items, bread, wheat flour, bakery goods, and milk products	[49]
7	Protein-glutamine: γ-glutamyltransferase (transglutaminase)	Production of cheese and other dairy products, animal processing, edible film production, and bakery goods manufacturing	[50]
8	1,4-α-D-glucan: amine 4-α-D(1,4 α-D-glucano)-transferase (cyclodextrin, glucanotransferase)	Extensively employed in the synthesis of cyclodextrin, which is utilized to create food-grade micro-encapsulants for vitamins, tastes, and colors	[51]
9	Triacyl glycerol acylhydrolase (lipase, tributyrase)	Extensively utilized in food processing, including taste enhancement, oil and fat alteration, and quality improvement	[52]
10	Phosphatidylcholine 2-acylhydrolase (phospholipase A, lechithinase A)	Phospholipase A and lechithinase A are examples of phosphatidylcholine 2-acylhydrolase	[53]
11	Tannin acyl hydrolase (tannase)	Making tea and cold beverages; making gallic acid and propylgallate, which are used as antioxidants and food preservatives	[54]
12	1,4-α-D-glucan glucohydrolase (amyloglucosidase, glucoamylase)	Making corn syrup, turning dextrins into glucose, making beer and malt liquor, and making fruit juice	[51]

(continued)

Table 5.4 (continued)

S. no	Systemic name (other names)	Application in food industry	References
13	1,4-(1,3;1,4)-β-D-glucan 4-glycanohydrolase (cellulase)	Production of fruit juice, solubilization of pentosan in baking, extractions of natural flavors and colors, application in the tea sector, and detergent additives	[55]
14	2,1-β-D-fructanfructanohydrolase (inulinase)	Inulin hydrolysis	[56]
15	Poly-(1,4-α-D-galacturonide) glycanohydrolase (pectinase, endopolygalacturonase)	Manufacturing of premium tomato ketchup and fruit pulps, making low-viscosity turbid vegetable juice, making fruit and vegetable juice, making coffee and tea, and retting and degumming fiber crops	[57]
16	β-D-galactosidegalactohydrolase (β-galactosidase, lactase)	Lactase-reducing enzyme, making ice cream	[51]
17	4-β-D-xylanxylohydrolase (pentosanase)	Production of nonstandard baked items, degerming, distillery mesh, vineger mesh, and beer manufacturing	[51]
18	Subtilisin (alkaline proteinase)	Meat tenderizer, bread, beer, milk coagulation for cheese manufacture, hydrolysate production for soups and savory dishes	[46, 51]
19	Ficain (ficin)	Tenderization of meat; hydrolysis of plant, animal, and milk proteins	[46, 51]
20	Fruit bromelain (juice bromelain)	Preparing beer, malt liquor, and tenderizing meat	[46, 51]

5.4 Opportunities and Challenges for the Biocatalysis in the Chemical Industry

5.4.1 *Challenges*

Despite its potential for efficient and sustainable manufacturing, biocatalysis faces a number of challenges to broad industrial adoption. For biocatalysts, regulatory obstacles including the EU's Classification, Labelling, and Packaging (CLP) Regulation and safety regulations in the US and the EU pose major obstacles, especially in food and feed applications where severe safety regulations are in place. Furthermore, biocatalysis is sometimes perceived as being more expensive and complicated than conventional chemical techniques, and biocatalysts frequently have stability, specificity, and scalability issues. While leftover DNA in enzymes obtained from genetically modified organisms (GMOs) creates societal worries despite the low dangers, contamination with poisons or infectious agents also causes safety concerns, particularly in the food and pharmaceutical industries. These issues are being watched by regulatory agencies such as the US FDA and EFSA, especially in industries like food processing, pharmaceuticals, cosmetics, and animal feed where the use of enzymes is expanding yet still requires intricate safety analyses. These technical, legal, and safety concerns must be resolved if biocatalysis is to reach its full industrial potential [58].

5.4.2 Opportunities

The chemical sector, which mostly uses oil-based resources to produce chemicals, is under pressure from competition due to the global instability of fossil energy prices. Less than 4% of biomass production is being employed for industrial reasons, compared to an estimated 170 billion tons of mineral oil reserves. The transition from fossil oil to biomass as a renewable raw material for manufacturing is being driven by worries about climate change and resource scarcity. This shift is becoming more possible thanks to developments in metabolic engineering and pathway design, but creating a new value chain may require large financial outlays and cooperation between business and academia. Concerns regarding the effects on food crops are also raised by the use of water and fertile acreage for biomass production. Despite these obstacles, businesses are moving away from biofuels and toward a wider variety of chemicals as metabolic engineering gains traction in the commodities and fine chemical industries. Businesses like Lonza are using their knowledge of chemistry and biotechnology to meet the increasing demand for bio-based products. Table 5.5 shows the opportunities and challenges in the applications of biocatalysis in the chemical industry.

In the 1990s, the US Environmental Protection Agency created the idea of "green chemistry," which highlights the need for safer, more environmentally friendly chemical manufacturing methods. By decreasing the demand for organic solvents and increasing process efficiency, biocatalysis—which employs enzymes as catalysts—is becoming more widely acknowledged for improving sustainability. For instance, Pfizer's transition to a lipase-catalyzed method for pregabalin synthesis greatly decreased the amount of solvent used and trash produced. By encouraging cooperation across industries, the Swiss Industrial Biocatalysis Consortium (SibC), which was established in 2004, has significantly advanced biocatalysis. Its members have developed a roadmap for quickening biotransformation in industrial applications [58–61].

Table 5.5 Opportunities and challenges in the applications of biocatalysis in the chemical industry

S. No	Perspectives	Opportunities	Challenges
1	Precursors	Fossil fuel price fluctuations and Abundant renewable resources	Restructuring of the value chain
2	Chemical production	Technological revolution in action	Innovation Gap and Competitiveness Loss
3	Public perception	Preemptive communication strategies	Lack of scientific literacy
4	Regulatory compliance	Evolving legal frameworks for applications	Multi-layered regulatory environment
5	Future prospects	Versatile chemical industry tool and Ultimate solution for the chemical industry	–
6	Education	Concerted education effort	Complexity, Awareness Deficit, and Tool Limitations
7	Green Sustainability	Waste minimization as well as Limiting hazardous material use	Production cost efficiency
8	Cost of development	Streamlined and cost-effective development	Enzyme Selection Constraints and Licensing Dependence

5.5 Conclusion

Currently, the global goal is green sustainability. Biocatalysis is a solution for green sustainability. The critical role that biocatalysis plays in industrial green chemistry has been illustrated in this chapter. According to how the chemical industry views biocatalysis, it is a sustainable method that makes chemical synthesis more energy-efficient (biocatalysis prefers to work only in mild conditions), environmentally acceptable (by lowering waste generation or E-factor, using greener solvents instead of organic toxic solvents, and having a biodegradable nature), and commercialized (due to high-yield product production due to the selective as well as specific nature of enzymes). According to the current scenario, the entire world shares a common goal: green sustainability. Biocatalysis is one of the solutions to achieve this green sustainability. The current state of environmental degradation has caused every chemical manufacturing industry to abandon traditional methods in favor of more environmentally friendly ones. These industries include biodiesel ion, agrochemicals (such as pesticides, herbicides, and insecticides), fine chemicals, pharmaceuticals, the food and beverage sector, and the textile sector. Biocatalysis plays a significant role in the environmentally friendly and sustainable production of these products. Although there are many advantages of biocatalysis over conventional techniques, there are also some drawbacks to using enzymes in the reaction, which will present further difficulties for proponents of the field. In comparison to the conventional method of using enzymes and their activity (which is highly sensitive to the acidity or basicity of the reaction mixture environment), biocatalysis using presents a number of difficulties, including cost and time.

References

1. Sheldon, R.A., Woodley, J.M.: Role of biocatalysis in sustainable chemistry. Chem. Rev. **118**(2), 801–838 (2018). https://doi.org/10.1021/acs.chemrev.7b00203
2. Anastas, P.T., Warner, J.C.: Green chemistry: theory and practice. Oxford University Press (2000). https://doi.org/10.1093/oso/9780198506980.001.0001
3. Ivanković, A., Dronjić, A., Bevanda, A.M., Talić, S.: Review of 12 principles of green chemistry in practice. Int. J. Sustain. Green Energy **6**(3), 39–48 (2017). https://doi.org/10.11648/j.ijrse.20170603.12
4. Margetić, D.: Mechanochemical organic reactions without the use of solvents. Kemija u industriji: časopiskemičaraitehnologaHrvatske **54**(7–8), 351–358 (2005)
5. Kakaei, K., Esrafili M.D., Ehsani, A.: Introduction to catalysis. Interface Sci. Technol. **27**, 1–21 (2019). https://doi.org/10.1016/B978-0-12-814523-4.00001-0
6. Richardson, J.T.: Principles of catalyst development. Springer (2013). https://doi.org/10.1007/978-1-4899-3725-4
7. Hagen, J.: Introduction. Industrial catalysis: a practical approach. Wiley, pp. 1–16 (2015). https://doi.org/10.1002/9783527684625.ch1
8. Bartholomew, C.H., Farrauto, R.J.: Fundamentals of industrial catalytic processes. Wiley (2011)
9. Paul, P.E.V., Sangeetha, V., Deepika, R.G.: Emerging trends in the industrial production of chemical products by microorganisms. In: Recent Developments in Applied Microbiology and Biochemistry, pp. 107–125 (2019). https://doi.org/10.1016/B978-0-12-816328-3.00009-X
10. Subin, S.R., Bhat, S.G.: Enzymes: concepts, nomenclature, mechanism of action and kinetics, characteristics and sources of food-grade enzymes. In: Enzymes in Food and Beverage Processing, pp. 3–38 (2015). https://doi.org/10.1201/b19408-3
11. Garcia-Quinto, E., Aranda-Cañada, R., García-García, P., Fernández-Lorente, G.: Use of potential immobilized enzymes for the modification of liquid foods in the food industry. Processes **11**(6), 1840 (2023). https://doi.org/10.3390/pr11061840
12. Fischer, E.: Einfluss der configuration auf die wirkung der enzyme. Berichte der deutschen-chemischen Gesellschaft **27**(3), 2985–2993 (1894).https://doi.org/10.1002/cber.18940270364
13. Prof. Dr. Koshland Jr, D.E.: The key–lock theory and the induced fit theory. AngewandteChemie Int. Ed. Engl. **33**(23–24), 2375–2378 (1995). https://doi.org/10.1002/anie.199423751
14. Koshland Jr, D.E.: Application of a theory of enzyme specificity to protein synthesis. Proc. Natl. Acad. Sci. **44**(2), 98–104 (1958). https://doi.org/10.1073/pnas.44.2.98
15. Parawira, W.: Biotechnological production of biodiesel fuel using biocatalysed transesterification: a review. Crit. Rev. Biotechnol. **29**(2), 82–93 (2009). https://doi.org/10.1080/07388550902823674
16. Devanesan, M.G., Viruthagiri, T., Sugumar, N.: Transesterification of Jatropha oil using immobilized Pseudomonas fluorescens. Afr. J. Biotechnol. **6**(21) (2007). https://doi.org/10.5897/AJB2007.000-2396
17. Marchetti, J.M., Miguel, V.U., Errazu, A.F.: Possible methods for biodiesel production. Renew. Sustain. Energy Rev. **11**(6), 1300–1311 (2007). https://doi.org/10.1016/j.rser.2005.08.006
18. Chen, J.W., Wu W.T.: Regeneration of immobilized Candida antarctica lipase for transesterification. J. Biosci. Bioeng. **95**(5), 466–469 (2003). https://doi.org/10.1016/S1389-1723(03)80046-4

19. Watanabe, Y., Shimada, Y., Sugihara, A., Noda, H., Fukuda, H., Tominaga, Y.: Continuous production of biodiesel fuel from vegetable oil using immobilized Candida antarctica lipase. J. Am. Oil Chemists' Soc. **77**(4), 355–360 (2000). https://doi.org/10.1007/s11746-000-0058-9
20. Crabbe, E., NolascoHipolito, C., Kobayashi, G., Sonomoto, K., Ishizaki, A.: Biodiesel production from crude palm oil and evaluation of butanol extraction and fuel properties. Process Biochem. **37**(1), 65–71 (2001). https://doi.org/10.1016/S0032-9592(01)00178-9
21. Prof. Dr. Liese, A., Dr. Seelbach, K., Prof. Dr. Wandrey, C.: Industrial biotransformations. Wiley **6**(12), 1044–1046 (2001). https://doi.org/10.3390/61201044
22. Yamada, H., Kobayashi, M.: Nitrile hydratase and its application to industrial production of acrylamide. Biosci. Biotechnol. Biochem. **60**(9), 1391–1400 (1996). https://doi.org/10.1271/bbb.60.1391
23. Mersinger, L.J., Hann, E.C., Cooling, F.B., Gavagan, J.E., Ben-Bassat, A., Wu, S., Petrillo, K.L., Payne, M.S., DiCosimo, R.: Production of acrylamide using alginate-immobilized E. coli expressing comamonastestosteroni 5-MGAM-4D nitrile hydratase. Adv. Synth. Catal. **347**(7–8), 1125–1131 (2005). https://doi.org/10.1002/adsc.200505039
24. Maksimov, A.Y., Maksimova, Y.G., Kuznetsova, M.V. et al.: Immobilization of Rhodococcusruber strain gt1, possessing nitrile hydratase activity, on carbon supports. Appl. Biochem. Microbiol. **43**, 173–177 (2007). https://doi.org/10.1134/S000368380702007X
25. Ran, N., Zhao, L., Chenb, Z., Tao, J.: Recent applications of biocatalysis in developing green chemistry for chemical synthesis at the industrial scale. Green Chem. **10**(4), 361–372 (2008). https://doi.org/10.1039/b716045c
26. Niu, W., Draths, K.M., Frost, J.W.: Benzene-free synthesis of adipic acid. Biotechnol. Prog. **18**(2), 201–211 (2002). https://doi.org/10.1021/bp010179x
27. Draths K.M., Frost, J.W.: Environmentally compatible synthesis of adipic acid from D-glucose. J. Am. Chem. Soc. **116**(1), 399–400 (1994). https://doi.org/10.1021/ja00080a057
28. Mabermann, C.E., Kroschwitz, J.I.: Encyclopedia of Chemical Technology. Wiley, New York **1**, 251 (1991)
29. Ling, I., Podányi, B., Hámori, T., Sólyom, S.: Asymmetric reduction of a carbon–nitrogen double bond: Enantioselective synthesis of 4, 5-dihydro-3 H-2, 3-benzodiazepines. J. Chem. Soc. Perkin Trans. **1**(11), 1423–1427 (1995). https://doi.org/10.1039/P19950001423
30. Zappalà, M., Grasso, S., Micale, N., Polimeni, S., De Micheli, C.: A simple and efficient synthesis of GYKI 52466 and GYKI 52895. Synth. Commun. **32**(4), 527–533 (2002). https://doi.org/10.1081/SCC-120002397
31. Anderson, B.A., Hansen, M.M., Harkness, A.R., Henry, C.L., Vicenzi, J.T., Zmijewski, M.J.: Application of a practical biocatalytic reduction to an enantioselective synthesis of the 5H-2, 3-benzodiazepine LY300164. J. Am. Chem. Soc. **117**(49), 12358–12359 (1995). https://doi.org/10.1021/ja00154a048
32. http://www.epa.gov/greenchemistry/pubs/docs
33. Ran, N., Zhao, L., Chen, Z., Tao, J.: Recent applications of biocatalysis in developing green chemistry for chemical synthesis at the industrial scale. Green Chem. **10**(4), 361–372 (2008). https://doi.org/10.1039/B716045C
34. Toushik, S.H., Lee, K.T., Lee, J.S., Kim, K.S.: Functional applications of lignocellulolytic enzymes in the fruit and vegetable processing industries. J. Food Sci. **82**(3), 585–593 (2017). https://doi.org/10.1111/1750-3841.13636
35. Kuhad, R.C., Singh, A.: Biotechnology for Environmental Management and Resource Recovery. Springer (2013). https://doi.org/10.1007/978-81-322-0876-1
36. HadjTaieb, N., Grati, N., Ayadi, M., Attia, I., Bensalem, H., Gargouri, A.: Optimisation of olive oil extraction and minor compounds content of Tunisian olive oil using enzymatic formulations during malaxation. Biochem. Eng. J. **62**, 79–85 (2012). https://doi.org/10.1016/j.bej.2011.04.003
37. Chiacchierini, E., Mele, G., Restuccia, D., Vinci, G.: Impact evaluation of innovative and sustainable extraction technologies on olive oil quality. Trends Food Sci. Technol. **18**(6), 299–305 (2007). https://doi.org/10.1016/j.tifs.2007.01.008

38. Chaudhary, S., Sagar, S., Kumar, M., Sengar, R.S., Tomar, A.: The use of enzymes in food processing: a review. South Asian J. Food Technol. Environ. **1**(3–4), 190–210 (2015)
39. Sharma, H.P., Patel, H., Sharma, S.: Enzymatic extraction and clarification of juice from various fruits–a review. Trends Post Harvest Technol. **2**(1), 1–14 (2014)
40. Gomes Gonçalves, P.: Síntesisheteróloga de la quimosina bovina en Escherichia coli para suusoenalimentación (2022)
41. Ferreira-Lazarte, A., Moreno, F.J., Villamiel,M.: Application of a commercial digestive supplement formulated with enzymes and probiotics in lactase non-persistence management. Food Funct. **9**(9), 4642–4650 (2018). https://doi.org/10.1039/C8FO01091A
42. Dekker, P.J., Koenders, D., Bruins, M.J.: Lactose-free dairy products: market developments, production, nutrition and health benefits. Nutrients **11**(3), 551 (2019). https://doi.org/10.3390/nu11030551
43. Chen, Z., Shen, Y., Xu, J.: A strategy for rapid acquisition of the β-D-fructofuranosidase gene through chemical synthesis and new function of its encoded enzyme to improve gel properties during yogurt processing. Foods **12**(8), 1704 (2023). https://doi.org/10.3390/foods12081704
44. Wong, C.M., Wong, K.H., Chen, X.D.: Glucose oxidase: natural occurrence, function, properties and industrial applications. Appl. Microbiol. Biotechnol. **78**(6), 927–938 (2008). https://doi.org/10.1007/s00253-008-1407-4
45. Faccio, G., Nivala, O., Kruus, K., Buchert, J., Saloheimo,M.: Sulfhydryl oxidases: sources, properties, production and applications. Appl. Microbiol. Biotechnol. **91**, 957–966 (2011). https://doi.org/10.1007/s00253-011-3440-y
46. Whitehurst, R.J., Van Oort, M., (eds.): Enzymes in food technology. Wiley-Blackwell, Singapore vol 388 (2010). https://doi.org/10.1002/9781444309935
47. Kirk, O., Borchert, T.V., Fuglsang, C.C.: Industrial enzyme applications. Curr. Opin. Biotechnol. **13**(4), 345–351 (2002). https://doi.org/10.1016/S0958-1669(02)00328-2
48. Gianfreda, L., Xu, F., Bollag, J.M.: Laccases: a useful group of oxidoreductive enzymes. Bioremediat. J. **3**(1), 1–26 (1999). https://doi.org/10.1080/10889869991219163
49. Olempska-Beer, Z., Hattan, D.: Hexose oxidase from chondrus crispus expressedin hansenula polymorpha. In: Safety evaluation of certain food additives, pp. 37 (2006)
50. Kieliszek, M., Misiewicz, A.: Microbial transglutaminase and its application in the food industry. a review. Folia Microbiol. **59**, 241–250 (2014). https://doi.org/10.1007/s12223-013-0287-x
51. Van Der Maarel, M.J., Van der Veen, B., Uitenhage, J.C., Leemhuis, H., Dijkhuizen.: Properties and applications of starch-converting enzymes of the α-amylase family. J. Biotechnol. **94**(2), 137–155 (2002). https://doi.org/10.1016/S0168-1656(01)00407-2
52. Hasan, F., Shah, A.A., Hameed, A.: Industrial applications of microbial lipases. Enzym. Microb. Technol. **39**(2), 235–251 (2006). https://doi.org/10.1016/j.enzmictec.2005.10.016
53. Casado, V., Martín, D., Torres, C., Reglero,G.: Phospholipases in food industry: a review. In: Lipases and Phospholipases: Methods and Protocols, pp. 495–523 (2012). https://doi.org/10.1007/978-1-61779-600-5_29
54. Chae, S.K., Yu, T.J.: Experimental manufacture of acorn wine by fungal tannase. Korean J. Food Sci. Technol. **15**(4), 326–332 (1983)
55. Uhlig, H.: Industrial enzymes and their applications. Wiley (1998)
56. Fernandes, M.R., Jiang, B.: Research article fungal inulinases as potential enzymes for application in the food industry. Adv. J. Food Sci. Technol. **5**(8), 1031–1042 (2013). https://doi.org/10.19026/ajfst.5.3201
57. Bhat, B.M.: MK14538100. Cellulases and related enzymes in biotechnology. Biotechnol. Adv. **18**(5), 355–383 (2000). https://doi.org/10.1016/S0734-9750(00)00041-0
58. Buller, R., Hecht, K., Mirata, M.A., Meyer, H.P.: An appreciation of biocatalysis in the Swiss manufacturing environment (2017). https://doi.org/10.1039/9781782629993-00001
59. Chiellini, E., Chiellini, F., Cinelli, P.: Polymers from renewable resources. In: Degradable Polymers: Principles and Applications, pp. 163–233 (2002). https://doi.org/10.1007/978-94-017-1217-0_7
60. United States Environmental Protection Agency. www.epa.gov. Last accessed Oct 2016

61. Wells, A.S., Finch, G.L., Michels, P.C., Wong, J.W.: Use of enzymes in the manufacture of active pharmaceutical ingredients a science and safety-based approach to ensure patient safety and drug quality. Org. Process. Res. Dev. **16**(12), 1986–1993 (2012). https://doi.org/10.1021/op300153b

Chapter 6
Integration of Renewable Energy Source into Chemical Production

Vaibhav Sharma, Piyush Gupta, and Pallavi Jain

Abstract Renewable energy sources are paving ways for sustainable chemical manufacturing, such as solar energy facilitates H_2 preparation, CO_2 reduction, and chemical synthesis, while wind energy aids in methanol production and hybrid methods promote biomethane and acetic acid generation. Nonetheless, obstacles persist in enhancing the integration of renewable energy sources while ensuring cost efficiency. Intermittent challenges necessitate intelligent grids, energy storage solutions, and demand-response mechanisms. Exorbitant expenses, catalyst degradation hinder large scale adoption. Ongoing technology progress, decrease in cost, and legislative support are essential for the integration of renewable energy sources in chemical manufacturing sector. This chapter discusses the synthesis of green hydrogen, ammonia, methanol, and acetic acid by the incorporation of solar, wind, and geothermal energy in their production method. It also highlights the technical challenges during the process of chemical production using renewable energy.

Keywords Renewable energy · Solar energy · Wind energy · Geothermal energy · Chemical production

6.1 Introduction

Currently, the escalating demand for energy is emerging as the foremost global concern. Fossil fuels, however, contribute to global warming through the emission of greenhouse gases, therefore, industries are required to find suitable solutions to overcome these challenges and reduce the impacts on the ecosystem [1]. Concerns

---

V. Sharma
Department of Chemistry, SRM Institute of Science and Technology, NCR Campus, Delhi-NCR Campus, Modinagar, Ghaziabad, Uttar Pradesh, India

P. Gupta (✉) · P. Jain
Department of Chemistry, Faculty of Engineering and Technology, SRM Institute of Science and Technology, NCR Campus, Delhi-NCR Campus, Modinagar, Ghaziabad, Uttar Pradesh, India
e-mail: piyushg.1977@gmail.com

P. Jain et al. (eds.), *Towards Green Chemical Processes: Strategies and Innovations*, Intelligent Systems Reference Library 277,
https://doi.org/10.1007/978-3-031-94090-3_6

regarding global warming, rising demand for energy, and the more stringent regulations on greenhouse gaseous emissions have facilitated the integration of renewable energy (RE) sources in the chemical industries. RE sources have gained significance due to their sustainability and reduced environmental impact. Hydropower, wind, solar, and biomass are the most often utilized sources. Solar energy, emerges as a viable solution for meeting global energy demands. Presently, battery technologies lack the capability to store significant amounts of energy. Consequently, to satisfy specific power and fuel demands while reducing the impact of renewable resource availability, various technological solutions exist, including the generation of thermal storage chemicals such as H_2 or methane, as well as hydro-storage solutions. The different types of RE sources were utilized to solve the issue of uneven flow of energy in the production of chemicals in chemical industries [2]. There is an increasing emphasis on developing environmental sustainable energy systems and expanding the integration of RE sources. Energy integration systems, capable of meeting diverse energy demands, are seen as a feasible strategy for enhancing the usage of RE resources and optimizing generation efficiency [3]. RE helped in the electrification of chemical sector. H_2 was prepared from water electrolysis utilizing solar or wind energy while methanol, ethanol and formic acid were prepared by using CO_2 and biomass fermentation using RE sources. NH_3 was prepared by using green H_2 in Haber–Bosch method with renewable electricity. The anticipated shift towards electrification and the utilization of RE, which mutually reinforce each other, is expected to foster a "green" sector and play a pivotal role in achieving environmental objectives [4]. This chapter highlights the manufacturing of various chemicals like green H_2, ammonia (NH_3), methanol and acetic acid by integration of solar, wind and geothermal energy in the manufacturing process of these chemicals.

6.2 Green H_2 Production using RE

H_2 underground storage was introduced to store renewable energy in chemical form at large scale. H_2 generated by using electrolysis of water was utilized as a substitute to natural gas, as a fuel for transportation and as a chemical feedstock for industry [5]. Mixed-integer linear programming was integrated with electricity generation and H_2 production methods by using RE sources, aimed at minimizing the total annual cost. The efficacy of the suggested approach was corroborated by a case study conducted on Jeju Island, Korea [6]. The integration of RE with electrolysis process could produce green H_2 which improved the environmental sustainability of energy systems and strengthens global initiatives to reduce the concentration of carbon in environment. Green H_2, produced by electrolyzing H_2O molecules into H_2 and O_2 with renewable electricity, holds significant potential to address global energy demands while furthering climate change objectives [7]. RE-refinery H_2 management system could decrease the overall yearly cost by 21.21% for abandoned wind power rates and 13.49% for regular wind power prices. As the cost of wind power decreases in the future, the overall yearly expenditure may decline by 15.09% in the 2035 scenario and 16.61% in the 2050 scenario. The price of wind power significantly influences the utilization rate, which rises from 57.67% in 2025 to 90.11% in 2050. Wind power utilized for H_2 compression exhibits cost benefits relative to green H_2 manufacturing [8]. H_2 generated using RE source, creates new opportunities for sustainable and low-carbon chemical manufacturing [9]

6.3 Solar Aided H_2 Production

The power generated from solar energy could supply the necessary heat for chemical reactions by heating the reactor or reactants [10]. Parabolic troughs are the predominant technology utilized for achieving the temperatures of 400 °C. Power towers provided 650 °C or higher temperature. This device employed mirrors arranged in a circular formation to focus solar rays onto towers, where molten salt or heating liquid captures the solar energy and raises the temperature. The molten salt is transferred to a reservoir for thermal energy storage, which can later be employed in the essential operations [11]. Spectrum expansion, enhancement, and assignment could enhance the efficiency of chemical H_2 synthesis by using solar photocatalysis, photothermal catalysis, and photoelectric catalysis methods. The overall efficacy of the solar spectrum utilization system was improved and facilitated mid- to low-temperature solar H_2 production method [12]. Solar Steam Methane Reforming utilizing a Volumetric Receiver Reactor (SSMR-VRR), Solar Steam Methane Reforming employing Molten Salt (SSMR-MS), and Solar Thermal Power generation integrated with Water Electrolysis (STP-WE) were the methods utilized for the production of H_2 (2577 kmol of H_2/hr.) SSMR-VRR had $2.5/kg cost of H_2 production, SSMR-MS had 68% of energy efficiency, while STP-WE was the green process with respect to causing environmental pollution [1].

6.3.1 Sunlight-driven Water Splitting Method for H_2 Generation

Sunlight-driven water splitting (WrS) with semiconductor photocatalysts was highly promising method for generating clean H_2 fuel. Solar H_2 was produced using photoelectrochemical WrS method [13]. WrS methods utilize solar energy as the main source for H_2 production. Photocatalytic WrS, photoelectrochemical WrS, and photovoltaic electrochemical WrS had been extensively used in the last five years [14]. The photocatalytic H_2 generation efficacy of doped-TiO_2 was significantly influenced by the nature and concentration of the dopants. The sol–gel process was used for the synthesis of TiO_2 doped photocatalysts [15]. The efficacy of WrS and H_2 production was contingent upon the characteristics of photocatalysts, which were influenced by their nature, composition, and various other parameters [16]. Fe_2O_3 based redox materials were created which could extract O_2 from H_2O and generate pure H_2 at 800 °C, and regenerated below 1300 °C. Ceramic honeycombs were produced and integrated into a specialized solar receiver/reactor. The operational parameters of the solar reactor were refined to achieve the temperatures of up to 1300 °C across the honeycomb, so enabling the entire cycle to be executed by a single solar energy converter [17]. A rotary-type solar reactor had been constructed for solar H_2 manufacturing by using two-step WrS method in which, CeO_2 and $Ni_{0.5}Mn_{0.5}Fe_2O_4$ were used. The elevated O_2-releasing reaction temperature over 1800 K was validated using the approximately tenfold scaled-up rotary-type solar reactor [18]. A sunlight-driven WrS reactor used Al-doped $SrTiO_3$ photocatalyst. Solar to H_2 conversion efficiency was 0.4% by utilising photocatalyst and WrS method. WrS panel is a feasible method for generation of solar H_2 [19].

6.4 Integration of RE Source in Different Chemical Production

The transformation of CO_2 into fuels and chemicals utilizing RE source presents a viable strategy to alleviate rising CO_2 levels in the atmosphere and ocean acidification. CO_2 could be electrochemically converted to hydrocarbons utilizing RE source like wind and solar, in a process that effectively reverses combustion and resembles photosynthesis. This technique will not compete with the direct utilization of RE as electricity, as its purpose is to store surplus energy for future use [20]. The substitution of combustion of fossil fuels in the transient Joule heating method with renewable electricity facilitated an environmentally friendly heating process and obviates the necessity for burners, thereby permitting a compact reactor design. Additionally, electrified heating is exceptionally versatile and controllable, enabling intermittent or dynamic heating that could align with the sporadic generation of renewable electricity. Electrified transient high-temperature synthesis helps in achieving green synthesis within the chemical industry [21]. The synergistic interaction of hot compressed water with energy derived from renewable sources is a novel approach to a greener carbon energy system for fuel generation, yielding advantages such as diminished energy consumption and reduced carbon dioxide emissions [22]. Utilizing electrical energy to empower microbes for the production of valuable compounds, including biofuels, has recently emerged as a sustainable approach to diminished reliance on oil. Figure 6.1 shows the utilization of different types of RE sources for the preparation of chemicals.

Microbial electrosynthesis (MES) is a bioelectrochemical technique developed which involves electroautotrophic microorganisms utilizing electrical current as an electron source to convert CO_2 into multicarbon organic compounds. Electricity required for MES could be obtained from solar energy and wind turbines [23]. Bioelectrochemical systems were used for the conversion of solar-to-chemical as

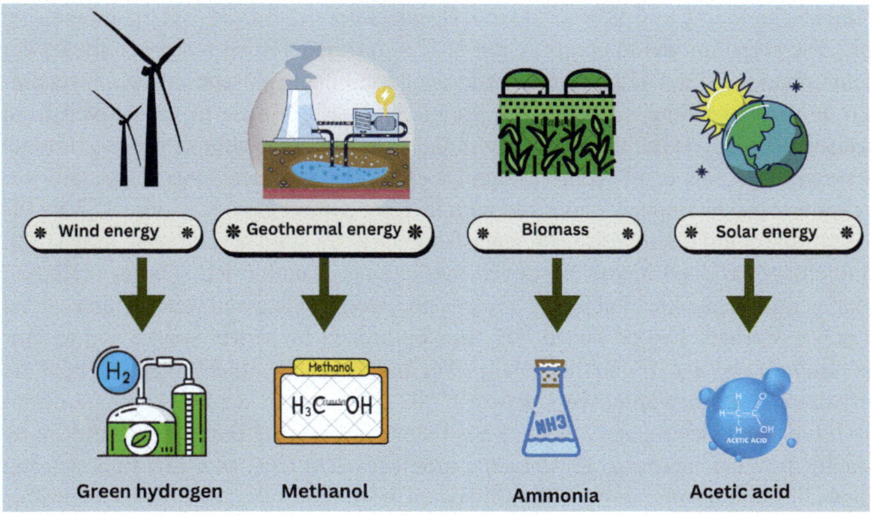

Fig. 6.1 RE sources in manufacturing different chemicals

they electrochemically integrate inorganic catalysts with microorganisms. Specifically, microbial electrosynthesis systems and biohybrid systems utilized CO_2 and electricity or light, respectively to produce organic acids [24]. Solar photons are optimal green reagents as they incur no expense and produce no residuals in the reaction mixture. In numerous instances, sun radiation can effectively substitute for harmful or costly chemical reagents to surmount the activation energy in organic synthesis [25]. CO_2 reduction combined with oxidative organic synthesis and solar energy in which CdSe/CdS quantum dots utilized the photocatalytic conversion of CO_2 to CO at an impressive rate of approximately 412.8 mmol/ g/ h exhibiting a high selectivity exceeding 96% in the presence of triethylamine [26]. Solar-driven chemistry represents a novel approach for low-carbon sustainable production. Electrocatalysis is an essential technology for facilitating the shift to solar-driven chemistry. Electrocatalytics helped in the enhancement of direct NH_3 synthesis and CO_2 reduction [27]. The dye-sensitized photooxygenation of 1,5-dihydroxy naphthalene was done under moderately concentrated sunshine. 79% yields of 5-hydroxy-1,4-naphthoquinone (Juglone) were attained after 4 h of illumination [28]. The Haber–Bosch process utilizing renewable sources for sustainable NH_3 production was modeled in Aspen. NH_3 may be generated efficiently from RE sources [29]. A solar and geothermal energy system (Rankine cycles). This system had an overall energy efficiency of 54.7%, but its exergy efficiency was 76.4% [30]. Two-stage solar-driven biogas reforming system was used to produce syngas. This system increased CO_2 conversion rate from 27.32 to 63.72%. The implementation of biogas dry reforming diminishes thermal energy and water usage [31]. The amalgamation of a methanol (CH_3OH) manufacturing process with oxy-combustion technology could improve the efficiency of synthesis methods. The integration encompasses multiple elements, including a 90 MW wind farm, H_2 generators, and a CH_3OH manufacturing unit. The preparation and compression of CO_2 for the CH_3OH reactor in the power plant enhanced CH_3OH production and decreased energy usage in the O_2 separation unit by providing additional O_2 via electrolysis. The use of O_2 derived from electrolysis reduced the use of energy in the oxidant compressor of the gas turbine. Integrated systems resulted in substantial efficiency enhancements, especially for low-power production systems,s and with increased input from RE sources. 7.5 MW Gas turbine could improve the efficiency from 0.8 to 6% with several wind farms [32]. Biomethane production was improved by the amalgamation RE sources including solar, geothermal, and wind [33]. NH_3 was prepared from air and water which served as power storage. The produced NH_3 was optimized for power source and reactor operation. It was also found that solar power and indirect synthesis reactors were the best alternatives for optimal production of NH_3 [34]. Acetic acid was prepared from fossil fuels and electrocatalytic synthesis helped in lowering the carbon footprint by a factor 5–6 and increased the faradaic efficiency and CO_2 as a raw material and renewable sources of energy were utilized. 50 € MW/hr. was the cost of renewable electricity, which was lower than the currently established route [35]. RE source-driven methods may achieve the manufacturing of NH_3 from N_2 and H_2O in a one step. This facilitates a substantial enhancement of energy efficiency, leading to a reduction of over 90% in greenhouse gas emissions and consequently in total energy consumption. The RE-driven synthesis of olefins led to a significant decrease in greenhouse gas emissions and total energy consumption [36].

6.5 NH_3 Synthesis Powered by Solar Energy

Solid-state synthesis, molten salt synthesis, thermochemical looping, and photocatalytic processes are the methods for the generation of NH_3 under mild conditions. In contrast to the Haber–Bosch process, which required 200–400 bar, these methods function at low pressures. Moreover, such pathways enable the direct synthesis of NH_3 from H_2O and N_2, eliminating the intermediate H_2 generation phase. These benefits facilitate the seamless incorporation of RE [37]. Ru-based catalyst utilized nanostructured CeO_2 support and Cs electronic promotion to remove H_2 and NH_3 inhibition, respectively. The absorbent used $MnCl_2$ to prevent the acid releasing decomposition of $MgCl_2$, and a support of SiO_2 to concurrently increase $MnCl_2$ dispersion and improve stabilization. This integrated catalyst-absorbent system consistently surpasses the equilibrium of single-pass NH_3 production. The results established a framework for integrating RE with NH_3 synthesis by reducing capital complexity which increased the fertilizer production with minimal CO_2 emissions and advance energy storage solutions [38]. An innovative method for producing NH_3 and fuel cell technology coupled with solar energy. The peak energetic efficiency performance attains 15.72%. The maximum exergetic efficiency was 16.55%. The daily discharge duration and energy capacity were determined to be 8.9 h and 8502.4 kw-h [39]. A novel integrated solar-powered NH_3 synthesis and fuel cell system had been created. The whole system energy efficiency ranged from 15.68 to 15.83%. The exergy efficiency of the entire system ranged from 16.44 to 16.67%. Discharge time capacities were seen to fluctuate between 2.6 and 8.3 h throughout the year [40]. NH_3-leaf technology would competitively perform in sunny regions compared to the conventional production scenario. Furthermore, a mere 8% improvement in energy efficiency would enable them to surpass the gHB for climate change-related effects in the sunniest regions. The cost of NH_3 production would be thrice the industrial benchmark, with the possibility of alignment contingent upon a significant reduction in investment expenses and exceptionally good selectivity for NH_3 in eN_2R processes [41]. A technique for NH_3 preparation employed solar thermochemical energy storage method. Enhancing heat transmission could minimize reactor size. An optimization technique was employed to build a system with minimal material volume. A proposed modular system design demonstrated a reduction in the volume of wall materials [42]. A solar tower system was employed to energize a thermo-catalytic cracking unit. The total energy and exergy efficiencies attained were 35% and 39%, respectively. NH_3 production approached 1000 tons/d at a cracking temperature of 900 °C [43]. A dual temperature zone catalysis was used which Illuminated by sunlight, TiO_2-xHy/Fe hybrid attained 495 °C temperature and there was 137 °C temperature difference between the heated region (Fe) and the "cooling" region (TiO_2-xHy) due to the plasmonic local heating effect. The heated Fe-bearing could effectively dissociate N_2, while TiO_2-xHy helped in the transfer of N_2 from Fe by using successive hydrogenation process, significantly reducing the reverse equilibrium shift and consequently achieving NH_3 concentrations of 1,939 (1 atm) and 19,620 ppm (10 atm) at 495 °C, which were 1.55 and 1.57 times more than the theoretical equilibrium limits of 1,249 and 12,459 ppm, respectively [44]. Ru/TiO_2-xHx demonstrated superior reactivity and stability under light than thermal heating. H atoms in TiO_2-xHx facilitated a low-energy barrier pathway for solar based NH_3 production. The activation energy of the reaction was reduced by light induced confined energy system [45]. Photovoltaic systems exhibited far greater efficiency in solar energy conversion compared to organic techniques. Even in the least favourable scenario assessed in Indiana, less than 1% of a cornfield converted to a Photovoltaic system was sufficient to generate adequate energy for NH_3 production necessary for fertilizing the remaining corn crops. The system was designed to supply NH_3 for fertilization in corn fields exceeding 1079 acres with the most adverse soil conditions [46].

6.6 Wind Energy-Driven Chemical Production

The integrated plant comprising wind energy-driven electrolytic H_2 generation, bioethanol-based CO_2 capture and compression, and direct CH_3OH synthesis. ASPEN Plus was employed to model a facility generating 97.01 metric tons of CH_3OH/d utilizing 138.37 metric tons of CO_2 and 18.56 metric tons of H_2 daily.

The total energy efficiency of renewable CH_3OH was approximately 58%. The fixation of carbon decreased CO_2 equivalent emissions by approximately −1.05 CO_2e/kg of CH_3OH. The cost of electrolytic H_2 generation was the primary factor influencing the economics of the entire facility [47]. NH_3 produced via wind energy was approximately twice as expensive as that generated through conventional fossil fuel methods. To decrease production costs, traditional NH_3 condensation method was replaced with a specific absorber comprising $CaCl_2$ which works at near-synthesis temperatures. This technique enabled the synthesis of NH_3 at 20 bar using air, water, and wind-generated electricity, while traditional process operated at 150–300 bar. The rate of NH_3 production in this reaction-absorption method was now governed not by the chemical reaction itself, but predominantly by the pump utilized for recycling the unreacted gases. The novel approach to decentralized NH_3 production that could locally provide nitrogen fertilizer and a technique for harnessing stranded wind energy as a carbon–neutral liquid fuel [48]. A mixed-integer linear programming formulation was created and alternative models derived from sophisticated natural language processing steady-state models. This methodology was utilized in the context of CH_3OH production from CO_2 and H_2 derived from solar or wind sources [49]. The efficiency of wind power for the CH_3OH conversion was approximately 50%. The price of CH_3OH was 300€/ton and × 800€/ton without and with wind turbine capital expenses. The cost of renewable CH_3OH generated from air-captured CO_2, H_2O, and renewable electricity was 750–800 €/ton [50]. Table 6.1 tabulates the key features of integrated RE source in different chemicals production.

6.7 Technical Challenges and Solutions

The shift from laboratory-scale research to industrial-scale manufacturing presents issues in sustaining efficiency and cost-effectiveness in the integration of RE source in chemical industries. Utilizing modern reactor designs and process optimization methods could enhance production scalability while preserving efficiency. Ongoing surveillance and enhancement of process parameters are necessary. RE source, like solar and wind, are intermittent, resulting in variations in energy supply for fuel production and could retain surplus RE for utilization during intervals of decreased production. Smart grid technologies and demand-response tactics can effectively regulate energy variations. Renewable NH_3 is two to three times more costly than conventional NH_3. The price of photovoltaic panels should decrease by 75%, while their efficiency should improve by 15% [34]. It is very important to determine the market demand of green H_2 before selecting anode reaction and there is need to improve the efficiency of electrocatalyst and electrodes. The separation and purification could be enhance by increasing the selectivity and faraday efficiency [51]. Catalysts employed in fuel and chemical production processes may deteriorate with time, diminishing their efficacy and lifespan. Enhancing catalyst stability and longevity can be achieved by developing more stable catalytic materials, including nanostructured

Table 6.1 Different types of chemicals are generated by various forms of RE source

Type of renewable energy/ methods	Key features	Chemical generated	References
Solar spectrum	Enhance the efficiency	H_2	[12]
SSMR-VRR	\$2.5/kg cost of production		[1]
Rotary-type solar reactor	Two-step WrS process was used		[18]
Sunlight-driven WrS reactor	Solar to H_2 energy conversion efficiency was 0.4%		[19]
Solar energy and wind turbines	MES	Organic acids	[24]
Solar energy	Selectivity more than 96%	CO_2 to CO	[26]
	Electrocatalytic method was used	NH_3 synthesis and CO_2 reduction	[27]
Two-stage solar-driven biogas reforming system	CO_2 conversion rate was increased from 27.32% to 63.72%	Syngas	[31]
Solar, geothermal, and wind	–	Biomethane production	[33]
RE	50 € MW/hr	Acetic acid	[35]
Wind power	750–800 €/ton	CH_3OH	[50]

and hybrid catalysts, and optimizing operational conditions. In-situ monitoring and regeneration strategies can assist in sustaining catalyst performance. The use of RE sources in chemical manufacturing necessitates significant technological and scientific transformations, as catalysts, reactors, and operations must be entirely reengineered. The integration of RE with chemical production process must be economically feasible to rival traditional fossil fuels. Ongoing enhancement of RE system efficiency and the reduction of catalyst and material costs through research and development can improve the economic viability of combining RE with chemical manufacturing. Government incentives and regulations that encourage the utilization of RE can enhance the economic viability of these technologies.

6.8 Conclusion

This chapter gives insights about the generation of H_2 by using solar energy and WrS method. NH_3 can also be prepared by solar based integrated process. Integration of RES with chemical synthesis method can generated chemical effectively. Wind energy can be used for the preparation of specific chemicals. Solar energy is crucial for H_2 preparation, CO_2 reduction, and chemical synthesis, with advancements such as solar spectrum, WrS reactors, and electrocatalysis enhancing efficiency and selectivity. Wind energy aids in CH_3OH manufacturing, whereas hybrid methods, incorporating solar, wind, and geothermal sources, promote the generation of biomethane and acetic acid. 750–800 €/ton and 50 €/MW·hr were the cost of CH_3OH and acetic acid produced by using wind energy and RE sources respectively. These technologies demonstrate the varied applications of RE in sustainable chemical production, improving efficiency and decreasing carbon emissions. The incorporation of RE source into the synthesis process of chemicals improves sustainability, although has problems related to efficiency and high cost in large-scale implementation. Ongoing technological advancement, cost minimization, and administration support are crucial for rendering RE source-driven chemical manufacturing cost-effective compared to chemicals prepared from fossil fuels.

References

1. Likkasit, C., Maroufmashat, A., Elkamel, A., Ming Ku, H., Fowler, M.: Solar-aided hydrogen production methods for the integration of renewable energies into oil and gas industries. Energy Convers. Manag. **168**, 395–406 (2018)
2. Martín, M., Grossmann, I.E.: Optimal integration of renewable-based processes for fuels and power production: Spain case study. Appl. Energy **213**, 595–610 (2018)
3. Bagherian, M.A., Mehranzamir, K.: A comprehensive review on renewable energy integration for combined heat and power production. Energy Convers. Manag. **24**, 113454 (2020)
4. Chen, C., Yang, A.: Power-to-methanol: the role of process flexibility in the integration of variable renewable energy into chemical production. Energy Convers. Manag. **228**, 113673 (2021)
5. Bünger, U., Michalski, J., Crotogino, F., Kruck, O.: Large-scale underground storage of hydrogen for the grid integration of renewable energy and other applications. Compend. Hydrog. Energy Hydrog. Use Saf. Hydrog. Econ. **4**, 133–163 (2016)
6. Won, W., Kwon, H., Han, J.H., Kim, J.: Design and operation of renewable energy sources based hydrogen supply system: technology integration and optimization. Renew. Energy **103**, 226–238 (2017)
7. Nnabuife, S.G., Hamzat, A.K., Whidborne, J., Kuang, B., Jenkins, K.W.: Integration of renewable energy sources in tandem with electrolysis: a technology review for green hydrogen production. Int. J. Hydrogen Energy **06**, 342 (2024)
8. Zhou, Y., Wang, Y., Yang, M.: Optimal integration of renewable energy in refinery hydrogen management systems: energy storage and direct utilization. Energy Convers. Manag. **304**, 118223 (2024)
9. Lanzafame, P., Centi, G., Perathoner, S.: Catalysis for biomass and CO_2 use through solar energy: opening new scenarios for a sustainable and low-carbon chemical production. Chem. Soc. Rev. **22**, 7562–7580 (2014)
10. González-Garay, A., Dowell, N.M., Shah, N.: A carbon neutral chemical industry powered by the sun. Discov. Chem. Eng. **1**, 1–22 (2021)
11. Naveenkumar, R., et al.: Comprehensive review on various parameters that influence the performance of parabolic trough collector. Environ. Sci. Pollut. Res. **18**, 22310–22333 (2021)
12. Zhao, N., Wang, J.: Solar full spectrum management in low and medium temperature light-driven chemical hydrogen synthesis—a review. Renew. Sustain. Energy Rev. **196**, 114368 (2024)

13. Xing, Z., Zong, X., Pan, J., Wang, L.: On the engineering part of solar hydrogen production from water splitting: photoreactor design. Chem. Eng. Sci. **104**, 125–146 (2013)
14. Imran, S., Hussain, M.: Emerging trends in water splitting innovations for solar hydrogen production: analysis, comparison, and economical insights. Int. J. Hydrogen Energy **77**, 975–996 (2024)
15. Ismael, M.: A review and recent advances in solar-to-hydrogen energy conversion based on photocatalytic water splitting over doped-TiO_2 nanoparticles. Sol. Energy **211**, 522–546 (2020)
16. Rafique, M., et al.: A comprehensive study on methods and materials for photocatalytic water splitting and hydrogen production as a renewable energy resource. J. Inorg. Organomet. Polym. Mater. **10**, 3837–3861 (2020)
17. Agrafiotis, C., et al.: Solar water splitting for hydrogen production with monolithic reactors. Sol. Energy **79**, 409–421 (2005)
18. Kaneko, H., et al.: Rotary-type solar reactor for solar hydrogen production with two-step water splitting process. Energy Fuels **21**(4), 2287–2293 (2007)
19. Goto, Y., et al.: A particulate photocatalyst water-splitting panel for large-scale solar hydrogen generation. Joule **2**, 509–520 (2018)
20. Ager, J.W., Lapkin, A.A.: Chemical storage of renewable energy. Science **1979**(360), 707–708 (2018)
21. Dong Y., et al. (2024) Highly efficient chemical production via electrified, transient high-temperature synthesis. eScience **4**, 100253 (2024)
22. Hosseinpour, M., Soltani, M., Nathwani, J.: Renewable energy integration with hot compressed water in heavy oil upgrading: a practice toward sustainability. J. Clean. Prod. **334**, 130268 (2022)
23. Tremblay, P.L., Zhang, T.: Electrifying microbes for the production of chemicals. Front. Microbiol. **6**, 134494 (2015)
24. Su, L., Ajo-Franklin, C.M.: Reaching full potential: bioelectrochemical systems for storing renewable energy in chemical bonds. Curr. Opin. Biotechnol. **57**, 66–72 (2019)
25. Protti, S., Fagnoni, M.: The sunny side of chemistry: green synthesis by solar light. Photochem. Photobiol. Sci. **8**, 1499–1516 (2009)
26. Guo, Q., et al.: Efficient and selective CO_2 reduction integrated with organic synthesis by solar energy. Chem **5**, 2605–2616 (2019)
27. Perathoner, S., Centi, G.: Catalysis for solar-driven chemistry: the role of electrocatalysis. Catal. Today **330**, 157–170 (2019)
28. Oelgemöller, M., Healy, N., de Oliveira, L., Jung, C., Mattay, J.: Green photochemistry: solar-chemical synthesis of Juglone with medium concentrated sunlight. Green Chem. **8**, 831–834 (2006)
29. Frattini, D., Cinti, G., Bidini, G., Desideri, U., Cioffi, R., Jannelli, E.: A system approach in energy evaluation of different renewable energies sources integration in ammonia production plants. Renew. Energy **99**, 472–482 (2016)
30. Suleman, F., Dincer, I., Agelin-Chaab, M.: Development of an integrated renewable energy system for multigeneration. Energy **78**, 196–204 (2014)
31. Su, B., et al.: System integration of multi-grade exploitation of biogas chemical energy driven by solar energy. Energy **241**, 122857 (2022)
32. Brzęczek, M., Kotowicz, J.: Integration of alternative fuel production and combined cycle power plant using renewable energy sources. Appl. Energy **371**, 123738 (2024)
33. Calise, F., Cappiello, F.L., Cimmino, L., D'accadia, M.D., Vicidomini, M.: A review of the state of the art of biomethane production: recent advancements and integration of renewable energies. Energies **14**, 4895 (2021)
34. Sánchez, A., Martín, M.: Optimal renewable production of ammonia from water and air. J. Clean. Prod. **178**, 325–342 (2018)
35. Lanzafame, P., et al.: Beyond solar fuels: renewable energy-driven chemistry. Chemsuschem **10**, 4409–4419 (2017)
36. Genovese, C., Ampelli, C., Perathoner, S., Centi, G.: Mechanism of C-C bond formation in the electrocatalytic reduction of CO_2 to acetic acid. A challenging reaction to use renewable energy with chemistry. Green Chem. **19**, 2406–2415 (2017)

37. Klaas, L., Guban, D., Roeb, M., Sattler, C.: Recent progress towards solar energy integration into low-pressure green ammonia production technologies. Int. J. Hydrogen Energy **49**, 25121–25136 (2021)
38. Smith, C., Torrente-Murciano, L., Smith, C., Torrente-Murciano, L.: Exceeding single-pass equilibrium with integrated absorption separation for ammonia synthesis using renewable energy—redefining the Haber-Bosch loop. Adv. Energy Mater. **13**, 2003845 (2021)
39. Siddiqui, O., Dincer, I.: Development and evaluation of a solar-based integrated ammonia synthesis and fuel cell system. J. Clean. Prod. **256**, 120393 (2020)
40. Siddiqui, O., Dincer, I.: A new solar energy system for ammonia production and utilization in fuel cells. Energy Convers. Manag. **208**, 112590 (2020)
41. D'Angelo, S.C., Martín, A.J., Cobo, S., Ordóñez, D.F., Guillén-Gosálbez, G., Pérez-Ramírez, J.: Environmental and economic potential of decentralised electrocatalytic ammonia synthesis powered by solar energy. Energy Environ. Sci. **16**, 3314–3330 (2023)
42. Chen, C., Lovegrove, K.M., Sepulveda, A., Lavine, A.S.: Design and optimization of an ammonia synthesis system for ammonia-based solar thermochemical energy storage. Sol. Energy **159**, 992–1002 (2018)
43. Mohamed, A.M.O., Bicer, Y.: Development and assessment of concentrated solar energy driven ammonia synthesis from liquefied natural gas. Int. J. Hydrogen Energy **16**, 10093–10103 (2021)
44. Mao, C., et al.: Beyond the thermal equilibrium limit of ammonia synthesis with dual temperature zone catalyst powered by solar light. Chem **10**, 2702–2717 (2019)
45. Mao, C., Yu, L., Li, J., Zhao, J., Zhang, L.: Energy-confined solar thermal ammonia synthesis with K/Ru/TiO_2-xHx. Appl. Catal. B **224**, 612–620 (2018)
46. Du, Z., Denkenberger, D., Pearce, J.M.: Solar photovoltaic powered on-site ammonia production for nitrogen fertilization. Sol. Energy **122**, 562–568 (2015)
47. Matzen, M., Alhajji, M., Demirel, Y.: Chemical storage of wind energy by renewable methanol production: feasibility analysis using a multi-criteria decision matrix. Energy **93**, 343–353 (2015)
48. Malmali, M., Reese, M., McCormick, A.V., Cussler, E.L.: Converting wind energy to ammonia at lower pressure. ACS Sustain Chem Eng **1**, 827–834 (2018)
49. Martín, M.: Methodology for solar and wind energy chemical storage facilities design under uncertainty: methanol production from CO_2 and hydrogen. Comput. Chem. Eng. **92**, 43–54 (2016)
50. Bos, M.J., Kersten, S.R.A., Brilman, D.W.F.: Wind power to methanol: renewable methanol production using electricity, electrolysis of water and CO_2 air capture. Appl. Energy **264**, 114672 (2020)
51. Wei, Z., et al.: Electrochemical synthesis in company with hydrogen production via renewable energy: opportunities and challenges. Chin. J. Catal. **58**, 1–6 (2024)

Chapter 7
Green Chemistry for Industrial Sustainable Water Use

Khushbu Upadhyaya, Ishika Taliyan, Shruti Shukla, and Rekha Sharma

Abstract Water is an integral part of many sectors and its misuse has created adverse effects such as contamination and water scarcity. To combat this problem, contactless solutions are required of which green chemistry appears to be a good target. Introducing its tenets in practice will allow industrialists to implement sustainable practices in terms of water usage and have minimal negative effects on the environment. An important aspect of these strategies is the substitution of hazardous security parameters of industrial processes with biodegradable polymeric containers to minimize pollution. New generation water reclamation systems permit industries to recycle processed water which in turn reduces the pressure on raw materials. Also, green chemistry offers several processes that require no solvent and energy-conserving ones that cut down on the number of necessary resources and emissions released into the environment. In addition, some new approaches have also been found to the problem of wastewater disposal and treatment: bio-based solvent, membrane filtration, catalysis, and electrochemical methods are powerful tools for the elimination of unwanted substances and for the recovery of useful substances. Thus, the integration of green chemistry in the industrial sphere is also the implementation of sustainable practices that are not only in compliance with international environmental strategies but also reduce the water footprint and pollution.

Abbreviations

FO	Forward Osmosis
$SsCO_2$	Supercritical carbon dioxide
RTILs	Room-temperature ILs
AOPs	Advanced Oxidation Processes
MT	Mixed Textile effluent
FST	Fabric Softening Treatment effluent

K. Upadhyaya · I. Taliyan · S. Shukla · R. Sharma (✉)
Department of Chemistry, Banasthali Vidyapith Niwai, Tonk 304022, Rajasthan, India
e-mail: sharma20rekha@gmail.com

P. Jain et al. (eds.), *Towards Green Chemical Processes: Strategies and Innovations*, Intelligent Systems Reference Library 277,
https://doi.org/10.1007/978-3-031-94090-3_7

ST	Standard Textile effluent
MBR	Membrane bioreactor
BPA	Bisphenol A

7.1 Introduction

Water is essential for many industries and along with different stages of any production process, it is also required for incorporation in multiple goods ranging from machinery manufacturing to cooling them [1]. The food and beverage, electronics, and medical industries all depend on ultra-pure water for the production of their goods. Industrial wastewater has to be managed with care as it poses threats of environmental pollution [2] and it also affects the economy. Any nation's economy comprises several industries and dairy, sugar, textiles, pulp and paper, oil and gas, breweries, and even the automobile sector [3] are all industries that consume significant quantities of water and require treatment.

The impact of climate change and ineffective water management strategies on water economics is further intensified due to the increased issues associated with water scarcity [4]. At the same time, water bodies and ecosystems targeted by these industries are also afflicted by pollutants mainly composed of chemical substances. Most of these substances can't be purified by conventional processes [5]. This quite often leads to a high level of water pollution as a result of industrial waste. For instance, as a result of the pharmaceutical industry activities the quality of untreated wastewater reaches a level where it stinks of raw materials, additives, solvents, etc. It is dangerous to use this water for swimming and even for drinking as it poses grave health risks. So treatment of this water is required. These pollutants are commonly represented by a collective parameter known as chemical oxygen demand (COD) [6]. The pathway of these contaminants along the food chain's trophic levels is harmful to population health and ecological systems. Recently, how ironic it may sound, there has been a growing concern about the amount of wastewater including raw water effluents, and even the water that has gone through treatment as it contains traces of pharmaceutical and textile substances [7]. The green approach is one of the most effective methods that aid in the prevention of water pollution.

Green chemistry, an innovative technique focusing on resource efficiency and environmental responsibility, is the only solution to these challenges. This design principle of avoiding pollution, moving towards less harmful chemicals, and new water management technology, all are the principles of green chemistry, which concepts can reduce industrial water use and harmful discharges [8].

In this paper, we mention the correlation between green chemistry and industrial water sustainability. It showcases state-of-the-art eco-friendly technologies such as membrane systems, catalytic treatments, and green solvents, along with case studies highlighting successful implementation in different industries [9]. Adopting these practices will influence water management systems industrially, so we use greener chemistry to advance resilience and sustainability in these systems for both environmental and economic benefits.

7.2 Principles of Green Chemistry in Industrial Water Management

Green chemistry aims to reduce pollution by designing chemical products and processes that minimize hazardous substances [10]. Key principles are included in Fig. 7.1.

Fig. 7.1 Principles of green chemistry

7.3 Role of Green Chemistry for Water Sustainability

Green chemistry offers solutions to water sustainability challenges. It promotes water-efficient processes, green solvents, and catalysts to reduce pollution [11]. By adopting green chemistry, we can conserve water, improve water quality, and mitigate environmental impact, ensuring a sustainable future for water resources.

7.3.1 *Wastewater Prevention*

Wastewater treatment in industries is a critical component of sustainable water management, particularly in sectors like textiles, chemicals, pharmaceuticals, and food processing, which are known for their high water consumption and effluent discharge [12]. In the textile sector, water use and pollutant output are interconnected with the widespread adoption of innovations, such as enzymatic processes, counter-current washing units, and waterless dyeing techniques using supercritical CO_2 [13]. Closed-loop systems also treated and reused dye baths and rinse water to reduce waste streams [14]. Likewise, the chemical industry can implement solutions to eliminate toxic wastewater and water-rich stages with solvent-free transformations, green catalysis, and integrated waste management operations.

Some techniques minimize wastewater such as dry cleaning, advanced filtration technologies, and the reuse of treated process water for cleaning [15]. In pharmaceutical manufacturing, solvent recovery systems, continuous flow processes, and membrane-based purification techniques reduce water waste without sacrificing production throughput. The steel, paper, and auto products use air-cooled dry cooling systems, zero-liquid discharge systems (ZLD), and pre-treatment optimization techniques to conserve water and eliminate wastewater discharge. Collectively, these moves enhance water efficiency, reduce costs, and lower environmental pollution. Such an integration of advanced technologies and green chemistry not only helps industries drastically reduce waste but also provides a roadmap for a greener and sustainable future.

7.3.2 *Use of Safer Chemicals*

It provides a solution for those reactions in which toxic and hazardous solvents are substituted by non-toxic, biodegradable, and ecologically friendly compounds [16]. This strategy boosts the sustainability of industrial operations while lowering the dangers to ecosystems, industrial workers, and human health. Safer substances like natural dyes and biodegradable surfactants are also substituted by harmful, synthetic dyes and wetting agents in the textile business, which lowers water pollution. Comparably, in the chemical manufacturing industry, less volatile and safer alternatives like supercritical CO_2, or ionic liquids are replacing conventional solvents like benzene and chloroform [17].

To reduce soil and water pollution, the agriculture sector also incorporates bio-based fertilizers and pesticides, which come from natural sources and break down in the environment without harming the ecosystem. The pharmaceutical industry also shifts from dangerous chemicals to biocatalyst enzymes for drug synthesis and provides safer production paths. The food and beverage industry also uses natural flavourings and preservatives that don't harm the environment or consumer safety. Modern technologies, such as safer reaction conditions and real-time monitoring are used to reduce the production and discharge of hazardous by-products in various sectors of the economy [18].

7.3.3 Energy and Resource Efficiency

The most sustainable industrial practices are energy and resource efficiency in water-using sectors. To conserve this essential resource, effective water management is applied in which the energy required for water, treatment, and disposal is low, which has major positive effects on the environment and the economy.

Closed-loop water recycling systems come under water-efficient technologies which are being adopted by industries. These technologies enable the reuse of water in various processes, including manufacturing, cooling, and washing [19]. As a result, less fresh water is required, and a smaller amount of energy is needed for pumping and treatment. Advanced water recovery systems, for instance, recycle dye bath water in the textile industry, reducing energy and water use. By replacing conventional cooling techniques, the dry cooling technique is used in which low energy and water consumption in industries is used [20]. In a similar vein, low-energy desalination techniques, such as forward and reverse osmosis, are being used to supply processed water with less energy. By combining these practices, it can save a significant amount of energy and water while promoting environmental stewardship and long-term sustainability.

7.3.4 Recycling and Reuse

To achieve water sustainability, recycling, and reuse are the most important strategies, particularly in industries where water consumption and wastewater generation are significant. Industries can drastically reduce their demand for freshwater, by treating and reusing water within processes, minimizing wastewater discharge, and enhancing overall resource efficiency [21].

It is a process of treating wastewater to remove impurities so that it can be used again for things like irrigation, cleaning, and cooling. For instance, water from dye baths is recovered and recycled in the textile sector using cutting-edge technologies like reverse osmosis and ultrafiltration. Similar to this, membrane technology sin the chemical and pharmaceutical industries cleans and reuses water used for washing and reactions, lowering water withdrawal and effluent outflow.

Using treated wastewater for boiler feed, industrial cooling, and landscaping is an example of reuse that goes beyond simple recycling. Industries are also using Zero Liquid Discharge (ZLD) systems more and more to recover almost all of the water from waste streams with little to no discharge [22]. Reusing treated wastewater in food processing ensures resource conservation without ignorance of hygiene by using it for non-potable uses like agricultural irrigation or equipment cleaning. Reusing and recycling also lowers operating expenses while also bringing companies into compliance with environmental and water conservation regulations. They are essential to sustainable water management methods because they lower the environmental impact of industrial activities, improve resource sustainability, and limit the hazards of water scarcity.

7.4 Sustainable Water Management for Industrial Water Use

Green technologies act as a game changer for industrial water use by promoting sustainability, reducing environmental impact, and enhancing resource efficiency. These technologies focus on minimizing water consumption, treating wastewater effectively, and enabling the recycling and reuse of water within industrial processes.

7.4.1 Solvent-free Reaction

The concept of conducting chemical reactions without water-based chemistry has become increasingly popular in the field of sustainable and green chemistry. These reactions exhibit unique features that make them well-suited for environmentally friendly and efficient chemical processes.

Key characteristics of reactions without water-based solvents include:

- *Independence from Solvents*: In this type of reaction only reactant is required, often aided by solid or heterogeneous catalysts and these catalysts are reusable and recover throughout the reaction.
- *Improved Atom Economy*: The absence of solvents allows most reactants to contribute to the desired product, reducing waste.
- *Reduced Environmental Impact*: Eliminating hazardous solvents from the reaction is beneficial for the environment and prevents wastewater production.

Various industries are increasingly using solvent-free reaction methods to minimize environmental impact and enhance process efficiency [23]. Some key industries and examples of solvent-free reactions are as follows.

7.4.1.1 Solid-Phase Peptide Synthesis (SPPS)

It is a technique for synthesizing peptides directly on solid supports, in which solvents are not required in the reaction and purification steps. By anchoring reactants to a solid matrix, this method allows for the stepwise addition of amino acids, resulting in high-purity peptides. The reduced solvent usage contributes to cost savings and environmental sustainability, making it a preferred method in pharmaceutical development [24].

7.4.1.2 Microwave-Assisted Organic Synthesis

This innovative technique involves microwave radiation to heat the reactants uniformly, and significantly increase or accelerate the reaction. Many organic reactions, including those critical for drug synthesis, can be conducted efficiently without solvents. The technique improves reaction yields and reduces reaction times, offering a greener and more cost-effective alternative [25].

7.4.1.3 Chemical Industry

- **Mechanochemical Synthesis**

Mechanochemical is one of the methods in which mechanical energy is utilized, such as grinding or milling, to facilitate chemical reactions between solid reactants. This approach eliminates the need for solvents and is particularly advantageous for producing high-purity compounds. Applications include pharmaceutical intermediates, catalysts, and specialty chemicals [26].

- **Ionic Liquid-Mediated Reaction**

Ionic liquids serve as dual-function agents, they act as solvents and catalysts in chemical reactions. These low-volatility liquids provide a cleaner or greener alternative to organic solvents, improving reaction efficiency while reducing emissions and waste. They are widely used in catalysis, separation processes, and organic transformations [27].

7.4.1.4 Materials Science

- **Polymer Synthesis**

In this method, solvent-free polymerization techniques, such as bulk polymerization and interfacial polymerization, are used to produce a variety of polymers with minimal solvent use. These methods are very efficient and offer better control over polymer properties, which makes them suitable for applications in coatings, adhesives, and packaging materials [28].

- **Nanomaterial Synthesis**

Solvent-free processes also allow nanometre-sized particle formation. These methods often use solid precursors like thermal, mechanical, or chemical activation to produce materials with size reduction at nanoscale. This sustainable approach for synthesis of nanoparticles and nanocomposite is widely applied in electronics, catalysis, and biomedicine for water treatment [29].

7.4.2 Close-loop Water System

Closed-loop water systems provide a newer approach for sustainable water management, particularly in water-dependent industries. In this method continuously water is recycled and reused within industrial processes, thereby reducing the need for freshwater intake and minimizing wastewater production [30]. This approach effectively addresses the water scarcity problem and helps to meet environmental compliance regulations. The main purpose of the closed-loop method is to reduce freshwater consumption by treating and reusing processed water.

An example of Closed-loop water systems which is an innovative incorporation of a novel design that combines Forward Osmosis (FO) with nanofilter, which results in highly efficient industrial wastewater treatment. This design can overcome typical challenges related to low-energy FO systems, such as concentration polarization. By reducing concentration polarization, this system can operate for a long duration without membrane clogging or significant flux reduction. For example, the system can remove over 97% of chemical oxygen demand (COD) from pharmaceutical wastewater, and deliver a water transporting rate that exceeds 50 LMH, and it suppresses all previous research findings. In addition, the continuous treatment process recovers and recycles 99% of contaminated water, with pure water and its transport rate is 58–60 LMH. The flat-sheet cross-flow module's horizontal membrane alignment within the FO loop minimizes concentration polarization and membrane clogging, due to the high-velocity flow across the membrane surfaces. This method greatly improves the water flow rate in both the FO and subsequent nanofiltration stages, and it ensures its sustainable, steady flux operation for long-term industrial applications [31].

Another example is related to greywater treatment. The system's main focus is to minimize freshwater utilization and decrease wastewater discharge by membrane filtration method for greywater treatment. This method is an excellent solution to solve water scarcity and reduce wastewater discharge. In this process a semi-permeable membrane is used to separate particles and impurities from a liquid. In this case, the membrane filters the greywater, by removing impurities, contaminants and improving the quality of water for reuse. The system offers economic benefits, and environmental advantages, and promotes responsible water treatment or freshwater consumption [32].

7.4.3 *Waterless Technology*

In today's world water scarcity and environmental concern are a big issue, so in this condition waterless technologies play a crucial role. By changing industrial processes, waterless technologies offer a sustainable future with reduced water consumption, lower costs, and a smaller ecological footprint [33].

This method indicates that traditional textile dyeing consumes a lot of fresh water which is released as effluent. As a result, freshwater level is reduced and surface water becomes contaminated. To solve this problem air-dyeing and supercritical carbon dioxide ($scCO_2$) aided dyeing are two waterless dyeing techniques which can be used. Supercritical Carbon Dioxide ($scCO_2$) dyeing in which supercritical CO_2 as a solvent to dissolve and transfer colors into the fabric, whereas in air-dyeing, applies dye particles directly onto the cloth, so there is no need for water-based dyeing. Compared to traditional methods, approaches provide several benefits, such as shorter working periods, lower energy usage, reduced waste and greenhouse gas emissions. Because of their high diffusivity, reusability, easy availability, non-flammability, and non-toxicity, they are both ecologically benign and economically viable. Air Dye is already being used by several textile industries in North America, Asia, and Europe [34].

Another technique is to pretreat microfiber cloths for waterless vehicle washing. In this process, particulate particles, metal ions, and organic matter are among the pollutants that are successfully removed by the pre-treated microfiber cloth. Ethanol, 2-propanol, and ethylene glycol were used to pre-treat the microfiber cloths. To get rid of impurities from the textiles, these solvents were combined with ultrasonic treatment. The reduction of chemical oxygen demand (COD), surfactant content, and total nitrogen in the extract were used to assess each solvent's efficacy.

The environmental impact of textile waste can be reduced and the idea of the circular economy can be promoted by reusing or recycling this cleaned material into new goods. The study emphasizes the value of environmentally friendly practices in the automobile sector as well as the potential of cutting-edge technologies to solve environmental issues [35].

7.5 Advanced Water Treatment Methods

Green chemistry promotes sustainable wastewater treatment by:

7.5.1 *Green Solvent*

Green solvents are environmentally friendly and alternative to conventional water treatment or hazardous solvents, offering a sustainable solution to reduce water consumption and environmental impact in industrial processes. Green solvents are widely used in industries like pharmaceuticals, textiles, and chemical manufacturing, which is like a transformative change.

Replacing water-intensive solvents with ionic liquids, supercritical CO_2, or bio-based solvents has lowered the demand for water use in industries.

7.5.1.1 Bio-based Solvents

Bio-based solvents are obtained from renewable resources such as plants, sugars, and biomass, offering an eco-friendly alternative to traditional water-intensive or hazardous solvents. These solvents are biodegradable, non-toxic, and renewable, making them a sustainable choice for industries seeking to reduce water consumption and environmental impact Table 7.1.

By using bio-based solvent industries may reduce wastewater output, improve the process of sustainability, and drastically less freshwater resources are required. This trade is beneficial for both economic and environmental promotion by bringing industrial operations into line with global sustainability goals and conserving water.

7.5.1.2 Ionic Liquids as Green Solvents

Ionic liquid is a salt in liquid form that has negligible vapor pressure and is highly recyclable. They are used as an alternative for volatile organic compounds (VOCs), which reduce both water and energy consumption in extraction, catalysis, and separation processes. For instance, in the pharmaceutical industry, ionic liquids promote efficient drug synthesis with minimum water use and waste generation [43]. Organic salts with low melting points are the best substitute for aprotic polar solvents such as $scCO_2$, dimethyl formamide, and dimethyl sulphoxide due to their remarkable solvating abilities for polar compounds. It has been also found that a subclass of imidazole-based solvents is especially helpful for polymeric polymers like cellulose [44].

At room temperature, ionic liquids (ILs), which are liquids made entirely of ions, have low vapor pressure and limited evaporative emission. Therefore, a solution of sodium chloride in water, a molecular solvent, is an ionic solution, but molten sodium chloride is an IL [45]. Water, ethanol, and benzene are common solvents that are typically molecular liquids, regardless of whether they are polar or non-polar. Ions make up ILs, which behave differently from molecular liquids when utilized as solvents. The majority of ILs are made up of inorganic anions and organic cations which also include nitrogen, Phosphorus, imidazolinium, or tricaprylmethyl

Table 7.1 Summarizes the bio-based solvent its sources, application, and advantage

Bio-based solvent	Source	Industrial use	Advantage	Ref
Ethanol	Fermentation of sugarcane, corn, or other biomass	Pharmaceutical, cosmetics, and chemical industries as a solvent for extraction, cleaning, and formulation	Cleaning and preparation	[36]
Ethyl Lactate	Fermentation of corn or other starch-rich crops	Paints, coatings, pharmaceuticals, and electronics for cleaning and as a solvent in chemical reactions	water-miscible, biodegradable, VOC-emitting solvents in cleaning and manufacturing processes	[37]
Glycerol (Glycerin)	Biodiesel byproducts, vegetable oils	Cosmetics, food processing, pharmaceuticals	Biodegradable, eco-friendly, replaces water in emulsions and formulations	[38]
d-Limonene	Citrus peels	Industrial degreasers, cleaning products, adhesives	Biodegradable, renewable, effective in removing grease and residues	[39]
Vegetable oil-based solvents	Soybean oil, castor oil	Inks, paints, coatings	Renewable, biodegradable, effective viscosity and solubility properties	[40]
Cyclopentyl methyl ether	Addition of methanol to the cyclopentene	Bio-refineries	Biotransformations, bioextractions, chromatography, solid-phase peptide synthesis	[41]
Bio-ethanol	Fermentation of sugar and vegetable residues	Power and chemical industry	Motor fuel, additive in gasoline	[42]

ammonium cations with different hetero-atom activities which are found in most found in research systems. There is an endless variety of potential ILs; by 2009, more than a thousand such molecules had been produced [46].

A fascinating new class of liquids known as room-temperature ILs (RTILs) has been accessible since the early 1980s. They are salts that melt below roughly 100°C and remain liquid throughout a broad temperature range. Although RTILs have received a lot of attention lately, the application of RTILs has been hindered by the absence of sustainable methods for product separation. In-depth research on the toxicity, biodegradation, safety, environmental risk, and impact of ILs is required, and RTILs must become more affordable, cleaner, and simpler to filter [47].

ILs hold a range of versatile and unique properties that make them valuable in various industrial and research applications. These are salts with relatively large cations and/or anions, typically having a freezing point below 100 °C, enabling them to remain liquid across a wide temperature range, from –96 to 200 °C. Their viscosity is generally below 100 cP, which offers a flexible process, and they exhibit high thermal stability up to 200 °C, along with electrochemical stability, which makes them suitable for extreme operational environments. ILs have exceptional electrical conductivity, and specific conductivity which is usually below 10 mS/cm and molar conductivity is less than 10 S/cm^2, while serving as excellent solvents for organic, inorganic, and polymeric materials. Their acidic compositions often function as superacids (pKa ≈ -20), which are highly effective in catalyzing reactions [48].

While some ILs demonstrate that they are highly water-sensitive and require handling in dry conditions, others are hydrophobic and air-stable, further expanding their utility. They also carry a very low or negligible vapor pressure at room temperature, which makes them environmentally friendly and non-flammable below their decomposition temperature, though some can be explosive under certain conditions. Their highly solvating nature reduces solvent volume requirements, enabling process intensification [49]. Moreover, they serve as dual-function materials, acting as both catalysts and solvents and also promote highly selective reactions. These features collectively establish ILs as a powerful tool in advancing sustainable, efficient, and innovative chemical processes. Because of all the above properties, ILs are considered to be green (environmentally benign) reaction media. They can contribute significantly to sustainable chemistry and the development of green technology (Table 7.2).

Compared to all other ILs, energetic ILs have shown their popularity over conventionally applied energetic compounds because of their higher thermal stability, easy synthesis methods, density, very low vapor pressure, and negligible toxicity.

Industries are using supercritical fluids (SCFs) as an effective and sustainable substitute for water in several applications. It is a substance that possesses temperature and pressure above its critical points and exhibits both liquid-like density and gas-like viscosity, which makes it unique and versatile. Supercritical carbon dioxide (SC-CO_2) is the most commonly utilized SCF because it is non-flammable, non-toxic, and environmentally benign [66]. In many industrial applications, supercritical fluids-in particular, supercritical CO_2 (SC-CO_2)—are being utilized as a sustainable substitute for water. SC-CO_2 takes the role of water and organic solvents for the extraction of bioactive substances, essential oils, and other sectors such as food, medicine, and cosmetics [67]. It is also an efficient substitute for water-based cleaning solutions. SC-CO_2 is used in the electronics, aerospace, and automotive industries to precisely clean delicate parts for cleaning and degreasing. It also improves the efficiency of materials like coatings and adhesives by substituting water in polymerization during polymer processing and synthesis.

SC-CO_2 is used in pharmaceuticals because it enhances drug formulations and regulates the size of active pharmaceutical ingredients (APIs), which lowers the amount of water required during production. It makes drying, preservation, and flavor extraction easier in food processing, providing a more effective and sustainable way than conventional water-based techniques. By removing pollutants from soil and water, SC-CO_2 also contributes to environmental remediation by using less water during clean-up procedures [68]. It is an alternative of water and organic solvents

Table 7.2 Summarizes the roles of various ionic liquids in chemical processes, environmental applications, and pharmaceuticals

Ionic liquid	Role in chemical processes/ environmental applications	Role in pharmaceuticals	Ref
[C_2C_1IM][Ac] (Reagent)	Pre-treatment for extraction of collagen biopolymer from waste fish	–	[50]
[EMIM][NO_3] (Media)	Cryopreservation, refolding, and dissolution of amyloid aggregates (insulin)	–	[51]
[BMIM][Cl] (Media)	Exceptional extraction for the desulfurization of liquid fuels	Used in the conversion of cellulose into biofuel precursor 5-hydroxymethylfurfural	[52]
[BMIM]Br (Media)	Replacement for volatile organic solvents in acetylsalicylic acid production	–	[53]
Acidic IL (Media)	Sustainable solvents for the extraction process	–	[54]
[C_4MIM] [Cl] (Media)	Solvent for catalytic transesterification of cellulose	Synthesis of isoxazoline derivatives, accelerating reaction rates and improving yields	[55]
Diimidazolium salts of gelling ILs (Media)	Adsorption of organic dyes from wastewater	–	[56]
[BMIM][Cl] and [EMIM][DEP]	Pretreatment of lignocellulosic oil palm fiber for enhanced biocomposite properties	–	[57]
[C_4MIM] [BF_4] (Media)	–	Synthesis of isoxazoline derivatives with improved yield	[58]
1,10-bis(3-methylimidazolium-1-yl) decane diphenylalanine (Media)	–	Enhances coating morphology on Ti surfaces	[59]
[EMIM][ESO_4] and [EMIM][Cl] (Media)	–	Refolding agent for denatured serum albumins	[60]

(continued)

Table 7.2 (continued)

Ionic liquid	Role in chemical processes/ environmental applications	Role in pharmaceuticals	Ref
$[EMIM]^+$, $[BMIM]^+$, $[HMIM]^+$ with anions$[BF_4]^-$, $[AcO]^-$, $[TfO]^-$ (Media)	–	High dissolving capacity for nonsteroidal anti-inflammatory drugs (e.g., ibuprofen)	[61]
$[C_4MIM][PF_6]$ (Catalyst)	–	Biotransformation of benzaldehyde to (R)-PAC with decreased compound toxicity	[62]
$[C_1C_1MIM]$ $[C1SO_4]$ Dehydrating agent	–	Dehydrating agent for synthesizing hydrazinylphthalazines	[63]
$[C_2MIM]$ [OAc] (Media)	–	Enhances solubility and permeability of drugs (e.g., Amphotericin B)	[64]
$[BDMIM][BF_4]$ (Media)	–	Used as an adjuvant in vaccine formulations	[65]

across a range of sectors and its further usage in the production of biofuel and supercritical drying (Table 7.3).

Compared to conventional water-based processes, the use of supercritical fluids—in particular, supercritical CO_2—offers several benefits to industries. Their solvent

Table 7.3 Summarizes the various applications of supercritical fluids in industry as a replacement for water

Supercritical fluid used	Application area	Industry	Ref
Supercritical CO_2 (SC-CO_2)	Extraction processes	Food, pharmaceuticals, cosmetics	[69]
Supercritical CO_2 (SC-CO_2)	Cleaning and degreasing	Electronics, aerospace, automotive	[70]
Supercritical CO_2, SCFs	Polymer processing and synthesis	Plastics, coatings, packaging	[71]
Supercritical CO_2 (SC-CO_2)	Pharmaceuticals and drug development	Pharmaceuticals, healthcare	[72]
Supercritical CO_2 (SC-CO_2)	Food processing	Food, beverages	[73]
Supercritical CO_2 (SC-CO_2)	Environmental remediation	Environmental services	[74]
Supercritical CO_2 (SC-CO_2)	Biofuel production	Energy, biofuels	[75]
Supercritical CO_2 (SC-CO_2)	Polymerization reactions	Chemical, pharmaceutical	[76]
Supercritical CO_2 (SC-CO_2)	Supercritical drying	Food, biotechnology	[77]

qualities are one of their main advantages; they can also dissolve non-polar molecules in a way that water cannot, enabling more accurate and effective extraction or reactions. Additionally, SCFs have low viscosity and high diffusion, which promote quicker reaction rates, cut down on processing time, and boost efficiency. They are environmentally friendly and do not leave hazardous residues, because they are non-toxic, non-flammable, and recyclable.

For example, it can be utilized in a variety of processes, such as environmental remediation, polymer synthesis, extraction, and cleaning. These fluids frequently reduce water usage by solvent recovery, drying, and cleaning. Further, SCFs improve process intensification, which uses only fewer amounts of solvent and enables more concentrated reactions and improved product yield.

The use of this fluid in a variety of industries which promotes sustainability, resource efficiency, and economic benefits.

7.5.2 Catalysis in Water Treatment

Catalysis plays an important role in the degradation, which is primarily important for sustainability, which aims to improve industrial water use. In this process complex compounds or molecules break down into simpler ones in water and prevent transforming harmful pollutants in water, which in turn reduces their toxicity and environmental impact. This process is more efficient and sustainable compared to other conventional methods and offers significant advantages in water treatment and pollution control.

7.5.2.1 Photo-Catalysis

This process uses light energy to operate the catalytic degradation of pollutants which is released by industrial effluents. This method is useful for breaking down the complex organic pollutants in wastewater and further this water is integrated into green water treatment technologies. A semiconductor catalyst is used, like titanium dioxide, tungsten trioxide, and zinc oxide are used to degrade organic pollutants released by effluent of industry into the exposure to light [78]. These semiconductor catalyses have stable and large surface areas, that is they generate highly reactive species which can oxidize and mineralize contaminants into harmless substances. They produce electron–hole pairs which help it to break down contaminants into innocuous by-products when exposed to light. To modify this degradation a metal nanocomposite is produced with light, this nanocomposite works well for organic compound degradation and makes it a viable technology for a clean environment.

To modify an efficient photocatalytic system, it is vital to ascertain the function of reactive species such as holes (h^+), electrons (e^-), hydroxyl (OH), and superoxide ion (O_2^-) radicals in the degradation process as shown in Fig. 7.2. The factors that impact the efficiency of photocatalysts are the selection of metal oxide, crystal structure,

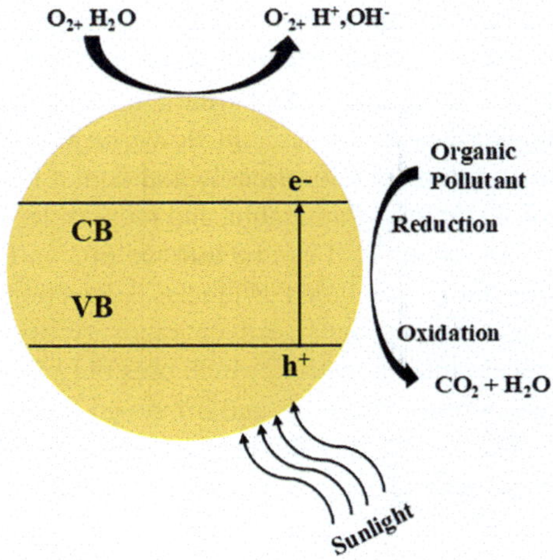

Fig. 7.2 Mechanism of photocatalytic degradation

light conditions, pH, catalyst loading, and pollutant concentration. To maximise degradation rates these parameters require optimising [79].

7.5.2.2 Advanced Oxidation Processes (AOPs)

This process involves the production of reactive species like hydroxyl radicals (OH) that efficiently degrade pollutants even at low concentrations. In AOPs, OH radicals are usually produced by oxidizing agents such as ozone (O_3), ultraviolet (UV) light, and hydrogen peroxide (H_2O_2) frequently in conjunction with catalysts [80]. Due to their high reactivity, these radicals target a variety of organic impurities, such as industrial dyes, chemicals, medications, and pesticides.

In the Ozonation process, injecting ozone (O_3) into water, which breaks down and produces hydroxyl radicals, is efficient by eliminating a variety of pollutants, such as dyes, chemicals, and other hazardous substances [81]. Furthermore, by utilizing UV radiation hydroxyl radicals produced from hydrogen peroxide (H_2O_2) or directly photolyzes organic molecules which are helpful for the disinfection of by-products and contaminated water [82]. In addition, Fenton's reagent also produces hydroxyl radicals by combining hydrogen peroxide (H_2O_2) with a ferrous ion catalyst (Fe^{2+}) which is very helpful for efficient cleaning of wastewater that contains organic contaminants and dyes [83]. High-frequency ultrasonography is used in sonolysis to produce oxidizing species and very reactive hydroxyl radicals which is efficient in purifying wastewater that contains organic contaminants that are difficult to remove.

In addition, they are more efficient and have quicker reaction times than conventional water treatment techniques. They may need careful optimization to strike a balance between cost and effectiveness, and they can be energy-intensive. AOPs can reduce the overall environmental effect of industrial effluents and often use less harmful chemicals, making them an eco-friendly alternative to conventional chemical treatments in the context of green technologies for industrial water consumption [84]. By integrating AOPs into wastewater management systems, industries may lessen their impact on the environment and provide cleaner water discharge.

7.5.3 Electrocatalysis

In electrocatalysis electrochemical reactions take place that are aided by catalysts, which break down dangerous pollutants into less toxic or inert substances. Electrolysis is of two types i.e., Direct electrolysis and indirect electrolysis which are used in electrocatalytic water treatment. In direct electrolysis, contaminants undergo direct oxidation or reduction at the electrode surface, depending on the applied voltage. This technique works well with basic organic substances. On the other hand, indirect electrolysis contaminants undergo oxidization, or reducing agents such as hydrogen peroxide or hydroxyl radicals are produced at the electrode surface [85].

This technology is also a beneficial technique for the water treatment process which uses semi-permeable membranes to separate and purify water. In this process water is passed through a membrane with specific pore sizes, which allow only certain molecules or particles to pass through and block unwanted impurities. This method offers an advantage that is eco-friendly industrial water treatment technique. Since it uses electric current rather than hazardous chemicals, which removes the chance of secondary contamination. This technique has a high degree of selectivity and allows it to specifically target certain pollutants, such as heavy metals, chlorinated compounds, and persistent organic contaminants, which are challenging to remove with other techniques. This energy efficiency process is also significantly increased by modern electrocatalysts, which reduces operating costs and makes large-scale industrial applications possible.

This method also allows for on-site breakdown of pollutants, reducing the requirement for transportation and storage of chemical reagents. Its scalability and adaptability make it suitable for use in a wide range of sectors, including pharmaceutical and textiles. Additionally, electrocatalysis ensures improved effluent discharge by preventing the addition of additional waste or residues to the water [86]. This approach is a viable way to address the rising problems of industrial water pollution since it can further align with sustainability goals by combining with renewable energy sources.

7.5.4 Membrane Technology

This technology is also a beneficial technique for the water treatment process which uses semi-permeable membranes to separate and purify water. In this process water is passed through a membrane with specific pore sizes, which allow only certain molecules or particles to pass through and block unwanted impurities [87].

In this membrane microfilter is also used in which process it removes large particles like pigments. It uses porous membranes with pore sizes around 0.1–1.0 mm, controlled at pressures up to 500 kPa. Ultrafiltration removes larger particles like bacteria, viruses, and macromolecules. It uses a membrane with pore sizes around 0.05 mm, operating at pressures up to 1000 kPa. Nanofiltration removes monovalent ions and larger molecules like dyes and sugars. It uses membranes operating on a solution-diffusion principle, with pressures up to 5000 kPa. Reverse osmosis uses the tightest membranes to remove dissolved salts and ions. It operates at high pressures (up to 7000 kPa) and is used for desalination, producing ultrapure water, and treating industrial wastewater [88].

Membrane technology is becoming more and more important in the water sector. Membranes are used in a variety of processes, including surface water treatment, wastewater treatment, reclamation, and seawater desalination. As the technology develops, better membranes and procedures are used. Numerous applications have been thoroughly evaluated; nevertheless, in order to improve their performance and lower their prices, new and improved materials and preparation procedures are needed to increase membrane life, chemical resistance, and the production of large membrane surfaces with uniform characteristics. Therefore, the character development strategies play a crucial role. Much work has to be undertaken to standardize some of the membrane production and testing processes.

7.5.5 Bioremediation

In the process in which biological organisms are used - such as bacteria, fungus, algae, and plants, to break down or eliminate pollutants from industrial effluent. It is both economical and environmentally beneficial. It is a sustainable method of controlling water pollution that works especially well for organic and certain inorganic contaminants, such as hydrocarbons, dyes, and heavy metals.

7.5.5.1 Microbial Degradation

Organic pollutants are broken down by microorganisms into less toxic forms like carbon dioxide, water, and biomass. For example, *Pseudomonas* and *Bacillus* species are frequently employed to break down phenols and hydrocarbons in industrial wastewater. An example in which *F. pinicola* IEBL-4 is used which is a suitable fungus for the bioremediation of industrial effluents. Optimization of growth conditions of this fungus resulted in significant biodegradation of various textile effluents, including a 72.06% reduction in MT effluent, a 75.15% reduction in FST effluent, and a 79.02% reduction in ST effluent. Additionally, the biodegradation process is enhanced by the addition of carbon and nitrogen sources. In this study, it was also found that ligninolytic enzymes, particularly lignin peroxidase, play a crucial role in the degradation of textile dyes. Overall, the findings suggest that *F. pinicola* IEBL-4 has the potential to degrade dye effectively and it is an eco-friendly solution for treating textile wastewater [89]. The study demonstrates the effectiveness of an MBR in removing ibuprofen from hospital wastewater but not diclofenac. Post-treatment is necessary for complete pharmaceutical removal. Aerobic biodegradation was effective for both pharmaceutical products, i.e., ibuprofen degrades faster than diclofenac. The presence of additional substrates did not enhance degradation, and nitrate was not a suitable electron acceptor. The study identified potential bacterial strains involved in the degradation process, with ibuprofen degrading completely and diclofenac degrading partially with intermediate formation [90].

Another study is conducted which has the potential of seaweed and algal biomass for efficient bioremediation of wastewater polluted with heavy metals. The benefits of employing these biomaterials, including their capacity to absorb and accumulate heavy metals, their quick development rate, and their potential for biomass production, are most likely highlighted in the study [91].

7.5.5.2 Enzymatic Degradation

During bioremediation, enzyme-catalyzed processes are essential for accelerating the biodegradation of contaminants. Certain enzymes produced by microorganisms like bacteria and fungi catalyze the conversion of complex and persistent pollutants into less hazardous, simpler compounds. These enzymes are perfect for treating industrial wastewater since they are extremely efficient, and selective, and work well in moderate environments [92]. The study comes to the conclusion that because of their effectiveness and environmental friendliness, enzyme-based technologies offer a lot of promise for treating wastewater. However, issues like the high cost of producing enzymes and immobilization methods must be resolved. The development of self-sustaining systems with regulated enzyme immobilization and on-demand substrate creation is the goal of future research [93]. Free enzymes and enzyme-loaded hydrogels comprising pectin, gelatin, and CMC were used in a study to evaluate the breakdown of BPA. In contrast to the other studied formulations, the results showed that pectin-based hydrogels considerably improved the breakdown of BPA. For pectin-based hydrogels, the maximum breakdown was 90%, whereas free enzymes could only accomplish 26%. This improvement is ascribed to the pectin hydrogel's ability to stabilize and shield the enzymes, preventing non-competitive inhibition-induced deactivation. Additionally, the molecular weight of the primary degradation product determined by LDTD-APCI analysis was 200.2 g/mol [94].

7.6 Conclusion

This chapter highlights that green chemistry offers a transformative approach to achieving industrial sustainable water use by emphasizing waste prevention, resource efficiency, and the adoption of environmentally friendly technologies. Industries face significant challenges related to water scarcity, pollution, and regulatory compliance. Through the principles of green chemistry, such as the use of safer chemicals, renewable feedstocks, and energy-efficient processes, industries can reduce their water dependency while minimizing environmental impact. This chapter also covered the innovative solutions like solvent-free reactions, advanced oxidation processes (AOPs), closed-loop water systems, and bioremediation which demonstrate that they have potential to clean and more sustainable industrial practices. In addition, technologies like membrane filtration in which microfilters are used, supercritical fluids, and enzyme-catalysed reactions further enhance water conservation and wastewater treatment efficiency. In this sense, we hope that new alternatives are also found that are more efficient and sustainable for industry in which water is used. By integrating these strategies, industries can transition towards circular water use models, promoting resource recovery and reducing their ecological footprint.

References

1. Scanlon, B.R., Fakhreddine, S., Rateb, A., de Graaf, I., Famiglietti, J., Gleeson, T., Grafton, R.Q., Jobbagy, E., Kebede, S., Kolusu, S.R., Konikow, L.F.: Global water resources and the role of groundwater in a resilient water future. Nat. Rev. Earth Environ. **4**(2), 87–101 (2023)
2. Englande Jr, A.J., Krenkel, P., Shamas, J.: Wastewater treatment and water reclamation. Ref. Modul. Earth Syst. Environ. Sci. B978–0 (2015)
3. Manasa, R.L., Mehta, A.: Wastewater: sources of pollutants and its remediation. Environ. Biotechnol. **2**, 197–219 (2020)
4. Taft, H.L.: Water scarcity: global challenges for agriculture. In: Food, Energy, and Water, pp. 395–429 (2015)
5. Pal, P., Kumar, R.: Treatment of coke wastewater: a critical review for developing sustainable management strategies. Sep. Purif. Rev. **43**(2), 89–123 (2014)
6. Kumar, R., Pal, P.: Membrane-integrated hybrid bioremediation of industrial wastewater: a continuous treatment and recycling approach. Desalin. Water Reuse. **1**(3), 26–38 (2013)
7. Wang, L., Rinklebe, J., Tack, F.M., Hou, D.: A review of green remediation strategies for heavy metal contaminated soil. Soil Use Manag. **37**(4), 936–963 (2021)
8. Khanam, Z., Sultana, F.M., Mushtaq, F.: Environmental pollution control measures and strategies: an overview of recent developments. In: Geospatial Analytics for Environmental Pollution Modeling: Analysis, Control and Management, pp. 385–414 (2023)
9. Bassey, K.E.: From waste to wonder: developing engineered nanomaterials for multifaceted applications. GSC Adv. Res. Rev. **20**(3), 109–123 (2024)
10. Chen, T.L., Kim, H., Pan, S.Y., Tseng, P.C., Lin, Y.P., Chiang, P.C.: Implementation of green chemistry principles in a circular economy system towards sustainable development goals: Challenges and perspectives. Sci. Total. Environ. **716**, 136998 (2020)
11. Anastas, P.T., Kirchhoff, M.M.: Origins, current status, and future challenges of green chemistry. Acc. Chem. Res. **35**(9), 686–694 (2002)

12. Muga, H.E., Mihelcic, J.R.: Sustainability of wastewater treatment technologies. J. Environ. Manage. **88**(3), 437–447 (2008)
13. Muhib, M.I., Rahman, M.M.: Water and energy efficiency in textile production. In: Dye Pollution from Textile Industry: Challenges and Opportunities for Sustainable Development, Springer Nature, Singapore, pp. 279–305 (2024)
14. Hai, F.I., Yamamoto, K., Fukushi, K.: Hybrid treatment systems for dye wastewater. Crit. Rev. Env. Sci. Tec. **37**(4), 315–377 (2007)
15. Bailone, R.L., Borra, R.C., Fukushima, H.C.S., Aguiar, L.K.: Water reuse in the food industry. Discov. Food. **2**(1), 5 (2022)
16. Clarke, C.J., Tu, W.C., Levers, O., Brohl, A., Hallett, J.P.: Green and sustainable solvents in chemical processes. Chem. Rev. **118**(2), 747–800 (2018)
17. Hessel, V., Tran, N.N., Asrami, M.R., Tran, Q.D., Long, N.V.D., Escribà-Gelonch, M., Tejada, J.O., Linke, S., Sundmacher, K.: Sustainability of green solvents–review and perspective. Green Chem. **24**(2), 410–437 (2022)
18. Bradu, P., Biswas, A., Nair, C., Sreevalsakumar, S., Patil, M., Kannampuzha, S., Mukherjee, A.G., Wanjari, U.R., Renu, K., Vellingiri, B., Gopalakrishnan, A.V.: Recent advances in green technology and industrial revolution 4.0 for a sustainable future. Environ. Sci. Pollut. Res. Int. **1** (2022)
19. Nižetić, S., Djilali, N., Papadopoulos, A., Rodrigues, J.J.: Smart technologies for promotion of energy efficiency, utilization of sustainable resources and waste management. J. Clean. Prod. **231**, 565–591 (2019)
20. Farhana, K., Kadirgama, K., Mahamude, A.S.F., Mica, M.T.: Energy consumption, environmental impact, and implementation of renewable energy resources in global textile industries: an overview towards circularity and sustainability. Mater. Circ. Econ. **4**(1), 15 (2022)
21. Azma, A., Kianfar, N., Zhang, Y.: Development on sustainable water recycle, reuse and management. In: 3rd Conference on Environment, Civil Architecture Urban Development, pp. 1–21 (2021)
22. Date, M., Patyal, V., Jaspal, D., Malviya, A., Khare, K.: Zero liquid discharge technology for recovery, reuse, and reclamation of wastewater: a critical review. J. Water Process Eng. **49**, 103129 (2022)
23. Gawande, M.B., Bonifacio, V.D., Luque, R., Branco, P.S., Varma, R.S.: Solvent-free and catalysts-free chemistry: a benign pathway to sustainability. Chem. Sus. Chem. **7**(1), 24–44 (2014)
24. Wegner, K., Barnes, D., Manzor, K., Jardine, A., Moran, D.: Evaluation of greener solvents for solid-phase peptide synthesis. Green Chem. Lett. Rev. **14**(1), 153–164 (2021)
25. Jun, C.H., Chung, J.H., Lee, D.Y., Loupy, A., Chatti, S.: Solvent-free chelation-assisted intermolecular hydroacylation: effect of microwave irradiation in the synthesis of ketone from aldehyde and 1-alkene by Rh (I) complex. Tetrahedron Lett. **42**(29), 4803–4805 (2001)
26. Sun, X., Shi, W., Zhou, X.Y., Ding, S.: Facile mechanochemical preparation of polyamide-derivatives via solid-state benzoxazine-isocyanide chemistry. Chin. J. Polym. Sci. **39**, 573–584 (2021)
27. Krusenbaum, A., Grätz, S., Tigineh, G.T., Borchardt, L., Kim, J.G.: The mechanochemical synthesis of polymers. Chem. Soc. Rev. **51**(7), 2873–2905 (2022)
28. Gonzalez, C.G., Choi, Y.H., Verpoorte, R.: Preanalytical treatments: extraction with deep eutectic solvents. In: Liquid-Phase Extraction, pp. 565–590 (2020)
29. Erdmenger, T., Guerrero-Sanchez, C., Vitz, J., Hoogenboom, R., Schubert, U.S.: Recent developments in the utilization of green solvents in polymer chemistry. Chem. Soc. Rev. **39**(8), 3317–3333 (2010)
30. Netskina, O.V., Mukha, S.A., Dmitruk, K.A., Ishchenko, A.V., Bulavchenko, O.A., Pochtar, A.A., Suknev, A.P., Komova, O.V.: Solvent-free method for nanoparticles synthesis by solid-state combustion using tetra (Imidazole) copper (II) nitrate. Inorg. **10**(2), 15 (2022)
31. Fathollahi-Fard, A.M., Ahmadi, A., Al-e-Hashem, S.M.: Sustainable closed-loop supply chain network for an integrated water supply and wastewater collection system under uncertainty. J. Environ. Manag. **275**, 111277 (2020)

32. Roy, M., Linnanen, L., Chakrabortty, S., Pal, P.: Developing a closed-loop water conservation system at micro level through circular economy approach. Sustain. Water Resour. Manag. **33**(12), 4157–4170 (2019)
33. Allafi, F.A.S., Hossain, M.S., Ab Kadir, M.O., Shaah, M.A.H., Lalung, J., Ahmad, M.I.: Waterless processing of sheep wool fiber in textile industry with supercritical CO_2: potential and challenges. J. Clean. Prod. **285**, 124819 (2021)
34. Mahmud, I., Kaiser, S.: Recent progress in waterless textile dyeing. J. Text. Sci. Eng **10**(6), 1–3 (2020)
35. Strikauska, S., Berzins, A., Arbidans, L., Kukela, A., Muter, O., Klavins, M.: Physicochemical pretreatment of contaminated microfibre cloths after their use in waterless car wash. Rigas Tehniskas Universitates Zinatniskie Raksti **32**, 85 (2015)
36. Edeh, I.: Bioethanol production: an overview. Bioethanol Technol. **1** (2021)
37. Pereira, C.S., Silva, V.M., Rodrigues, A.E.: Ethyl lactate as a solvent: properties, applications and production processes–a review. Green Chem. **13**(10), 2658–2671 (2011)
38. Azelee, N.I.W., Ramli, A.N.M., Manas, N.H.A., Salamun, N., Man, R.C., El Enshasy, H.: Glycerol in food, cosmetics and pharmaceutical industries: basics and new applications. Int. J. Sci. Technol. Res. **8**(12), 553–558 (2019)
39. Anandakumar, P., Kamaraj, S., Vanitha, M.K.: D-limonene: a multifunctional compound with potent therapeutic effects. J. Food Biochem. **45**(1), e13566 (2021)
40. Cheikhyoussef, N., Cheikhyoussef, A.: Vegetable oils as green solvents in the pharmaceutical industry. In: Green Sustainable Process for Chemical and Environmental Engineering and Science, pp. 1–11 (2021)
41. de Gonzalo, G., Alcántara, A.R., Domínguez de María, P.: Cyclopentyl methyl ether (CPME): a versatile eco-friendly solvent for applications in biotechnology and biorefineries. Chem. Sus. Chem. **12**(10), 2083–2097 (2019)
42. Tekin, K., Hao, N., Karagoz, S., Ragauskas, A.J.: Ethanol: a promising green solvent for the deconstruction of lignocellulose. Chem. Sus. Chem. **11**(20), 3559–3575 (2018)
43. Broughton, R.M.: Investigation of organic liquids for fiber extrusion. Natl. Text. Cent. Annu. Rep. **9**, 1–8 (2006)
44. Scammells, P.J., Scott, J.L., Singer, R.D.: Ionic liquids: the neglected issues. Aust. J. Chem. **58**(3), 155–169 (2005)
45. Hough-Troutman, W.L., Smiglak, M., Griffin, S., Reichert, W.M., Mirska, I., Jodynis-Liebert, J., Adamska, T., Nawrot, J., Stasiewicz, M., Rogers, R.D., Pernak, J.: Ionic liquids with dual biological function: sweet and anti-microbial, hydrophobic quaternary ammonium-based salts. New J. Chem. **33**(1), 26–33 (2009)
46. Lei, Z., Chen, B., Koo, Y.M., MacFarlane, D.R.: Introduction: ionic liquids. Chem. Rev. **117**(10), 6633–6635 (2017)
47. Wasserscheid, P., Welton, T., (eds.): Ionic liquids in synthesis. Weinheim: Wiley-Vch. **1**, 145 (2008)
48. Singh, S.K., Savoy, A.W.: Ionic liquids synthesis and applications: an overview. J. Mol. Liq. **297**, 112038 (2020)
49. Mallakpour, S., Dinari, M.: Ionic liquids as green solvents: progress and prospects. Green solvents II: properties and applications of ionic liquids, pp. 1–32 (2012)
50. Muhammad, N., Gonfa, G., Rahim, A., Ahmad, P., Iqbal, F., Sharif, F., Khan, A.S., Khan, F.U., Khan, Z.U.H., Rehman, F., Rehman, I.U.: Investigation of ionic liquids as a pretreatment solvent for extraction of collagen biopolymer from waste fish scales using COSMO-RS and experiment. J. Mol. Liq. **232**, 258–264 (2017)
51. Ishikawa, Y., Takekiyo, T., Yoshimura, Y.: Recovery and cryopreservation of insulin amyloid using ionic liquids. J. Mol. Liq. **272**, 1019–1024 (2018)
52. Nanda, B., Sailaja, M., Mohapatra, P., Pradhan, R.K., Nanda, B.B.: Green solvents: a suitable alternative for sustainable chemistry. Mater. Today: Proc. **47**, 1234–1240 (2021)
53. Alviz, P.L.A., Alvarez, A.J.: Comparative life cycle assessment of the use of an ionic liquid ([Bmim] Br) versus a volatile organic solvent in the production of acetylsalicylic acid. J. Clean. Prod. **168**, 1614–1624 (2017)

54. Pena-Pereira, F., Namieśnik, J.: Ionic liquids and deep eutectic mixtures: sustainable solvents for extraction processes. Chem Sus Chem. **7**(7), 1784–1800 (2014)
55. Schenzel, A., Hufendiek, A., Barner-Kowollik, C., Meier, M.A.: Catalytic transesterification of cellulose in ionic liquids: sustainable access to cellulose esters. Green Chem. **16**(6), 3266–3271 (2014)
56. Marullo, S., Rizzo, C., Dintcheva, N.T., Giannici, F., D'Anna, F.: Ionic liquids gels: soft materials for environmental remediation. J. Colloid Interface Sci. **517**, 182–193 (2018)
57. Mahmood, H., Moniruzzaman, M., Yusup, S., Akil, H.M.: Pretreatment of oil palm biomass with ionic liquids: a new approach for fabrication of green composite board. J. Clean. Prod. **126**, 677–685 (2016)
58. Chakraborty, B., Sharma, C.D.: A new route to the synthesis of isoxazoline derivatives from dihydropyran via cycloaddition reaction in ionic liquid. Tetrahedron Lett. **54**(40), 5532–5536 (2013)
59. Wheelis, S.E., Biguetti, C.C., Natarajan, S., Guida, L., Hedden, B., Garlet, G.P., Rodrigues, D.C.: Investigation of the early healing response to dicationic imidazolium-based ionic liquids: a biocompatible coating for titanium implants. ACS Biomater. Sci. Eng. **6**(2), 984–994 (2020)
60. Sindhu, A., Bhakuni, K., Sankaranarayanan, K., Venkatesu, P.: Implications of imidazolium-based ionic liquids as refolding additives for urea-induced denatured serum albumins. ACS Sustain. Chem. Eng. **8**(1), 604–612 (2019)
61. Huang, Y., Ji, Y., Zhang, M., Ouyang, D.: How imidazolium-based ionic liquids solubilize the poorly soluble ibuprofen? A theoretical study. AIChE J. **66**(5), e16940 (2020)
62. Kandar, S., Suresh, A.K., Noronha, S.B.: (R)-PAC biosynthesis in [BMIM][PF 6]/aqueous biphasic system using Saccharomyces cerevisiae BY4741 cells. Appl. Biochem. Biotechnol. **175**, 1771–1788 (2015)
63. Subramanian, G., Rajeev, C.B., Mohan, C.D., Sinha, A., Chu, T.T., Anusha, S., Ximei, H., Fuchs, J.E., Bender, A., Rangappa, K.S., Chandramohanadas, R.: Synthesis and in vitro evaluation of hydrazinyl phthalazines against malaria parasite, plasmodium falciparum. Bioorg. Med. Chem. Lett. **26**(14), 3300–3306 (2016)
64. Huang, W., Wu, X., Qi, J., Zhu, Q., Wu, W., Lu, Y., Chen, Z.: Ionic liquids: green and tailor-made solvents in drug delivery. Drug Discov. Today **25**(5), 901–908 (2020)
65. Uddin, M.N., Basak, D., Hopefl, R., Minofar, B.: Potential application of ionic liquids in pharmaceutical dosage forms for small molecule drug and vaccine delivery system. J. Pharm. Pharm. Sci. **23**, 158–176 (2020)
66. Gahtani, R.M., Talath, S., Hani, U., Rahamathulla, M., Khalid, A.: Pregabalin solubility in supercritical green solvent: a comprehensive experimental and theoretical-intelligent assessment. J. Mol. Liq. **408**, 125339 (2024)
67. Voutsas, E.: 7 Supercritical fluid. In: Food Engineering Handbook: Food Process Engineering, pp. 287 (2014)
68. Cvjetko Bubalo, M., Vidović, S., Radojčić Redovniković, I., Jokić, S.: Green solvents for green technologies. J. Chem. Technol. Biotechnol. **90**(9), 1631–1639 (2015)
69. Beckman, E.J.: Supercritical and near-critical CO_2 in green chemical synthesis and processing. J. Supercrit. Fluids **28**(2–3), 121–191 (2004)
70. Li, M.: Analysis of a New Cleaning Technology Based on Supercritical Carbon Dioxide Pre-treatment for Diesel Engine Remanufacturing. The University of Liverpool (United Kingdom) (2018)
71. Goodship, V., Ogur, E.: Polymer Processing with Supercritical Fluids. iSmithers Rapra Publishing. **15** (2005)
72. Zakharchenko, A.V., Bazarnova, N.G.: Influence of supercritical fluid extraction on lignocellulosic plant matrix and extractive substance yield. Acta biol. Sib. **8**, 107–113 (2022)
73. Bertran-Llorens, S., Perondi, F., de Freitas, A.L.S., Chen, J., van Erven, G., Deuss, P.J.: Supercritical CO_2 as effective wheat straw pretreatment for subsequent mild fractionation strategies. J. Chem. Eng. **497**, 154491 (2024)
74. Bhattacharjee, P., Paul, K., Dutta, S.: Supercritical carbon dioxide. In: Biotechnical Processing in the Food Industry: New Methods, Techniques, and Applications, **137** (2021)

75. Adeoti, I.A.: Extraction of Oil from Fish Processing Waste for Fuel Applications: Process Development, Analysis and Feasibility (Doctoral dissertation, Memorial University of Newfoundland) (2015)
76. Kendall, J.L., Canelas, D.A., Young, J.L., DeSimone, J.M.: Polymerizations in supercritical carbon dioxide. Chem. Rev. **99**, 543–564 (1999)
77. Braga, M.E., Gaspar, M.C., de Sousa, H.C.: Supercritical fluid technology for agrifood materials processing. Curr. Opin. Food Sci. **50**, 100983 (2023)
78. Eidsvåg, H., Bentouba, S., Vajeeston, P., Yohi, S., Velauthapillai, D.: TiO_2 as a photocatalyst for water splitting—an experimental and theoretical review. Mol. **26**(6), 1687 (2021)
79. Chong, M.N., Jin, B., Chow, C.W., Saint, C.: Recent developments in photocatalytic water treatment technology: a review. Water res. **44**(10), 2997–3027 (2010)
80. O'Shea, K.E., Dionysiou, D.D.: Advanced oxidation processes for water treatment. J. Phys. Chem. Lett. **3**(15), 2112–2113 (2012)
81. Venkatesh, S., Venkatesh, K., Quaff, A.R.: Dye decomposition by combined ozonation and anaerobic treatment: cost effective technology. J. Appl. Res. Technol. **15**(4), 340–345 (2017)
82. Zoschke, K., Börnick, H., Worch, E.: Vacuum-UV radiation at 185 nm in water treatment–a review. Water Res. **52**, 131–145 (2014)
83. Peralta-Hernández, J.M., Meas-Vong, Y., Rodríguez, F.J., Chapman, T.W., Maldonado, M.I., Godínez, L.A.: In situ electrochemical and photo-electrochemical generation of the fenton reagent: a potentially important new water treatment technology. Water Res. **40**(9), 1754–1762 (2006)
84. Mahamuni, N.N., Adewuyi, Y.G.: Advanced oxidation processes (AOPs) involving ultrasound for wastewater treatment: a review with emphasis on cost estimation. Ultrason. Sonochem. **17**(6), 990–1003 (2010)
85. Martínez-Huitle, C.A., Andrade, L.S.: Electrocatalysis in wastewater treatment: recent mechanism advances. Quim. Nova **34**, 850–858 (2011)
86. Singh, N., Goldsmith, B.R.: Role of electrocatalysis in the remediation of water pollutants. ACS Catal. **10**(5), 3365–3371 (2020)
87. Peters, T.: Membrane technology for water treatment. Chem. Eng. Technol. **33**(8), 1233–1240 (2010)
88. Ji, K., Liu, C., He, H., Mao, X., Wei, L., Wang, H., Zhang, M., Shen, Y., Sun, R., Zhou, F.: Research progress of water treatment technology based on nanofiber membranes. Polym. **15**(3), 741 (2023)
89. Mahmood, R.T., Asad, M.J., Hadri, S.H., El-Shorbagy, M.A., Mousa, A.A.A., Dara, R.N., Awais, M., Tlili, I.: Bioremediation of textile industrial effluents by Fomitopsis pinicola IEBL-4 for environmental sustainability. Hum. Ecol. Risk Assess. Int. J. **29**(2), 285–302 (2023)
90. Langenhoff, A., Inderfurth, N., Veuskens, T., Schraa, G., Blokland, M., Kujawa-Roeleveld, K., Rijnaarts, H.: Microbial removal of the pharmaceutical compound's ibuprofen and diclofenac from wastewater. Biomed. Res. Int. (1), 325806 (2013)
91. Znad, H., Awual, M.R., Martini, S.: The utilization of algae and seaweed biomass for bioremediation of heavy metal-contaminated wastewater. Mol. **27**(4), 1275 (2022)
92. Joshni, T.C., Subramaniam, K.: Enzymatic degradation of azo dyes-a review. Int. J. Environ. Sci. **1**(6), 1250–1260 (2011)
93. Maloney, A.J., Dong, C., Campbell, A.S., Dinu, C.Z.: Emerging enzyme-based technologies for wastewater treatment. In: Green Polymer Chemistry: Biobased Materials and Biocatalysis, pp. 69–85 (2015)
94. Gassara, F., Brar, S.K., Verma, M., Tyagi, R.D.: Bisphenol A degradation in water by ligninolytic enzymes. Chemosphere **92**(10), 1356–1360 (2013)

Chapter 8
Green Chemistry Strategies for Industrial Product Formulation

Gandhi Mansi, Vishwajit Chavda, Shashi Ranga, and Sapna Raghav

Abstract This chapter delves into the applications of green chemistry concepts in industrial product formulations, aiming for efficiency, sustainability, and environmental protection. It addresses strategies for minimizing waste, atomic economy, and replacing hazardous materials via greener alternatives. The twelve principles of green chemistry have established the framework, and this is of immense importance to the atom economy and the elimination of waste, including potential applications of renewable feedstocks. Green Metrics and Life Cycle Assessment (LCA) assist industries in assessing environmental impact and enhancing process efficiency. One of the main advancements is the availability of green solvents based on supercritical CO_2, ionic liquids, and water-based solvents instead of volatile organic compounds (VOCs). This chapter also discusses the importance of using renewable feedstocks such as biodegradable polymers, lignin, cellulose, and plant-based materials to enhance the durability of protective coatings, adhesives, and personal care products. Modern methods like biocatalysis, continuous manufacturing, and process intensification offer environmental benefits. In addition, researchers studied the applications of green additives, emulsifiers, and surfactants in industrial formulations. Case studies from the pharmaceutical, agrochemical, and specialty chemical industries demonstrate the successful application of green chemistry principles. Even though there are problems like limited resources, rules to follow, and limited technology, progress in AI-driven optimization, bio-based materials, and circular

G. Mansi
Department of Chemistry, Gujarat Technological University, Ahmedabad 382424, Gujarat, India

V. Chavda (✉)
Department of Chemistry, Institute of Science, Humanities and Liberal Studies, Indus University, Ahmedabad 382115, Gujarat, India
e-mail: vishwajitchavda1998@gmail.com

S. Ranga
Chemical Engineering Department, Government Engineering College-Valsad, Gujarat Technological University, Valsad 396001, Gujarat, India

S. Raghav
Department of Chemistry, Shri Jagdishprasad Jhabarmal Tibbrewala (JJT) University, Jhunjhunu 333001, Rajasthan, India

P. Jain et al. (eds.), *Towards Green Chemical Processes: Strategies and Innovations*, Intelligent Systems Reference Library 277,
https://doi.org/10.1007/978-3-031-94090-3_8

economy models shows promise for solutions. This chapter offers a comprehensive guide for researchers, industries, and policymakers, offering insights into sustainable industrial formulations aligned with global environmental and economic goals.

Keywords Green chemistry · Industrial formulations · Sustainable manufacturing · Green solvents · Process intensification · Renewable feedstocks · Biocatalysis · Life cycle assessment · Circular economy

8.1 Introduction

Green Chemistry: "hurt not the earth, neither the sea, nor the trees."—Ronald C.D. Breslow

The advancement of sustainable industrial procedures and product formulations is growing increasingly dependent upon green chemistry approaches. These approaches aim to reduce the impact on the environment and health risks by lowering the usage of harmful substances, eliminating waste production, and enhancing the use of resources. Green chemistry has found applications in a variety of industries, including personal care, cosmetics, and pharmaceuticals [1]. Improving production efficiency, considering the advantage of the full life cycle of materials, and developing environmentally friendly materials are key areas of focus [2]. The multiple benefits of green chemistry include less expensive production costs, cheaper disposal costs, and fewer resources needed while compromising the quality of the final product. Researchers and industry may develop innovative, sustainable procedures and products that promote sustainable development targets as the field improves.

Building on these principles, researchers in the field associated with green chemistry and the chemical industry are collaborating on developing environmentally compatible and greener alternatives to significant harmful procedures or materials. Those within the academic community who aim to tackle these sorts of problems should prioritize conducting studies to pinpoint the basic requirements currently determined by industry stakeholders. A consortium of pharmaceutical corporations stated the demands of the pharmaceutical industry in an annual report for the year 2007 [3]. This work has been applied in multiple articles in journals and, more importantly, in academic proposals as a basis for choosing an area of study. These interactions between academia and industry allow the application of the academic community's resources and resourcefulness to actual green chemistry demands [4] (Fig. 8.1).

In parallel, the formulated product operations search for additional environmentally conscious products and procedures. This industry offers approximately eighty billion dollars' worth of household products every year, comprising air fresheners, polishing agents, laundry detergents, and insecticides, along with approximately two hundred billion dollars in personal care goods annually, like toothpaste, shampoo, perfumes, cosmetics, and scented candles. The industry also emphasizes specific demands, and these are laid out here in an approach that is straightforward for learners

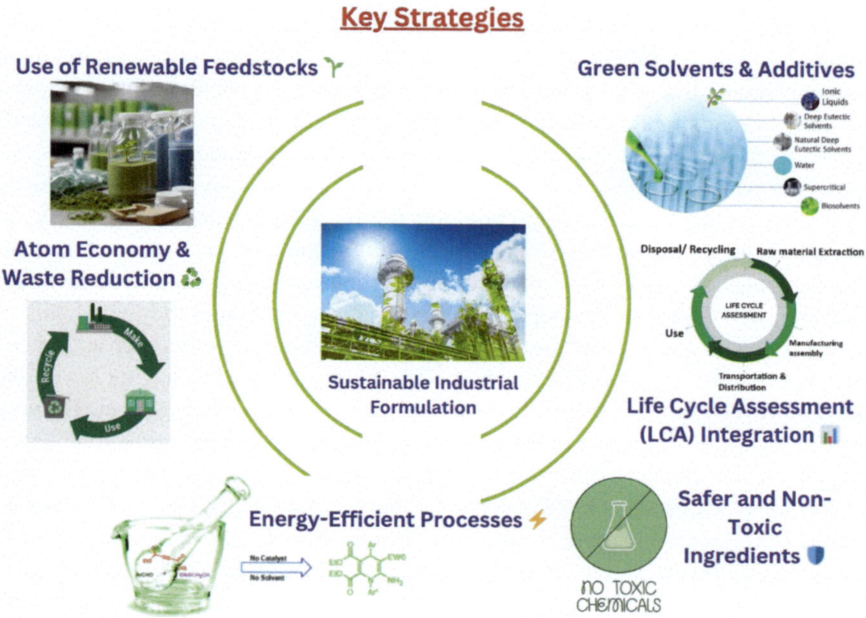

Fig. 8.1 Highlighting key strategies for sustainable industrial formulation

in academia to follow. Materials that are deemed required for their intended application currently have been shown or assumed to have detrimental effects on human beings or the natural world and play a role in each of these concerns. Because of the method in which they operate, numerous amounts of these materials finish in the sewage system or in the natural world. Consequently, more environmentally friendly alternatives are being researched [5].

In the formulation industry, hundreds of chemicals are incorporated to manufacture consumer personal care and cleaning products. Every component in the mixture has been chosen to fulfill a specific objective or offer a particular advantage. As further environmental, health, and safety data is available, a few parts must be removed or modified because they fail to satisfy the field's necessary requirements for these purposes. Identifying alternatives with significantly greater prominence is an issue this sector is interested in; substitutions must perform better than the original ingredients. The total cost of alternative chemicals must remain comparable. Before considering replacing a present chemical, the formulation industry now demands a complete set of environmental, safety, and health data. The chemicals sector must be sure that the new chemical has a good environmental, health, and safety profile before the chemical may be employed. A company never wants to switch out a chemical with one that they believe is superior, only to discover later that it has certain drawbacks that could affect consumers, the environment, or the company itself [6, 7].

The Green Chemistry Institute (GCI) Formulators' Roundtable collaborates with the GCI and 13 companies associated with the formulated products industry. They aim to encourage and promote the incorporation of green chemistry via the preparation of new products that will be less hazardous to consume and grow and more environmentally friendly through the duration of every phase of their life period. The previous Formulator's Roundtable contributors have discovered 10 specific needs promoting greener solutions to minimize the adverse environmental impacts of developed materials [2]. These findings will catalyze further study in these domains. According to the Roundtable, pinpointing the most significant opportunities for greener alternative potential from a formulator's point of view was critical for commencing the process of communicating with and convincing suppliers and higher education institutions to develop greener alternatives. Although the ingredients of presently produced goods are thought to be safe and effective, words such as "green" and "sustainable" are most effectively used as a comparison (i.e., having lower risk than known alternatives); therefore, there is always scope for advancement. Through the development of novel, eco-friendly materials, the Roundtable aims to enhance the general sustainability of formulation products [8].

This chapter examines key concepts, including green metrics, life cycle assessment (LCA), green solvent selection, renewable feedstocks, process intensification, sustainable additives, green catalysis, and real-world industrial case studies.

8.2 Principles of Green Chemistry in Industrial Formulations

Green chemistry, which first began in the early 1990s, has grown into a major, globally active focus area within chemistry. Green chemistry is an initiative-taking approach that aims to design a synthesis in a sustainable way from the start. Green chemistry's primary goals are to establish atom-efficient methods based on renewable resources that have limited energy consumption and inherently less hazardous chemicals, prevent the application of volatile organic solvents and substitute them with safer and less harmful alternatives, establish sustainable processes by replacing catalytic reagents with stoichiometric reagents, ensure that final products are biodegradable, and allow real-time surveillance during manufacturing. While it is not always possible to incorporate all the principles in each process, efforts should be made to follow as many of them as possible [9, 10] (Fig. 8.2).

A number of principles have been suggested by Hutchins [11] for sustainable chemistry:

(1) Waste management: removing or reducing waste.
(2) Atom economy: little or no atoms are wasted.
(3) Catalysis: Balanced reagents are used excessively and only operate once, whereas catalysts are recovered, recycled, and reused for the majority of components.

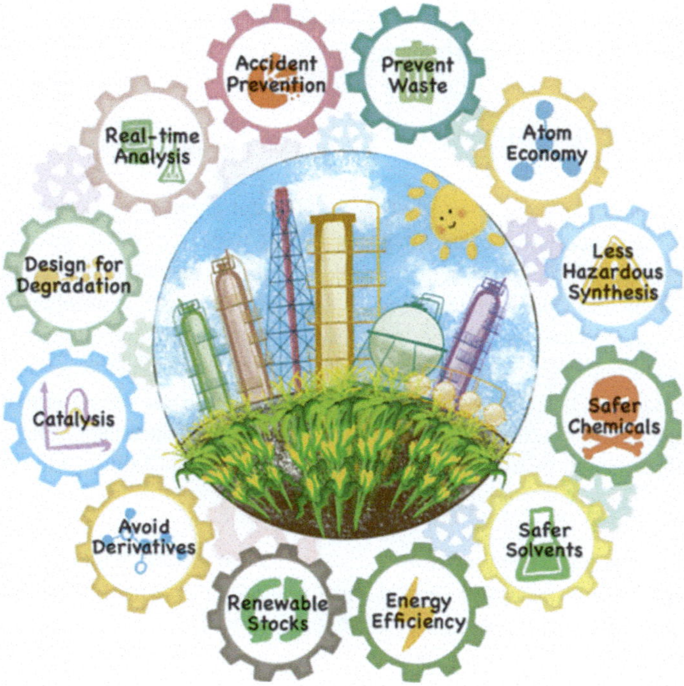

Fig. 8.2 Twelve principles of the green approach. Reproduced with permission Ref. [12]. Copyright 2024. American Chemical Society

(4) Derivatives or intermediate processes need more reagents and may produce more waste than direct reactions, which employ the fewest or most reaction steps.
(5) Safer reactions: synthetic processes must be developed to use and produce materials that are less or completely hazardous to the environment and human health.
(6) Applying non-depleting and renewable feedstocks is known as renewable raw resources.
(7) Safer products: maintaining functionality while lowering toxicity.
(8) Biodegradability: using substances that break down quickly and safely without building up in the environment.
(9) Green auxiliaries: If feasible, auxiliary chemicals (such as solvents and separation agents) should be avoided; if they are used, they should ideally be safe.
(10) Energy economy: Reactions that occur at ambient temperature and pressure conditions are the best way to save energy.
(11) Safer by-products: real-time by-product monitoring, management, and reuse.
(12) Hazard control: avoiding dangerous substances to reduce the possibility of devastating releases, fires, and explosions.

8.3 Green Metrics and Life Cycle Assessment (LCA) in Industrial Formulations

8.3.1 Green Chemistry Metrics

Using metrics to encourage researchers to implement greener methods has emerged as an increasing concern in the past ten years. People widely accept this as a successful metric that is simple to comprehend, straightforward to measure, unprejudiced, and capable of inspiring appropriate action. Under this premise, a number of metrics have been suggested and recommended to motivate engineers and chemists to design safer, more efficient, and greener processes. For instance, certain mass-based metrics include the E factor, atom economy, space-time yield, process mass intensity (PMI), mass efficiency, and reaction mass efficiency. The E factor, atom economy, and PMI constitute three of the most often-used indications [13, 14] (Table 8.1).

Environmental factor (E factor)

The E factor, which Sheldon recommended in 1992, constitutes an exception to the principle that the most straightforward ideas frequently prove the most effective. The E factor can be defined as the ratio of waste to product [15].

$$\text{E factor} = \frac{\text{Total waste (Kg)}}{\text{Product (Kg)}}$$

Atom economy

Trost first introduced the idea of the "atom economy" in 1991. Around the same time, Sheldon came up with a remarkably similar idea of atom use, although the term "atom economy" has gained wider usage [16]. The simplest definition of the atom economy is:

$$\text{Atom economy} = \frac{\text{Molecular weight of the desired product } \times 100\%}{\text{Molecular weight of all incoming chemicals}}$$

Process mass intensity (PMI)

The total mass of materials required for making a given mass product is referred to as the process mass intensity. Catalysts, reagents, reactants, and solvents for reaction and purification include instances of these materials. This should ideally equal unity when there is no waste generated and all products that incorporate materials [17].

$$\text{PMI} = \frac{\text{Total mass in a process or process step (kg)}}{\text{Mass of Product}}$$

Table 8.1 Green chemistry strategies for industrial formulations

Strategy	Description	Examples in industry	References
Implementing sustainable feedstocks	Make use of sustainable, biobased raw materials	Surfactants derived from plants in detergents	[20]
Atom economy	Design procedures that optimize the use of materials	Stoichiometric reductions instead of catalytic hydrogenation	[21]
Alternative solvents	Replace toxic solvents with green alternatives	Substitution of ionic liquids or supercritical CO_2 for organic solvents	[22]
Energy efficiency	Reduce energy consumption by optimizing reaction conditions	Pharmaceutical synthesis aided by microwaves	[23]
Catalysis	Use catalysts to improve selectivity and efficiency	Enzyme-based catalysts in biofuel generation	[24]
Waste reduction & valorization	Minimize byproducts and recycle waste	Conversion of glycerol waste from biodiesel into valuable chemicals	[15]
Benign by design	Develop formulations and chemicals that are safer by nature	Solvent-based paints being replaced by water-based paints	[25]
Process intensification	Merge several steps into one effective unit	Fine chemical synthesis using continuous flow chemistry	[17]
Biodegradable & eco-friendly products	Formulate products that degrade safely in the environment	Biodegradable cleansers and packaging	[26]
Minimal toxicity & safer alternatives	Minimize dangerous chemicals and wastes	Electronics solders without lead	[27]
Water efficiency	Use fewer resources and clean wastewater efficiently	Closed-loop water dyeing systems for textiles	[15]
Use of green nanotechnology	Make use of nanoparticles to improve sustainability	Nano-catalysts for reducing pollution	[28]

It has been shown that metrics like the E factor could promote waste reduction; while this is an excellent point to start, it is not the most effective. It could potentially claim that there is only a single difference between PMI and E factor.

$$\text{E factor} = \frac{\text{Total waste (Kg)}}{\text{Product (Kg)}} = \frac{\text{Total mass used in process or process step - mass of product (Kg)}}{\text{Product (Kg)}}$$

$$\boxed{\text{E factor} = \text{PMI - 1}}$$

However, that direct difference applies to a company's marketable product; in simple terms, it is the factor that provides income for any company operating in the chemical and related industries. The optimal PMI state, for example, is when every material used in the process is incorporated into the final product, actively helping to ensure the generation of income (this corresponds to a PMI value of 1 and thus an E factor of zero). However, the waste from the production process is only a portion of the equation. The cost of high-value materials has historically attracted management attention to a greater degree than waste and waste reduction, so in a broad sense, it makes no difference which of these metrics is used.

8.3.2 *Life Cycle Assessment (LCA)*

Since the late 1990s, some work has been conducted on the Life Cycle Assessment of pharmaceutical processes or LCA. An approach called LCA makes it possible to calculate the total environmental effects of production with greater accuracy. All the chemicals, supplies, and machinery needed to produce a good or provide a service, giving an extensive understanding of the possible trade-offs in environmental effects related to a certain process or good across its whole life cycle. It is desirable to have a range of measures available to assess a process's "greenness," including LCA measurements, that most accurately capture the sustainability of a process or product in its entirety (Fig. 8.3) [18].

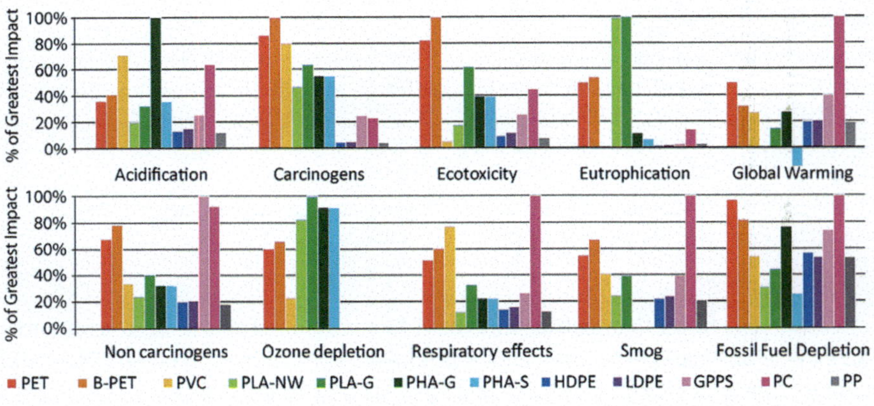

Fig. 8.3 Cradle-to-gate life cycle assessment (LCA) findings for different polymers are shown, normalized to the largest effect observed. In five categories such as ozone depletion, acidification, and ecotoxicity biopolymers have the greatest impact because of fertilizer usage and ethanol production. The overall effects of polyolefins (HDPE, LDPE, PP) are lower than those of complicated petro-polymers (PET, PVC, PC), which rank lowest because of extra synthetic stages that raise emissions and environmental costs. Reproduced with permission Ref. [19]. Copyright 2024. Royal Society of Chemistry

However, compared to determining PMI or even the E factor, the LCA requires a lot more time and work. Green chemistry and engineering must be really included in chemical processes by focusing on the inputs rather than the outputs, which leads to measurements that enable us to adjust. In contrast, the routes and processes are being developed and evaluated. In this case, the first-line leading metrics for chemists and engineers creating chemical and pharmacological pathways are mass and energy inputs (how much and what kinds). The synthetic procedures of some of the most popular FDA-approved medications that use green chemistry concepts in their synthesis are highlighted in this analysis [18].

8.4 Green Solvent Selection and Sustainable Alternatives

A wide range of chemical classes is included in the word "solvent," including terpenes, alcohols, hydrocarbons, amines, glycol ethers, glycols, esters, amides, and oxygenated hydrocarbons. Due to their diverse applications, petroleum distillates and other traditional solvents are crucial parts of many product applications. For a great deal of formulators, these components are necessary to make concentrated cleaning solutions with satisfactory performance [29–32]. Their diverse range of qualities makes them essential in formulations for personal care and cleaning. The disintegration of raw materials (such as waxes and resins), the removal of different sorts of soil (such as glue, grease, and inks), and the distribution of essential oils are merely a few of the possible uses for solvents. Both biobased and fossil-based solvents can dissolve in water or oil.

Petroleum distillates are dangerous for a variety of reasons, including the fact that most of them induce dermatitis, skin defatting, and other skin responses, and others have respiratory or inhalant side effects. Fossil-based solvents are mostly volatile organic compounds (VOCs) or low vapor-pressure VOCs (LVP VOCs), which typically have a $T_b > 216$ °C. Some of these solvents are more dangerous, such as flammable. Their non-renewable source makes them less sustainable and less resource-depleting than biobased solvents. Biobased solvents [33] derived from sustainable feedstocks are highly sought after by formulators. Foods may ferment to produce ethanol and ethyl lactate (Fig. 8.4).

Within the compounds, the fact that can be made chemically from bio-derived compounds have been N-methylpyrrolidone [34], glycerol [35], glycerol derivatives [36], propanediols, soy methyl esters ("methyl soyate") [33], fusel oil esters [37], levulinic acid derivatives (levulinic ketal esters, 2-methyltetrahydrofuran [38], and γ-valerolactone [39]. The synthesis of these biobased solvents can, at times, impact food crops, but depending on the method by which they originate, they may have less of a potential impact on global warming than their fossil-based replacements. By pressing or steam-distilling waste biomass, citrus oils [38] and coniferous (like pine) derivatives can be obtained without triggering a chemical reaction or harming food crops. This means that they might have less of an effect on the environment.

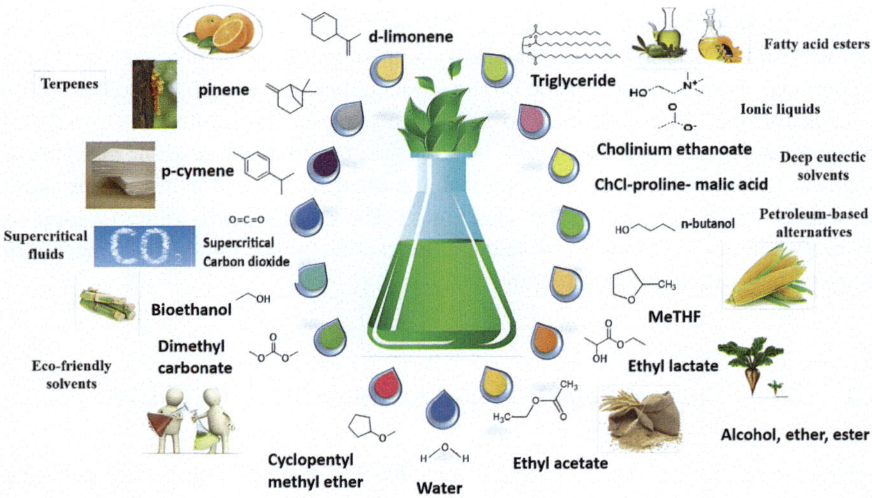

Fig. 8.4 Various green solvents are used in sustainable chemistry, including water, supercritical CO_2, ionic liquids, and deep eutectic solvents. Adapted from [44], open access under CC BY license, copyright 2024, MDPI

Life cycle assessments have been published for citrus oils and methyl soyate, but for pine derivatives, a partial life cycle assessment (energy only) has been published. A further drawback of biobased solvents is that they may reduce local biodiversity by reducing natural areas or monoculture and then replanting them with the preferred crop. Both energy and reagents may be needed to convert a biomass feedstock into a biobased solvent, which would increase the solvent's GWP and environmental effect.

Surface water eutrophication is an unfavorable outcome of the production of diverse biobased compounds. Researchers advise both an environmental impact assessment and a life cycle evaluation when using renewable feedstocks. Based on petroleum, new solvents are likely to be substituted with bio-based or renewable solvents; however, if LCA shows that its products are greener than the solvents currently employed for certain formulations, then the formulator's industry continues to use them. The effects of the solvent's make, usage, and disposal or recycling ought to be included within the life cycle assessment. A number of reviews and perspective articles have recently focused on the design and selection of biobased and sustainable solvents. In contrast, many academic studies focus on solvents as reaction media instead of formulations. Among researchers concerned with the issue of green solvents, we recommend starting with Kerton and Marriott's publication [40].

The Roundtable of ACS GCI Formulators is looking for more environmentally friendly substitutes for conventional solvents. In addition to satisfying any of the 12 Principles of Green Chemistry as feasible, the following identifies some of the key characteristics of suitable alternatives. These substitutes are derived from renewable resources, avoiding petroleum feedstocks wherever possible.

- They should have minimal toxicity to people, be non-irritating, and not cause sensitization. Solvent toxicity requirements are often higher than for other substances due to their prevalence in various applications. A solvent's oral and dermal mammalian LD_{50} should be greater than 2000 mg kg^{-1}, and its inhalation LC_{50} should be greater than 5000 ppm, according to the EPA DfE.
- The solvent should not show any toxicity to reproduction. The OECD 415 and 416 tests are standard. They possess little color and smell, which has a negligible effect on the final product's appearance. There are important and useful studies that compare the effects of solvents without looking at how they are made. Cradle-to-cradle, cradle-to-grave, or cradle-to-gate life cycle assessments (LCAs) are terms that are employed to refer to it [41, 42].
- In the unlikely scenario that a scholar lacks the necessary skills for conducting an LCA, researchers can at least employ mine raw materials in order to map out the entire manufacturing procedure and observe if there are any obvious issues [25] or if these materials have cleansing characteristics like dissolving and reducing grease. In general, finished formulations have been employed for standard cleaning efficacy tests, such as CSPA DCC-17, ASTM D5343, and ASTM G122-96, instead of the pure solvent. These tests consider the specific needs of each application.
- The components that must be dissolved also differ based on the application: "greasy kitchen soil," considered a standard test material for cleaning kitchen surfaces, is a combination of bacon grease, Wesson® cooking oil, and Crisco® shortening; for laundry, test materials include motor oil, bacon grease, and lipstick. Therefore, the physical properties of finished formulations undergo modifications, including decreased viscosity, freeze-thaw recovery, reduced freezing point in wintertime and high-temperature stability in the summertime.
- The solvent's impact on the formulation's qualities is usually more significant than the solvent's characteristics. The solvent stabilizes formulations by avoiding precipitates and maintaining solids in the solution, all while satisfying the EPA DfE requirements for permissible formulation ingredients [43].

8.5 Renewable Feedstocks and Green Raw Materials

Currently, just 3–4% of the 200 billion tons of biomass produced annually by natural photosynthesis are utilized by humans in the food and non-food sectors. In two key areas of development, biomass can be crucial to sustainable development and addressing the problems of global warming: using biomass directly or indirectly for energy purposes and using it as a base resource for chemicals, materials, and goods. We can identify three primary categories in the chemical company specifically:

- The replacement of fossil fuel feedstocks with renewable raw materials, where biomass is converted into fermentable sugars and subsequent downstream products by novel enzymes and entire cell systems.

- Many organic and other compounds are being made using bioprocesses rather than conventional chemical methods.
- Novel bioproducts, including chemicals, high-performance polymers, and novel plastics, are being developed.

The utilization of biomass compared to fossil fuels is among the most important collections of technologies to decrease CO_2 emissions. Particularly liquid biofuels for greener transportation, such as biodiesel (produced from vegetable oil), bioethanol, and its derivative ETBE (derived from plant sugars). Various liquid biofuels have been produced from lignocellulosic material, including biomethanol and its derivative MTBE. The methyl ester of C16–18 fatty acids is what biodiesel is. Approximately 1.5 million tons (tpa) are produced annually in Europe. Bioethanol is frequently manufactured by fermenting simple sugars derived from the starches of corn, wheat, and potatoes. Companies also use other feedstocks, like waste from the forest, food, and beverage industries.

Rice straw, municipal solid waste, corn stover, sugar cane bagasse, and energy crops are among the agricultural wastes used in research on ethanol manufacturing. Bioethanol and gasoline are combined to create "gasohol," which is utilized as an automotive fuel. In the United States, 12% of all vehicle fuels are ethanol-blended. Natural oils and fats derived from animals, marine life, or plants are the building blocks of oleochemicals used in detergents, cosmetics, plastics, textile additives, rubber, paint, surface coatings, and various other applications in the pharmaceutical and food industries.

Renewable raw resources, including coconut, palm, and palm kernel, are already widely used in the oleochemical sector. The Dow Chemical Company and Cargill established a 50/50 joint venture in 1997 to create a new synthesis of polylactic acid (PLA) [26], a polymer made from dextrose derived from corn, in the field of biopolymers. Nature Works Produces from renewable resources yearly; PLA is the first class of synthetic polymers. It is recyclable back to the monomer or completely compostable in commercial settings. Lactide synthesis and effective vacuum distillation are two important technological developments for low-cost, large-scale production.

Compared to conventional plastics, the Cargill–Dow method consumes 30–50% fewer fossil resources, yields greater than 95%, and lowers greenhouse gas emissions by 30–60%. Cargill–Dow has contributed over $300 million to the project. With a capacity of 140,000 tons per year, the first plant was constructed in Nebraska, USA [45]. Biomass feedstocks, including agricultural waste, will be accessible to future resources. 1,3-propanediol (PDO) is produced by metabolic engineering of an Escherichia coli K12 microbe using a straightforward sugar fermentation method created by DuPont and GENENCOR. In a Tate & Lyle pilot plant, the PDO production increases to 135 g l^{-1} at a rate of 4 g l^{-1} h^{-1}. PDO is used to manufacture PTT (polytrimethylene-terephtalate), a new polymer for high-end fibers marketed under the Sorona brand.

By 2010, production should reach 500 ktpa. Xanthane, a fermentation-derived polysaccharide used as a viscosifier in paints, sauces, food, and even cement, is an example of a biopolymer. Biopol is a natural polyester produced by ICI. Wheat polysaccharides are fermented using Alcaligenes eutrophius to yield the copolymer poly-3-hydroxybutyrate/-3-hydroxyvalerate. Metabolics' recent advancements utilizing modified enzymes should enhance product qualities and yields. Scientists have successfully modified the genetic codes of microbes and plants to create polyhydroxybutyric acid. Genencor and Dow Corning are working on a large-scale project worth $35 million USD to create bio-based organosilicon. This so-called "Silicon Biotechnology" promises to provide novel physiologically mediated organosilicon-based materials for usage in diagnostics, biosensors, regulated active ingredient distribution, and personal care products [46].

One of the earliest examples of biotechnology used in the chemical industry to produce acrylamide was Mitsubishi Rayon's use of the enzyme "nitrile hydratase." The traditional method used copper catalysts and/or sulfuric acid to hydrate the nitrile. Using a batch reactor at 5 C to prevent polymerization, the enzymatic reaction yields an output, conversion, and selectivity of 99.99% and a productivity of about 2 kg l^{-1} d^{-1}. In Slovakia, Degussa has a plant that is growing to produce 30,000 tons per year of L-threonine, an amino acid used in animal nutrition. The improved process eliminates up to 90% of the fermentation broth. It employs a "repeated fed-batch procedure" to inoculate the following batch after scientists genetically modified a new strain of Escherichia coli bacteria to reduce sensitivity to fermentation conditions and products, increase productivity, and produce stability. One crystallization stage is sufficient to obtain the pure product since the broth is purer [47].

Numerous industrial bioprocesses have been created in the pharmaceutical industry. Beginning in the early 1980s, DSM investigated biocatalysis for antibiotics like cephalosporins. The bio-coupling of the 7-ADCA synthon reduces the overall waste by 67%, steam by 60%, organic reagents and solvents by 80%, and liquid N_2 refrigerant by 100%, compared to the prior chemical approach. Fermentation has replaced the conventional multi-step chemical process to produce vitamin B2 (riboflavin) at BASF and Hoffman Laroche, resulting in a 40% cost reduction and 95% less waste. Biotechnology also produces vitamin C (ascorbic acid) [48]. The manufacturing of several crop protection agents, such as glyphosate, has improved because of biotechnology. DuPont has disclosed a new technique for this broad-spectrum herbicide that oxidizes glycolic acid using enzymes derived from spinach and yeast. This decreases the number of steps in the process and reduces product loss to waste streams [48]. To produce chiral molecules, white biotechnology is particularly helpful.

- Avecia uses a dehalogenase enzyme to produce 2,000 tpa of 99% pure chloro-2 propionic acid.
- L-DOPA is produced by Ajinomoto using ammonia, pyruvate, and catechol.
- Decussa: hollow-fibered modules of chiral amino acids.

The International Monetary report titled "Application of Biotechnology to Industrial Sustainability" from 2001 contains 21 biotechnology case studies. A 2001 South Korean review titled "Biotechnology for Industrial Sustainability" includes more examples [49].

8.6 Process Intensification and Sustainable Manufacturing Technologies

"Doing more with less" is the primary driving force behind industrial process intensification. Initially, Process Intensification (PI) was intended to decrease capital intensity by shrinking chemical plants. In the late 1970s, ICI pioneered this approach, industrializing the Higee technology, an innovative distillation method that functioned in a radial gravitational field. A more modern example of PI acting as a "size reducer" is the DSM Urea 2000 Plus process, which is only 20 m higher than the prior 50 m. PI takes a broader approach by lowering diffusional constraints to realize the full promise of physicochemical transformation. The physicochemical transformation requires adapting the process to the chemical reaction, selecting the most appropriate technology for each step, and modifying the equipment size to the reaction rather than adapting it to already existing and occasionally inadequate equipment. Smaller, more cost-effective, and more efficient plants are intended to replace large, costly, and inefficient equipment. It also involves combining several functions into smaller equipment [50, 51].

PI's engineering approach, in conjunction with the development of equipment, techniques, and sensors, significantly improves manufacturing and processing. Reductions in size/production ratio, energy usage, and waste generated are among these improvements, along with the adoption of smaller, more compact, safer, and better-controlled (in addition to consequently more economical) solutions. "PI is a potentially profitable way to generate products in a more condensed, safer, cleaner, and less expensive environment." By using better tools of varied sizes and streamlining the process, "sustainable manufacturing" solutions can be found that combine built-in safety, environmental friendliness (low energy and mass), and better product quality. Three factors indicate that PI will produce positive outcomes: In the upcoming five years, it will be the main tool for improving process safety. It provides (groundbreaking) technology innovation to solve societal limitations in the medium run. It also allows for more control over product quality assurance and improvement, distinguishing end-use properties [52].

Despite an inherent conservatism frequently rooted in common sense, the industry is creating more sophisticated machinery and procedures. Like other novel ideas, there are numerous obstacles to PI's adoption: Since plant managers are inherently risk-adverse and careful, they prefer comprehensive demonstrations of novel concepts to reduce risk; in periods of slow development, gradual expansion and debottlenecking are favored; and a unique process is costly. Standard batch processes must be managed very differently for PI, and any changes might result in significant regulatory costs. Technology plays a role, too. The current functions for process intensification units (reactive separations, reaction, and heat exchange) are insufficient. We require advanced integrated equipment. Many features remain unknown, including pressure drop, microchannel clogging, and microdevice fouling. In most synthetic chemistry, exothermic bond formation requires sacrifice to establish molecule order [53]. The reaction must be "cooled down" on an industrial scale by reducing the inherent chemical possibilities if removing heat is the limiting step. This can be achieved by deliberate diffusional limits, partial catalyst deactivation, delayed reagent input, dilution in a solvent, and lower concentrations. The transition is, therefore, "slow [54]." The goal of intensified equipment is to fully utilize the transformation's inherent capacity. compact reactor exchangers, spinning disc reactors, spinning tube-in-tube reactors, static mixers, intensified static mixers, and rotating packed bed reactors are some examples of these new parts [55].

Using microtechnology and structured devices, we have an unprecedented opportunity to explore innovative concepts for a new generation of high-precision devices. This opens the door to even more process design innovation. Particularly in continuous plug flow, new structured components made at low cost from a variety of materials (corrosion-resistant metals, plastics, ceramics, and also convenient, economical, and efficient glass) visible formidable opportunities for intensifying sustainable processes due to their highly effective mass and heat transfer and increased control over the process. Shorter equipment is more efficient for mass and energy transmission for two reasons: larger gradients of physical characteristics and stronger driving forces for mass transport or heat transfer. Surface-to-volume ratios may reach up to 200,000 $m^2\ m^{-3}$ in microreactors, and as these ratios rise, the net exchange surface consequently increases. Reduced volume results in better safety conditions and less material hold-up [56].

8.7 Greener Surfactants

Surfactants are utilized in a broad range of industries, such as mining, textiles, pesticides, paints, inks, polishes, fuels, detergents, lubricants, personal care, pharmaceutical dosage forms (i.e., inclusion in formulations to ensure delivery of the pharmaceutical medication to the target organ), and environmental remediation. Surfactants have various uses in various applications, including decreasing static, cleaning, emulsifying, solubilizing, foaming, and conditioning hair [57–63]. Their numerous applications explain the enormous volume of surfactant usage and their consequent widespread diffusion in the environment. So, formulators are searching for more environmentally friendly surfactants. The best green surfactant should be environmentally friendly and ideally derived from sustainable sources rather than petroleum. Preferably, the source should not be harmful to biodiversity or have any food value. We can make surfactants from renewable resources using plants, animal fats, or even microbes (Fig. 8.5).

Despite coming from a renewable resource, plant-based surfactants can still harm the environment. Reduced biodiversity in the region is usually a consequence of monoculture or the removal of natural areas and replacing them with the desired crop. The food supply may also suffer if we place more importance on the crop's chemical content than its nutritional value. A life cycle assessment should be considered to decide if using renewable feedstocks results in a less negative impact on the

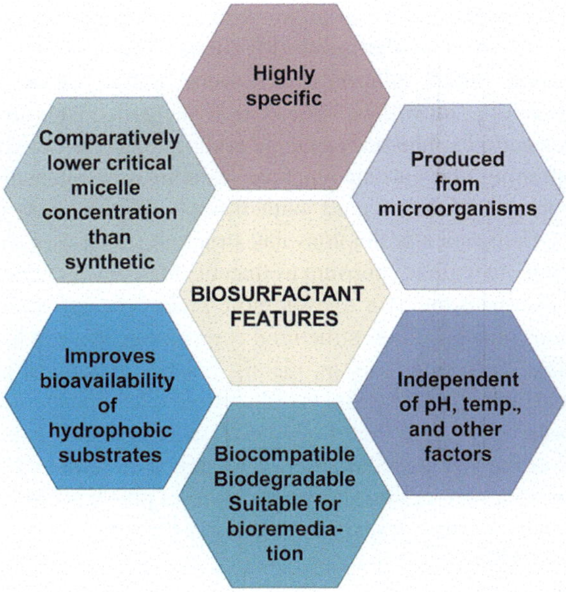

Fig. 8.5 General properties of biosurfactants. Adapted from [64], open access under CC BY license licensed under CC-BY-NC-ND 4.0, copyright 2023, Published by American Chemical Society

environment. Microorganisms secrete biosurfactants [65] on live surfaces, such as the surfaces of their cells.

Some biosurfactants have low critical micelle concentration (CMC) and high surface activity, but they are limited by how much it costs to make them. This makes them promising alternatives to synthetic surfactants. Compared to surfactants made from petroleum, the variety of surfactants derived from renewable feedstocks has fewer physical features. Formulators need a broad variety of surfactants that satisfy the various requirements of disinfecting and personal care products. A large number of surfactants derived from renewable feedstocks are anionic, such as sulfates, carboxylates, phosphate esters, sulfonates, and esters. As a result, formulators are searching for cationic and nonionic surfactants derived from renewables.

A surfactant's toxicity and environmental fate might vary depending on its chemical structure, particularly its hydrophobic moiety. Most surfactants used in wastewater treatment today are easily biodegradable, resulting in extremely low quantities in the environment. Surfactants' environmental concentration and aquatic toxicity levels should be compared when assessing their environmental impact. If the concentrations of surfactants are much lower than the aquatic toxic levels because of their rapid biodegradation, then they are not a risk to the environment. The products of surfactant biodegradation should also be considered into account when assessing their environmental destiny.

Alkyl phenol ethoxylates, and other surfactant biodegradation products may be more hazardous than the surfactant themselves. These biodegradation products are water pollutants and need to be stopped using them. According to Annex XIV of the REACH regulation, nonyl phenol ethoxylates are candidates for EU prohibition as of 2014. Due to their ability to attach to organic substances, surfactants can also find their way into aquatic sediments in tiny concentrations. Therefore, to reduce their environmental effect, surfactants are advised to have rapid anaerobic biodegradability. The European "Flower" Eco-label is one of the green certification programs that currently requires suitable anaerobic biodegradability in order to deal with this issue [66]. This further reduces the number of green surfactants. Users should not experience any health issues with green surfactants. They must not provide a danger of developmental toxicity nor be mutagenic or carcinogenic.

Diverse uses dictate a wide range of physical properties for the ideal surfactant. Therefore, without analyzing the exact use of a new surfactant, it is important to clarify what physical characteristics it should feature. Therefore, we recommend examining the physical characteristics of novel surfactants. It is then possible to determine which applications would benefit from a new surfactant with this data. Thus, in addition to sustainability and environmental impact concerns, each novel surfactant must undergo testing to ascertain its physical characteristics, which should include the following, to evaluate its applicability for a range of applications:

- Cloud Point: the temperature at which precipitation will begin to form in a 1 weight percent solution of a nonionic surfactant in water, or the Krafft point for ionic surfactants.
- The hydrophile-lipophile balance is commonly referred to as HLB.
- The "pour point" is the lowest temperature at which a clear liquid surfactant continues to flow.
- EO moles (the quantity of ethylene oxide units in the surfactant molecule's structure),
- CMC in water at 25 °C, in parts per million,
- The viscosity (cP) of the plain liquid surfactant at 25 °C [67].
- The density of the clean liquid surfactant at 20 °C (g ml^{-1})
- ASTM D93 Flash Point, Closed Cup [18]
- Surface tension (dynes cm^{-1}) at 1% at 25 °C, as well as the Ross-Miles foam heights in millimeters at 0.1% actives at 25 °C, both at the start and after five minutes.

Preferable properties for environmentally friendly surfactants include:

- Ready biodegradability in anaerobic (soil), freshwater, and marine environments.
- Low aquatic toxicity to invertebrates, fish, and algae ($LC_{50} > 10$ mg l^{-1})

It is made using a procedure that considers the 12 Principles of Green Chemistry and is derived from a feedstock that has no food value and will not harm biodiversity.

8.8 Green Catalysis and Biocatalysis

The pharmaceutical industry has made a major effort to transition from stoichiometric, non-selective reagents. Complete catalyst removal and, if feasible, recycling is necessary for homogeneous processes to guarantee product quality and reduce expenses. While heterogeneous catalysis can have some benefits, it has been difficult to immobilize catalysts without causing much leaching. Catalytic procedures, whether homogeneous, heterogeneous, biological, or organocatalytic, are becoming increasingly popular because of the increased need for greener API manufacture [68]. Their integration in processes has strong business motivations since it can boost yield while reducing waste development, time, and energy use.

The alternative of biocatalysis is becoming increasingly popular. Most enzymes are biodegradable and frequently produced by fermenting renewable feedstocks, resulting in an increasingly sustainable catalytic choice. The cost of these catalysts is decreasing in accordance with the rise in the number of commercial enzymes and their related capabilities [69]. From an environmental and commercial standpoint, businesses have benefited from this methodology's high-yield production of chiral APIs or intermediates. Biocatalysis-produced intermediates are found in various commercial APIs, such as paclitaxel (Bristol-Myers Squibb), aliskiren (Novartis), and sitagliptin (Merck).

Merck researchers were able to eliminate the necessity for high-pressure rhodium-catalyzed hydrogenation and boost yield by 10–13% and total waste by 19% by using a recombinant ω-transaminase enzyme. In 2010, this synthesis was awarded the Presidential Green Chemistry Challenge's "Greener Reaction Conditions" prize. As with their metal-based counterparts, complete enzyme extraction or process monitoring is necessary to ensure minimum leaching into the reaction stream. The International Council on Harmonization (ICH) should approve these catalytic processes, and they should meet regulatory requirements when the appropriate risk assessments and control mechanisms are assigned to them. Incorporating biocatalytic processes into telescoped flow syntheses would be the next obvious step. Recently, Heidlindemann et al. showed that it is possible to create hybrid sequences with each catalyst immobilized in its compartment [70]. Efficient in-line separations/solvent swaps or compatibility with reaction solvents will be necessary for mainstream use. The application of catalytic approaches, either as a stand-alone step or as a component of an end-to-end train, will increase as industries work to maintain their viability on both an ecological and economic level.

8.9 Industrial Case Studies in Green Chemistry

8.9.1 Green Chemistry in the Pharmaceutical Industry

Historically, the pharmaceutical industry has been attributed to extensive waste production, energy-intensive procedures, and excessive solvent consumption. Companies such as Amgen and Bachem have responded by using green chemistry concepts to enhance process effectiveness and environmental impact [4, 71]. Amgen and Bachem, for instance, collaborated to optimize the solid-phase peptide synthesis (SPPS) process for the medication ParsabivTM, which treats secondary hyperparathyroidism. Due to the extensive use of toxic solvents in traditional SPPS methods, manufacturing costs and environmental effects are increased. Green solvent alternatives and process intensification techniques helped Amgen cut production time by 56% and solvent use by 71%. These enhancements maintained excellent product quality while drastically reducing the production's carbon impact [72–74]. This case study illustrates how green chemistry concepts may be applied in the pharmaceutical industry to maximize productivity and minimize environmental damage.

8.9.2 Sustainable Textile Industry Practices

The textile business is one of the most resource-intensive industries, especially regarding energy and water use, chemical treatments, and dyeing procedures. In terms of green chemistry applications, Sweden's textile industry has advanced significantly via the use of:

- Biodegradable detergents are used in washing procedures to reduce toxic discharge.
- Enzyme-based cotton pre-treatment substitutes harsh chemicals and uses less water and energy.
- Advanced water recycling systems enable businesses to collect and repurpose as much as 60% of their processed water.

With the implementation of ISO 14000 environmental management standards, textile producers are now even more motivated to adhere to global sustainability laws and ensure that their manufacturing methods meet environmentally friendly requirements. These efforts have significantly decreased water contamination and the textile industry's overall environmental effects [75, 76].

8.9.3 Green Solvent Use in Chemical Manufacturing

The manufacture of chemicals has been under growing pressure to cut emissions of volatile organic compounds (VOCs), which contribute to environmental deterioration and air pollution.

Pfizer and GlaxoSmithKline (GSK), two significant corporations, have achieved groundbreaking breakthroughs in the reduction of solvent waste:

- Pfizer has demonstrated dedication to removing dangerous solvents from its production by not scaling any process employing chlorinated hydrocarbons for more than eight years.
- GSK achieved a 50% decrease in solvent waste by investing in solvent recovery and closed-loop recycling methods.

By reconsidering solvent use and moving toward eco-friendly substitutes, these initiatives help to lessen the environmental impact of chemical manufacture [77, 78].

8.9.4 Circular Economy in the Electronics Industry

To improve industrial sustainability, the Japanese electronics industry has adopted the circular economy concept. The Matsushita Eco Technology Center of Panasonic, which specializes in recycling and reusing parts from household appliances, is an excellent example.

Among the initiative's major accomplishments are:

- A 50 and 60% recycling rate for items such as air conditioners, refrigerators, washing machines, and television sets.
- The incorporation of green chemical solutions into product design and disassembly to reduce electronic waste.
- Notice Japan's stringent legal requirements, which call for increased recycling and resource recovery goals.

This case study shows how resource efficiency and the application of design-for-recycling principles may help industrial formulations meet sustainability objectives [79].

8.9.5 Chemical Leasing for Industrial Efficiency

In European businesses, a new business model called Chemical Leasing (ChL) is becoming increasingly popular to maximize chemical use and reduce hazardous waste.

How Chemical Leasing Operates:

- Businesses offer chemical services rather than selling bulk chemicals, guaranteeing that only the required quantity is utilized.
- This method improves process efficiency while lowering waste production, excessive consumption, and environmental contamination.

Important Impact: Businesses using this technique have noted a 30–50% drop in chemical usage. Enhanced safety and adherence to environmental standards; decreased emissions and expenses because of better chemical utilization. According to this case study, there can be significant sustainability gains from switching from a product-based to a service-based chemical supply [80, 81].

These industry case studies demonstrate the practical implementation of green chemistry concepts in electronics, textiles, pharmaceuticals, and chemical manufacturing. By implementing circular economy concepts, green solvents, and process intensification strategies, enterprises may minimize their environmental impact while preserving operational and financial effectiveness.

8.10 Challenges in Adopting Green Chemistry in Industrial Formulations

Green cannot be sustainable enough to make a method effective. Industries struggle implementing green processes due to regulatory, economic, political, and technical difficulties. Green chemistry reduces inherent risk by reducing hazard, while current laws reduce risk by reducing exposure. In the U.S., manufacturers must recertify with the FDA after changing production processes. It is expensive and time-consuming, so companies do not invest in atom-efficient chemistry that reduces waste [82]. Cost-intensive, control-oriented regulation prevents gentler process changes. Stakeholder ignorance hinders green process implementation. To create a successful green process, one must understand green chemistry, engineering, biotechnology, economics, and toxicology. Industrial green chemistry is not improving since chemists are not trained in such fields [83]. Even if all other factors favor it, a green process may face commercial rejection if it lacks economic attractiveness.

Green industrial processes should cost a similar amount to those of their traditional counterparts. Business concerns led to abandoning some technically adequate and environmentally acceptable methods. In the chemical industry, environmentally friendly practices may not be profitable. Thomas Swan and Company built the first continuous-flow reactor using supercritical carbon dioxide in Consett, UK, using Martyn Poliakoff's work from Nottingham University [84]. Isophorone may be selectively hydrogenated to 3,3,5-trimethylcyclohexanone using the Sc-CO method without forming any byproducts. In conventional approaches, this prevented a costly and energy-intensive separation procedure. A lack of government subsidies prevented the plant from producing chemicals cheaper than traditional methods. After operating commercially from 2002 to 2009, the facility shut down in 2009. Similarly, the phosphonium iodide from Eastman Chemical Company isomerizes 3,4-epoxybut-1-ene to 2,5-dihydrofuran [85]. The commercialization of green technologies is also hampered by financial investment. In the Difasol procedure, IFP (France) used ionic liquids as a co-catalyst and solvent for nickel-catalyzed alkene dimerization [86]. Because of this biphasic process, the product has a layer above the ionic liquid layer, which facilitates separation. This method's improved catalytic activity, product separation, dimer selectivity, and reactor space-time yields make it preferable to the conventional Dimersol process. Commercial implementation was hindered by the prohibitively prohibitive cost of capital equipment. Green process commercialization needs several adjustments throughout the global supply chain [87].

Eden Organic Foods' non-BPA food packaging was suitable for beans but highly acidic tomato sauce. Shifting coatings for different foods reduces market size, changes manufacturing processes, and raises costs. Every value chain member must agree to and approve such initiatives. Since financial rewards drive most industries, voluntary adoption of sustainable practices is less expected [88]. Legislation that is balanced, appealing, and strong is required to encourage greener practices. On the registration, evaluation, licensing, and limitation of chemical substances, REACH, the 2007 EU regulation, is the most important and promising. REACH requires chemical manufacturers to disclose additional information about their products' health and environmental hazards, but it also permits sustainable innovation to gain five-year registration exemptions [89]. This European Union move has inspired other governments to develop sustainable chemical industry laws.

8.11 Conclusion and Outlook

Building sustainable development demands green chemistry because it offers responses that balance environmental responsibility and scientific advancement. The emergence of safer, more effective, and less harmful chemical processes has the potential to close the gap between science and society. Green chemistry's broad acceptance requires significant technical developments and robust legal and policy frameworks that encourage companies to promote sustainable practices. The establishment of incentives and regulations that encourage greener alternatives is a critical function of governments and regulatory agencies. The next generation of chemists must be exposed to green chemistry concepts early to promote long-term change. Including sustainability in chemistry classes will inspire students to consider the effects on the environment and create safer, more creative methods. Multidisciplinary and interdisciplinary cooperation is also necessary to overcome technological obstacles and make green technology commercialization possible. Researchers can produce scalable and profitable solutions by integrating knowledge from material science, biotechnology, engineering, and chemistry. Financial incentives, including grants, tax exemptions, and subsidies for green initiatives, will further accelerate the shift to sustainable industrial practices. Industries should understand that making an initial investment in green chemistry, whether through new machinery, procedures, or regulatory compliance, has long-term advantages for society and industry. Green practices can safeguard regulatory compliance, increase operational efficiency, boost corporate reputation, and lower environmental footprints. Researchers, engineers, industrialists, and legislators must work together and sincerely to successfully apply green chemistry. Using consistent innovation, policy backing, and industry dedication, chemistry has the potential to develop into a sustainable field that promotes a healthier and cleaner future for society.

References

1. Qudus Onagun.: Abdulrahman Abdulmumin Gbenga: Green chemistry in medicinal chemistry: a review on sustainable approaches to the synthesis of biologically active compounds. World J. Adv. Res. Rev. **24**, 1371–1382 (2024). https://doi.org/10.30574/wjarr.2024.24.2.3417
2. MSingh, R., Pramanik, R., Hazra, S.: Role of green chemistry in pharmaceutical industry: a review. J. Univ. Shanghai Sci. Technol. **23**, 291–299 (2021). https://doi.org/10.51201/JUSST/21/121018
3. Abdel-Moety, E.M., Rezk, M.R., Wadie, M., Tantawy, M.A.: A combined approach of green chemistry and quality-by-design for sustainable and robust analysis of two newly introduced pharmaceutical formulations treating benign prostate hyperplasia. Microchem. J. **160**, 105711 (2021). https://doi.org/10.1016/j.microc.2020.105711
4. Kar, S., Sanderson, H., Roy, K., Benfenati, E., Leszczynski, J.: Green chemistry in the synthesis of pharmaceuticals. Chem. Rev. **122**, 3637–3710 (2022). https://doi.org/10.1021/acs.chemrev.1c00631
5. Mukherjee, S., Das, A., Chowdhury, A., Biswas, A., Roy, S., Majumdar, S., Paul, S.: Harmful effect of personal care products on ecosystem and the possible alternative approach. Biocatal. Agric. Biotechnol. **57**, 103065 (2024). https://doi.org/10.1016/j.bcab.2024.103065
6. Syeda, S.R., Khan, E.A., Padungwatanaroj, O., Kuprasertwong, N., Tula, A.K.: A perspective on hazardous chemical substitution in consumer products. Curr. Opin. Chem. Eng. **36**, 100748 (2022). https://doi.org/10.1016/j.coche.2021.100748
7. Maertens, A., Golden, E., Hartung, T.: Avoiding regrettable substitutions: green toxicology for sustainable chemistry. ACS Sustain. Chem. Eng. **9**, 7749–7758 (2021). https://doi.org/10.1021/acssuschemeng.0c09435
8. Eckelman, M.J., Moroney, M.S., Zimmerman, J.B., Anastas, P.T., Thompson, E., Scott, P., McKeever-Alfieri, M., Cavanaugh, P.F., Daher, G.: Applying green chemistry to raw material selection and product formulation at the Estée Lauder Companies. Green Chem. **24**, 2397–2408 (2022). https://doi.org/10.1039/D1GC03081G
9. Pobłocki, K., Drzeżdżon, J., Gawdzik, B., Jacewicz, D.: Latest trends in the large-scale production of MOFs in accordance with the principles of green chemistry. Green Chem. **24**, 9402–9427 (2022). https://doi.org/10.1039/D2GC03264C
10. Schaub, T.: Efficient industrial organic synthesis and the principles of green chemistry. Chem. Eur. J. **27**, 1865–1869 (2021). https://doi.org/10.1002/chem.202003544
11. Hutchings, G.J.: A golden future for green chemistry. Catal. Today **122**, 196–200 (2007). https://doi.org/10.1016/j.cattod.2007.01.018
12. Whiteker, G.T.: Applications of the 12 principles of green chemistry in the crop protection industry. Org. Process Res. Dev. **23**, 2109–2121 (2019). https://doi.org/10.1021/acs.oprd.9b00305
13. Sherer, E.C., Bagchi, A., Kosjek, B., Maloney, K.M., Peng, Z., Robaire, S.A., Sheridan, R.P., Metwally, E., Campeau, L.-C.: Driving aspirational process mass intensity using simple structure-based prediction. Org. Process Res. Dev. **26**, 1405–1410 (2022). https://doi.org/10.1021/acs.oprd.1c00477

14. Sajid, M., Płotka-Wasylka, J.: Green analytical chemistry metrics: a review. Talanta **238**, 123046 (2022). https://doi.org/10.1016/j.talanta.2021.123046
15. Sheldon, R.A., Bode, M.L., Akakios, S.G.: Metrics of green chemistry: waste minimization. Curr. Opin. Green Sustain. Chem. **33**, 100569 (2022). https://doi.org/10.1016/j.cogsc.2021.100569
16. Murphy, M.A.: Professors Trost and Sheldon's promotion of catalytic technologies, atom economy, and the e-factor metrics in synthetic organic chemistry and the fine chemical and pharmaceutical industries, to speed the early evolution of "Green Chemistry." Substantia **7**, 41–55 (2023). https://doi.org/10.36253/Substantia-2140
17. Madabhushi, S.R., Pinto, N.D.S., Lin, H.: Comparison of process mass intensity (PMI) of continuous and batch manufacturing processes for biologics. New Biotechnol. **72**, 122–127 (2022). https://doi.org/10.1016/j.nbt.2022.11.002
18. Mondello, G., Salomone, R.: Assessing green processes through life cycle assessment and other LCA-related methods (2020). https://doi.org/10.1016/B978-0-444-64337-7.00010-0
19. Rogers, L., Jensen, K.F.: Continuous manufacturing—The Green chemistry promise? Green Chem. **21**, 3481–3498 (2019). https://doi.org/10.1039/C9GC00773C
20. Kjellin, M., Johansson, I., (eds.).: Surfactants from Renewable Resources. Wiley (2010). https://doi.org/10.1002/9780470686607
21. Lim, F.P.L., Dolzhenko, A.V.: Atom economy in green organic synthesis. In: Green Sustainable Process for Chemical and Environmental Engineering and Science, pp. 1–12. Elsevier (2020). https://doi.org/10.1016/B978-0-12-819539-0.00001-4
22. Jessop, P.G.: Searching for green solvents. Green Chem. **13**, 1391 (2011). https://doi.org/10.1039/c0gc00797h
23. Sheldon, R.A.: Atom efficiency and catalysis in organic synthesis. Pure Appl. Chem. **72**, 1233–1246 (2000). https://doi.org/10.1351/pac200072071233
24. Sheldon, R.A., Woodley, J.M.: Role of biocatalysis in sustainable chemistry. Chem. Rev. **118**, 801–838 (2018). https://doi.org/10.1021/acs.chemrev.7b00203
25. Kerton, F., Marriott, R.: Alternative solvents for green chemistry. R. Soc. Chem. (2013). https://doi.org/10.1039/9781849736824
26. Brandl, H., Gross, R.A., Lenz, R.W., Fuller, R.C.: Plastics from bacteria and for bacteria: Poly(β-hydroxyalkanoates) as natural, biocompatible, and biodegradable polyesters. In: Microbial Bioproducts, pp. 77–93. Springer-Verlag, Berlin/Heidelberg. https://doi.org/10.1007/BFb0010232
27. Brausch, J.M., Rand, G.M.: A review of personal care products in the aquatic environment: environmental concentrations and toxicity. Chemosphere **82**, 1518–1532 (2011). https://doi.org/10.1016/j.chemosphere.2010.11.018
28. Rose, H.B., Kosjek, B., Armstrong, B.M., Robaire, S.A.: Green and sustainable metrics: charting the course for green-by-design small molecule API synthesis. Curr. Res. Green Sustain. Chem. **5**, 100324 (2022). https://doi.org/10.1016/j.crgsc.2022.100324
29. Patel, B., Singh, S., Parikh, K., Chavda, V., Ray, D., Aswal, V.K., Kumar, S.: Micro-environment mapping of mole fraction inspired contrasting charged aqueous Gemini micelles: a drug solubilization/release study. J. Mol. Liq. **363**, 119885 (2022). https://doi.org/10.1016/j.molliq.2022.119885
30. Shah, P.A., Chavda, V., Hirpara, D., Pokar, R., Kumar, S., Shrivastav, P.S.: Advances in deep eutectic solvent-based synthesis of nanomaterials for environmental remediation. J. Mol. Liq. **416**, 126482 (2024). https://doi.org/10.1016/j.molliq.2024.126482
31. Shah, P.A., Shrivastav, P.S., Sharma, V.S., Chavda, V.: Uncovering the green frontier: harnessing deep eutectic solvents for sustainable bioanalysis. Bioanalysis **15**, 815–821 (2023). https://doi.org/10.4155/bio-2023-0088
32. Shah, P.A., Chavda, V., Hirpara, D., Sharma, V.S., Shrivastav, P.S., Kumar, S.: Exploring the potential of deep eutectic solvents in pharmaceuticals: challenges and opportunities. J. Mol. Liq. **390**, 123171 (2023). https://doi.org/10.1016/j.molliq.2023.123171
33. Shin, H.-M., McKone, T.E., Bennett, D.H.: Contribution of low vapor pressure-volatile organic compounds (LVP-VOCs) from consumer products to ozone formation in urban atmospheres. Atmos. Environ. **108**, 98–106 (2015). https://doi.org/10.1016/j.atmosenv.2015.02.067

34. Horváth, I.T., Mehdi, H., Fábos, V., Boda, L., Mika, L.T.: γ-Valerolactone—A sustainable liquid for energy and carbon-based chemicals. Green Chem. **10**, 238–242 (2008). https://doi.org/10.1039/B712863K
35. Lammens, T.M., Potting, J., Sanders, J.P.M., De Boer, I.J.M.: Environmental comparison of biobased chemicals from glutamic acid with their petrochemical equivalents. Environ. Sci. Technol. **45**, 8521–8528 (2011). https://doi.org/10.1021/es201869e
36. García, J.I., García-Marín, H., Pires, E.: Glycerol based solvents: synthesis, properties and applications. Green Chem. **16**, 1007–1033 (2014). https://doi.org/10.1039/C3GC41857J
37. Srinivas, K., Potts, T.M., King, J.W.: Characterization of solvent properties of methyl Soyate by inverse gas chromatography and solubility parameters. Green Chem. **11**, 1581 (2009). https://doi.org/10.1039/b913050k
38. Taddeo, F., Tesser, R., Serio, M., Di Russo, V.: Ethyl levulinate ketalization with glycerol promoted by H,Y-Zeolite: kinetic analysis in batch and continuous milli-reactor scale. Chem. Eng. Process. Process Intensif. **197**, 109712 (2024). https://doi.org/10.1016/j.cep.2024.109712
39. Pace, V., Hoyos, P., Castoldi, L., Domínguez de María, P., Alcántara, A.R.: 2-Methyltetrahydrofuran (2-MeTHF): a biomass-derived solvent with broad application in organic chemistry. ChemSusChem **5**, 1369–1379 (2012). https://doi.org/10.1002/cssc.201100780
40. Zhang, D., del Rio-Chanona, E.A., Shah, N.: Life cycle assessments for biomass derived sustainable biopolymer & energy co-generation. Sustain. Prod. Consum. **15**, 109–118 (2018). https://doi.org/10.1016/j.spc.2018.05.002
41. DeVierno Kreuder, A., House-Knight, T., Whitford, J., Ponnusamy, E., Miller, P., Jesse, N., Rodenborn, R., Sayag, S., Gebel, M., Aped, I., Sharfstein, I., Manaster, E., Ergaz, I., Harris, A., Nelowet Grice, L.: A method for assessing greener alternatives between chemical products following the 12 principles of green chemistry. ACS Sustain. Chem. Eng. **5**, 2927–2935 (2017). https://doi.org/10.1021/acssuschemeng.6b02399
42. Jenck, J.F., Agterberg, F., Droescher, M.J.: Products and processes for a sustainable chemical industry: a review of achievements and prospects. Green Chem. **6**, 544 (2004). https://doi.org/10.1039/b406854h
43. Estevan, C., Pamies, D., Sogorb, M.A., Vilanova, E.: Reproductive toxicity: in vivo testing guidelines from OECD. In: Reproductive and Developmental Toxicology, pp. 165–178. Elsevier (2022). https://doi.org/10.1016/B978-0-323-89773-0.00010-2
44. Chemat, F., Abert Vian, M., Ravi, H.K., Khadhraoui, B., Hilali, S., Perino, S., Fabiano Tixier, A.-S.: Review of alternative solvents for green extraction of food and natural products: panorama, principles. Appl. Prospect. Mol. **24**, 3007 (2019). https://doi.org/10.3390/molecules24163007
45. Killoran, J., Allen, L., Gallagher, J.F., Gallagher, W.M., O'Shea, D.F.: Synthesis of BF_2 chelates of tetraarylazadipyrromethenes and evidence for their photodynamic therapeutic behaviour. Chem. Commun. 1862–1863 (2002). https://doi.org/10.1039/B204317C
46. Koutinas, A.A., Vlysidis, A., Pleissner, D., Kopsahelis, N., Lopez Garcia, I., Kookos, I.K., Papanikolaou, S., Kwan, T.H., Lin, C.S.K.: Valorization of industrial waste and by-product streams via fermentation for the production of chemicals and biopolymers. Chem. Soc. Rev. **43**, 2587 (2014). https://doi.org/10.1039/c3cs60293a
47. Ioannidou, S.M., Pateraki, C., Ladakis, D., Papapostolou, H., Tsakona, M., Vlysidis, A., Kookos, I.K., Koutinas, A.: Sustainable production of bio-based chemicals and polymers via integrated biomass refining and bioprocessing in a circular bioeconomy context. Bioresour. Technol. **307**, 123093 (2020). https://doi.org/10.1016/j.biortech.2020.123093
48. Nielsen, J., Tillegreen, C.B., Petranovic, D.: Innovation trends in industrial biotechnology. Trends Biotechnol. **40**, 1160–1172 (2022). https://doi.org/10.1016/j.tibtech.2022.03.007
49. Lee, J.W., Kim, H.U., Choi, S., Yi, J., Lee, S.Y.: Microbial production of building block chemicals and polymers. Curr. Opin. Biotechnol. **22**, 758–767 (2011). https://doi.org/10.1016/j.copbio.2011.02.011
50. Saleem, A., Farooq, U., Riaz, A., Ahmed, F., Hussain, A., Lee, M.: Understanding the impact of reactive holdup on process intensification in the design of reactive distillation column. Chem. Eng. Process. Process Intensif. **191**, 109440 (2023). https://doi.org/10.1016/j.cep.2023.109440

51. Srivastava, S., Pandey, V.K., Fatima, A., V, M.S., Pandey, S., Singh, R., Dar, A.H., Dhillon, B.: A literature review on process intensification: An innovative and sustainable food processing method. Appl. Food Res. 4, 100363 (2024). https://doi.org/10.1016/j.afres.2023.100363
52. Yue, J.: Green process intensification using microreactor technology for the synthesis of biobased chemicals and fuels. Chem. Eng. Process. Process Intensif. **177**, 109002 (2022). https://doi.org/10.1016/j.cep.2022.109002
53. Chanthon, N., Ngaosuwan, K., Kiatkittipong, W., Wongsawaeng, D., Appamana, W., Quitain, A.T., Assabumrungrat, S.: High-efficiency biodiesel production using rotating tube reactor: new insight of operating parameters on hydrodynamic regime and biodiesel yield. Renew. Sustain. Energy Rev. **151**, 111430 (2021). https://doi.org/10.1016/j.rser.2021.111430
54. van der Wielen, L.A.M., Mussatto, S.I., van Breugel, J.: Bioprocess intensification: cases that (don't) work. New Biotechnol. **61**, 108–115 (2021). https://doi.org/10.1016/j.nbt.2020.11.007
55. Li, H., Wu, C., Hao, Z., Li, X., Gao, X.: Process intensification in vapor–liquid mass transfer: the state-of-the-art. Chin. J. Chem. Eng. **27**, 1236–1246 (2019). https://doi.org/10.1016/j.cjche.2018.08.005
56. Domínguez, M.I., Centeno, M.A., Martínez T., M., Bobadilla, L.F., Laguna, Ó.H., Odriozola, J.A.: Current scenario and prospects in manufacture strategies for glass, quartz, polymers and metallic microreactors: a comprehensive review. Chem. Eng. Res. Des. **171**, 13–35 (2021). https://doi.org/10.1016/j.cherd.2021.05.001
57. Hirpara, D., Chavda, V., Hirapara, N., Kumar, S.: Inorganic salt-induced micellar morphologies in deep eutectic solvent: structure and curcumin solubilization. J. Mol. Liq. **411**, 125761 (2024). https://doi.org/10.1016/j.molliq.2024.125761
58. Chavda, V., Patel, B., Singh, S., Hirpara, D., Rajeswari, V.D., Kumar, S.: A sustainable approach for the adsorption of methylene blue from an aqueous background: an adsorbent based on DES/CGS modified GO@ZrO_2. RSC Sustain. **1**, 2038–2057 (2023). https://doi.org/10.1039/D3SU00236E
59. Patel, B., Singh, S., Parikh, K., Chavda, V., Hirpara, D., Ray, D., Aswal, V.K., Kumar, S.: Composition triggered Aggregation/Solubilization behaviour of mixed counter charged gemini surfactants: a multi-technique investigations. J. Mol. Liq. **359**, 119242 (2022). https://doi.org/10.1016/j.molliq.2022.119242
60. Chavda, V., Hirpara, D., Kumar, S.: GO/ionic surfactant inspired photophysical modulation of rhodamine B in Reline with or without additives. J. Mol. Liq. **368**, 120614 (2022). https://doi.org/10.1016/j.molliq.2022.120614
61. Chavda, V., Hirpara, D., Rao, V., Kumar, S.: Deep Eutectic solvent/surfactant-engineered GO@TiO_2 advanced material for sustained and ultrafast adsorption of hazardous dye from contaminated aqueous environment. ACS Appl. Eng. Mater. **2**, 604–617 (2024). https://doi.org/10.1021/acsaenm.3c00696
62. Hirpara, D., Patel, B., Chavda, V., Kumar, S.: Micellization of conventional and gemini surfactants in aquoline: a case of exclusively water based deep eutectic solvent. J. Mol. Liq. **362**, 119672 (2022). https://doi.org/10.1016/j.molliq.2022.119672
63. Hirpara, D., Patel, B., Chavda, V., Desai, A., Kumar, S.: Micellization and clouding behaviour of an ionic surfactant in a deep eutectic solvent: a case of the reline-water mixture. J. Mol. Liq. **364**, 119991 (2022). https://doi.org/10.1016/j.molliq.2022.119991
64. Nagtode, V.S., Cardoza, C., Yasin, H.K.A., Mali, S.N., Tambe, S.M., Roy, P., Singh, K., Goel, A., Amin, P.D., Thorat, B.R., Cruz, J.N., Pratap, A.P.: Green surfactants (Biosurfactants): a petroleum-free substitute for sustainability-comparison, applications, market, and future prospects. ACS Omega **8**, 11674–11699 (2023). https://doi.org/10.1021/acsomega.3c00591
65. Liu, X., Yang, J., Zou, Q., Hu, Y., Li, P., Tan, L., Miljkovic, N., Yang, R.: Enhancing liquid-vapor phase-change heat transfer with micro/nano-structured surfaces. ACS Nano (2025). https://doi.org/10.1021/acsnano.4c15277
66. Nimrod, A.C., Benson, W.H.: Environmental estrogenic effects of Alkylphenol Ethoxylates. Crit. Rev. Toxicol. **26**, 335–364 (1996). https://doi.org/10.3109/10408449609012527
67. Burcik, E.J., Vaughn, C.R.: The effect of pH on the rate of surface tension lowering. J. Colloid Sci. **6**, 522–527 (1951). https://doi.org/10.1016/0095-8522(51)90050-5

68. Gu, Y., Jérôme, F.: Bio-based solvents: an emerging generation of fluids for the design of eco-efficient processes in catalysis and organic chemistry. Chem. Soc. Rev. **42**, 9550 (2013). https://doi.org/10.1039/c3cs60241a
69. Aguillón, A.R., Avelar, M.N., Gotardo, L.E., de Souza, S.P., Leão, R.A.C., Itabaiana, I., Miranda, L.S.M., de Souza, R.O.M.A.: Immobilized lipase screening towards continuous-flow kinetic resolution of (±)-1,2-propanediol. Mol. Catal. **467**, 128–134 (2019). https://doi.org/10.1016/j.mcat.2019.01.034
70. Heidlindemann, M., Rulli, G., Berkessel, A., Hummel, W., Gröger, H.: Combination of asymmetric organo- and biocatalytic reactions in organic media using immobilized catalysts in different compartments. ACS Catal. **4**, 1099–1103 (2014). https://doi.org/10.1021/cs4010387
71. Sharma, S., Das, J., Braje, W.M., Dash, A.K., Handa, S.: A glimpse into green chemistry practices in the pharmaceutical industry. Chemsuschem **13**, 2859–2875 (2020). https://doi.org/10.1002/cssc.202000317
72. Armenta, S., Esteve-Turrillas, F.A., Garrigues, S., de la Guardia, M.: Alternative green solvents in sample preparation. Green Anal. Chem. **1**, 100007 (2022). https://doi.org/10.1016/j.greeac.2022.100007
73. Xiouras, C., Kuijpers, K., Fanfair, D., Dorbec, M., Gielen, B.: Enabling technologies for process intensification in pharmaceutical research and manufacturing. Curr. Opin. Chem. Eng. **41**, 100920 (2023). https://doi.org/10.1016/j.coche.2023.100920
74. Jaspers, M., Tegel, F., Roelofs, T.P., Starsich, F., Song, Y.L., Meir, B., Elkes, R., Dickhoff, B.H.J.: Process intensification of pharmaceutical powder blending at commercial throughputs by utilizing semi-continuous mini-blending. Int. J. Pharm X **8**, 100264 (2024). https://doi.org/10.1016/j.ijpx.2024.100264
75. Panda, S.K.B.C., Sen, K., Mukhopadhyay, S.: Sustainable pretreatments in textile wet processing. J. Clean. Prod. **329**, 129725 (2021). https://doi.org/10.1016/j.jclepro.2021.129725
76. Periyasamy, A.P., Ramamoorthy, S.K., Rwawiire, S., Zhao, Y.: Sustainable wastewater treatment methods for textile industry (2018). https://doi.org/10.1007/978-981-10-8591-8_2
77. Dutta, P., McGranaghan, A., Keller, I., Patil, Y., Mulholland, N., Murudi, V., Prescher, H., Smith, A., Carson, N., Martin, C., Cox, P., Stierli, D., Boussemghoune, M., Barreteau, F., Cassayre, J., Godineau, E.: A case study in green chemistry: the reduction of hazardous solvents in an industrial R&D environment. Green Chem. **24**, 3943–3956 (2022). https://doi.org/10.1039/D2GC00698G
78. Idoko, F.A., Ezeamii, G.C., Ojochogwu, O.J.: Green chemistry in manufacturing: innovations in reducing environmental impact. World J. Adv. Res. Rev. **23**, 2826–2841 (2024). https://doi.org/10.30574/wjarr.2024.23.3.2938
79. Trivedi G (2025) Zero waste manufacturing: a case study of circular and sustainable economy practices. J. Inform. Educ. Res. **5** (2025). https://doi.org/10.52783/jier.v5i1.2346
80. Joas, R., Abraham, V., Joas, A.: Chemical leasing: a business model to drive resource efficiency in the supply chain (2018). https://doi.org/10.1007/978-3-319-50079-9_28
81. Weerakkody, M.P., Edirisinghe, L.G.L.M., Sivashankar, P.: Farmers' attitude towards chemical leasing for sustainability and environmental protection. Curr. Res. Environ. Sustain. **4**, 100175 (2022). https://doi.org/10.1016/j.crsust.2022.100175
82. Clark, J.H.: Green chemistry: challenges and opportunities. Green Chem. **1**, 1–8 (1999). https://doi.org/10.1039/a807961g
83. Poliakoff, M., Licence, P.: Supercritical fluids: green solvents for green chemistry? Philosophical transactions of the royal society a: mathematical. Phys. Eng. Sci. **373**, 20150018 (2015). https://doi.org/10.1098/rsta.2015.0018
84. Harry, D., Horton, D., Durham, D., Constable, D.J.C., Gaffney, S., Moore, J., Todd, B., Martinez, I.: Grand challenges and opportunities for greener chemical alternatives in hydraulic fracturing: a perspective from the ACS Green Chemistry Institute Oilfield Chemistry Roundtable. Energy Fuels **34**, 7837–7846 (2020). https://doi.org/10.1021/acs.energyfuels.0c00933
85. Taifouris, M., Martín, M., Martínez, A., Esquejo, N.: Challenges in the design of formulated products: multiscale process and product design. Curr. Opin. Chem. Eng. **27**, 1–9 (2020). https://doi.org/10.1016/j.coche.2019.10.001

86. Olivier-Bourbigou, H., Favre, F., Forestière, A., Hugues, F.: Ionic liquids and catalysis: the IFP Biphasic Difasol process. In: Handbook of Green Chemistry, pp. 101–126. Wiley (2010). https://doi.org/10.1002/9783527628698.hgc005
87. Ahmed, S.F., Mofijur, M., Rafa, N., Chowdhury, A.T., Chowdhury, S., Nahrin, M., Islam, A.B.M.S., Ong, H.C.: Green approaches in synthesising nanomaterials for environmental nanobioremediation: technological advancements, applications, benefits and challenges. Environ. Res. **204**, 111967 (2022). https://doi.org/10.1016/j.envres.2021.111967
88. Kapoor, A., Mehendale, S.: Challenges in implementation of green chemistry in Indian pharmaceutical sector. J. Pharm. Res. Int. 160–168 (2021). https://doi.org/10.9734/jpri/2021/v33i34A31836
89. Estévez, C.: The challenge of green chemistry. Green Chem. Lett. Rev. **1**, 5–5 (2007). https://doi.org/10.1080/17518250701642860

Chapter 9
Lifecycle Assessment of Green Chemistry Practices in Industry

Amita Tripathy and Abhishek Satpathy

Abstract Any industry's worldwide growth presents challenges, highlighting the necessity of thorough analysis and long-term management. As an evidence-based technique, lifecycle assessment (LCA) has become more and more popular in research. This tool supports decision making at broad and specific levels. On a larger scale, this aids in developing strategies, exploring scenarios, and evaluating policy options. At a more focused level, they typically address issues related to specific products or processes. Applications of LCA to a broad range of conservation initiatives in industries are examined in this chapter. Based on principles from green chemistry, sustainable development, this study seeks to establish guidelines for greenness. It considers the complexity of research problems as well as socioeconomic and environmental sustainability.

Keywords Green chemistry · Sustainable development · Environmental sustainability · Lifecycle assessment

9.1 Introduction

Advancements in industrial complexes and the aspiration to improve living standards, safety, affluence, and efficiency have elevated the significance of chemistry and its applications. The assumption that naturally occurring chemicals are inherently "environmentally friendly" is erroneous. Around the 1990s, the notion of green chemistry was established in the chemical industry and is widely implemented by institutions and companies to contribute to sustainability goals [1].

A. Tripathy (✉)
Department of Applied Sciences & Humanities, ABES Engineering College, Ghaziabad, UP, India
e-mail: tripathyamita21@gmail.com

A. Satpathy
Department of Electronics & Instrumentation, Birla Institute of Technology and Science, Pilani, Rajasthan, India

P. Jain et al. (eds.), *Towards Green Chemical Processes: Strategies and Innovations*, Intelligent Systems Reference Library 277,
https://doi.org/10.1007/978-3-031-94090-3_9

Green chemistry is a transformative strategy that incorporates environmental factors into design, development, application of products and processes. The Environmental Protection Agency (EPA) United States, emphasises green chemistry as an approach that considers the comprehensive lifecycle among chemical processes, with the objective of mitigating their adverse impacts on both environmental and human health [2]. By prioritizing sustainability, it reduces hazardous substances, minimizes waste, and promotes energy efficiency. The environmental benefits include significant reductions in greenhouse gas emissions, improved air and water quality, and reduced reliance on non-renewable resources. Economically, green chemistry fosters cost savings through enhanced resource efficiency, decreased waste disposal costs, and the creation of new markets for sustainable products. In 1998, green chemistry was formally established by12 principles [3] (Fig. 9.1).

Each stage in a product or process life cycle involves unique materials and energy inputs with environmental impacts. Owing to practical, economic, and institutional constraints, improving existing unsustainable products, processes, and systems will be necessary. This optimization serves as a crucial short-term measure for which green engineering principles provide an effective framework. Since these principles only represent a qualitative frame-work of best practices [4] without defining clear metrics [5], indicators to measure the greenness of chemical products & processes are needed. Some key limitations include the significant effort required, difficulty in

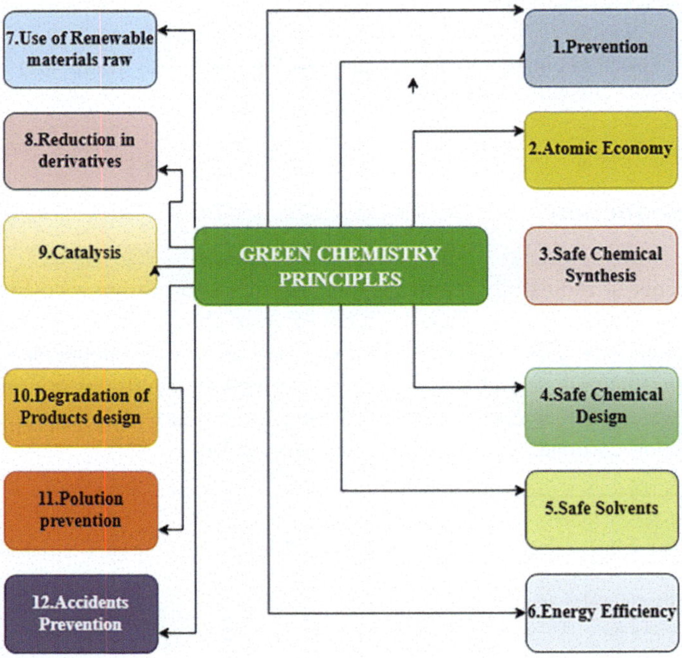

Fig. 9.1 Green chemistry principles

accessing necessary data, and reliance on specialized or proprietary datasets. Additionally, the results may lack transparency, making them harder to interpret and communicate effectively.

LCA (ISO 14040/44), is a widely utilized evaluation technique for impact assessment of products and processes via their life cycle [6] by providing a strong framework of practices with the value chain, from extraction to dispersal. This chapter explores the interplay between green chemistry and LCA, demonstrating their synergistic potential to foster sustainable industrial practices.

9.2 Assessment Methods

9.2.1 Green Chemistry Metrics

Green chemistry plays a crucial role in transforming traditional industrial practices by embedding sustainability into the core of chemical production and use. The core concepts of 12 green chemistry principles serve as a basis for incorporating environmentally friendly practices into industrial operations [7]. These principles can be categorized according to their goals of improving resource use, increasing energy efficiency, and reducing hazards to humans and the environment [8]. They can be seen as a conceptual framework describing what green chemistry means and how to design chemical products or processes in accordance with green chemistry. The aim is not to completely model a complex technical system, but to be able to derive information about the greenness, particularly in early development phases, based on individual process- or product-specific parameters.

Metrics have been developed to evaluate the eco-friendliness of a process. These measurements aim to determine how well the procedures adhere to the fundamentals of green chemistry. Currently, there is no unified metric or standardized definition on how to determine the level of greenness of a chemical product or process [9–11]. They have become crucial with an emphasis on mass-based metrics. The comprehensive mass-related green metrics (GMs) are presented in Table 9.1. Green chemistry mass metrics consider two key parameters: atom economy & the E-factor. AE increases the reactant atoms. E-factor underscores waste management in chemical production, particularly in the fine chemical industry [12–14]. Weight-based measurements specifically target dangers to the environment and to human health. Furthermore, it mentions methodologies that aim to concurrently evaluate all 12 principles, sometimes condensing them into a single comprehensive metric.

A comprehensive, multilevel metric with diverse indicators was proposed to assess the sustainability and workflows of chemical processes & production. This metric serves two objectives: (1) to enable comparisons between alternative technological approaches, and (2) to offer a broad framework across chemical sectors. Indicator selection is specific to each industry and is determined by aspects of their products that align with green chemistry principles, including climate impact, energy

Table. 9.1 Comprehensive mass-related GMs

Mass metrics	Abbreviation
Environmental impact factor	E-factor
Process mass intensity	PMI
Process mass efficiency	PME
Atom economy	AE
Reaction mass intensity	RMI
Reaction mass efficiency	RME
Carbon efficiency	CE
Greener atomic level, transformation green aspiration level	tGAL™
Green aspiration level TM	GAL™
Relative process greenness	RPG
Renewable percentage	RP
Wastewater intensity	WWI

sustainability, resource utilization footprint, product toxicity, minimization of solvent use, and recycling potential [15].

In recent years, there has been a proliferation of software applications [16–18] designed to assist chemists in GM calculations and greenness evaluation based on the 12 principles. Researchers have used spreadsheet templates and software to improve understanding of GM and enhance mass and environmental measurements. These tools calculate the RME of chemical processes in laboratories. The pentagonal diagram shows parameters for RME calculation, including Atom Economy, stoichiometric factor, recovery parameter. An Excel radar graph assesses environmental compatibility of a reaction by considering health and environmental risks. A Green Motion metric tool can evaluate impact on flavour and fragrance sectors. Chem Pager and Tibco Spotfire platforms can be used for data visualization.

The use of "big data" [19] in assessing chemical hazards and risks is crucial for determining toxicological properties. These applications utilize specific parameters related to processes or products, including material requirements, energy consumption, waste production, and potential hazards of associated chemicals. The results of the tools cannot be directly compared because the underlying concepts differ. The chapter suggests employing advanced environmental assessment methods, such as life cycle analyses and carbon footprint measurements, to assess the ecological impact of scientific work.

9.2.2 Lifecycle Assessment (LCA)

LCA [20–22] is a process for assessing how a product, procedure, or service will have an environmental impact over its whole life cycle. It identifies hotspots, informs

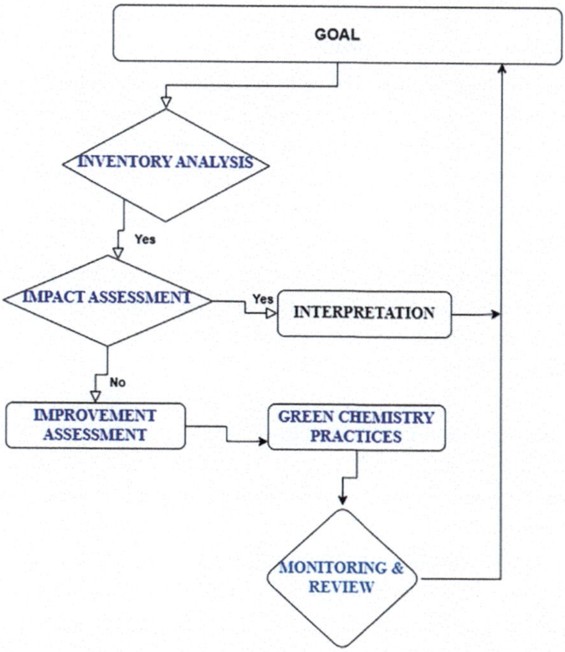

Fig. 9.2 Methodological framework of LCA

decision-making, enhances sustainability, ensures transparency, and drives innovation in materials, processes, and products through a baseline for measuring progress, setting targets, and monitoring environmental performance over time. LCA supports product differentiation and marketing by substantiating environmental claims and communicating sustainability efforts to stakeholders (Fig. 9.2).

LCA generally encompasses four stages:

- Goal & scope,
- Inventory.
- Impact analysis.
- Result.

In the initial stage of goal & scope, the objectives with functional units are established. This phase sets assessment boundaries and specifies the product or system under examination. System boundaries, determined based on the objectives, specify the unit processes to be incorporated. The inventory analysis phase focuses on gathering and measuring data related to the functional units. This comprises inputs of energy and resources, waste and emission, and environmental discharges for each life cycle. LCA inventories typically use specialized software to model various aspects, encompassing life cycle expenses and social effects, and environmental factors. This phase evaluates the energy consumption, material usage, and emissions linked to each stage of a product's lifecycle. In the context of sustainable chemistry, this involves examining the raw materials employed in synthesis, energy utilization, and the production of waste.

During the impact assessment stage, researchers utilize data to evaluate the effects of products, measures, or processes on both the environment and human populations. This evaluation encompasses various factors, such as ozone depletion, acidification, smog formation, resource depletion, and fossil fuel consumption. The impact assessment process comprises two essential steps: first, selecting and classifying impact categories, and second, characterizing the impacts. In the characterization step, researchers quantitatively model the impact (e.g., of emissions) using a specific impact score. This score is expressed in a unit that can be applied uniformly to all contributions within a particular impact category, such as kilograms of carbon dioxide equivalent ($kgCO_2$).

Two optional processes are also incorporated in the impact assessment: (i) normalization, which compares the various characterized impact scores to a standard, and (ii) weighting, which applies a hierarchical structure and/or quantitative valuation of the diverse types of environmental impact to reflect the relative significance of the impacts.

Intermediate environmental impact measures, called midpoint indicators, reflect changes in the natural environment from emissions or resource consumption. These indicators quantify effects of environmental stressors. For instance, potential for global warming of gases can be stated as carbon dioxide equivalents (CO_2e), serving as midpoint indicators for greenhouse gas emissions. Compared to endpoint indicators, midpoint indicators are typically more straightforward to compute and comprehend, being more closely linked to impact sources and less influenced by uncertainties. Nevertheless, midpoint indicators do not sufficiently account for all the impacts of total modification of the environment on human welfare or on ecosystem functions, or communicate the relative importance of individual impact categories. The rationale for mirroring the midpoint indicators in the paper is based on their higher maturity and strong linkage to inventory data and environmental impact sources compared to midpoints [23].

Endpoints indicators quantify the impact on human well-being, natural resources and the supply of ecosystem diversity due to pollutant discharge or resource consumption. Decision in LCA [24, 25] on selection of midpoint or endpoint indicators is driven to a great extent by the above stated objectives, scope, data, methodological availability and quality, meeting of the target audience needs, and stakeholder values. Integrity in the form of transparency and critical analysis of the assumptions and implications of different choice of indicators should be held when addressing the sensitivity and uncertainty of results.

9.3 Intersection of LCA and Green Chemistry Practices

Several evaluation methods for green chemistry practices exist which include LCA, chemical footprint, process mass intensity, atom economy and RME. The different evaluation approaches show unique methods to study environmental effects of chemical processes which leads to a complete sustainability examination.

When we perform LCA we measure environmental effects by tracking raw materials together with end products and waste emissions from product birth to death. This approach determines the carbon emission reduction effect of renewable resources while identifying advantages from waste prevention methods. The adherence to green chemistry principles through LCA helps industries determine sustainable choices and secure safer chemicals and enhanced energy performance and produce environmental advantages during manufacturing and utilization processes. This research investigates the connection between green chemistry principles and LCA by checking how implementation of guidelines supports environmental indicators tracked by LCA.

Chemical footprint serves as a quantitative approach to assess environmental effects of single chemical substances and chemical groups through ecological assessment methods. The method assesses a substance according to toxicological measures together with its persistence status and accumulation potential and human contact likelihood. A chemical process efficiency assessment requires different metrics for evaluation. The process mass intensity (PMI) measures how much material goes into making one unit of product therefore it helps identify ways to reduce waste and enhance efficiency. Reaction efficiency can be measured with atom economy through determining the percentage of initial starting materials which become integrated within the final product. RME serves to evaluate process efficiency because it calculates the amount of product created from each unit of starting materials. The evaluation tools serve as essential indicators for waste reduction assessment and chemical process efficiency enhancement.

Association between waste reduction strategies produces lower environmental emissions and reduced resource utilization in inventory analysis processes and atomic-level efficiency leads to minimized material requirements and lifecycle impact reduction. The extensive network demonstrates how green chemistry approaches fuel sustainability by boosting functionalities from material selection through final product use. Collaboration between chemists and engineers needs sustainability experts to guarantee that green chemical processes fulfil performance and environmental requirements from production through disposal stages.

9.4 Case Studies

9.4.1 *Pharmaceuticals*

The complex synthesis operations in pharmaceutical manufacturing units uses heavy energy that leads to substantial environmental harm. Also, numerous dangerous chemicals, experimental substances and reaction boosters, during the process of drug manufacturing inflict detrimental effects on human health. Stringent regulations pertaining to pharmaceutical industries pose challenges for complex manufacturing procedures. Green chemistry significantly contributes towards pharmaceutical research and development. Eventually, the scientists use eco- friendly solvents and reactants that lowers toxicity levels and decreases negative effects on health. Further, the implementation of modern techniques and computational methods in drug molecules has led to the decrease of experimental burdens and has eventually upgraded the quality of drugs. By the use of modern technologies, LCA of drugs is instrumental to perform environmental impact assessment of pharmaceutical manufacturing operations. The evaluation process includes four sections that includes active pharmaceutical ingredients production (APIs), drug preparation, packaging & distribution and disposal methods. By the application of LCA, drugs can be produced with precision synthetic processes and it would help to reduce unwanted side-products during production by increasing atom economy. Process optimization for pharmaceutical synthesis achieved better atom economy through waste reduction, though preliminary expenses are involved to redesign the manufacturing methods [26].

The LCA operates as a certified methodology to determine environmental effects that occur in pharmaceutical manufacturing, beginning with API production through packaging, distribution and finally the disposal phase. The method applies to biopharmaceutical processes to identify various vital environmental zones, during the phases of production.

According to recent LCA studies, power consumption, particularly electricity, chemical use of raw materials and solvents are the main factors that affect drug production and packaging. Pharmaceutical manufacturers should adopt alternative renewable power systems and advanced chemical production techniques to substantially decrease environmental strains. By adopting LCA operational strategy & manufacturing guidelines, the pharmaceuticals can undergo effective transformation by decreasing waste and enhancing efficiency.

9.4.2 *Material Industry*

The adoption of principles of green chemistry would help the materials industries to improve product development processes for plastics, polymers and composites. The material sector accumulates environmental hazards due to resource depleting, waste production as well as dangerous chemicals emissions. The material industry needs to use sustainable raw materials as traditional sources like petroleum and natural gas sources are depleting. The scarcity of these resources creates an immediate requirement to execute changes. Waste appears as a crucial issue for materials manufacturing companies in production and product expiration. Material waste poses problems for proper disposal leading to environmental risks and health-related dangers as they contain potentially dangerous and carcinogenic substances like heavy metals along with halogens and phthalates. The lifecycle journey of material products creates major environmental and human health risks assessment, until they reach their disposal stage.

9.4.3 Bioplastics Production

Biodegradable plastics would be a wise alternative for traditional synthetic plastics, that remain in the environment for centuries, posing a threat to the ecological balance. These environmentally friendly, biodegradable plastics, derived from natural corn starch or cellulose are produced using the green chemistry methodologies.

In the present day, LCA is used to contrast the environmental gains of biodegradable polymers versus synthetic plastics, focusing on metrics, like final result of life biodegradability and decreased application of non-renewable assets. Although there are challenges, such as feedstock availability, the LCA confirms that utilizing renewable feedstocks would reduce greenhouse gases.

9.4.4 Eco-Friendly Solvents

Production Plants conventionally use solvents, like benzene and toluene, that pose significant risk to environment and human health. To mitigate this challenge, benign solvents like water and supercritical CO_2 are used. Using these alternatives has reduced VOC emissions and improved workplace safety conditions. Evaluating and contrasting, the lifecycle of traditional solvents & sustainable alternatives, reduces the VOC emissions [27, 28].

9.4.5 Sustainable Materials Sourcing

The purchase of raw materials for production & design significantly affects the environment, safety, and performance of a plant. Safety concerns regarding material production affect both workers and end users. Using the concepts of green chemistry & LCA, plants can utilize recycled materials or materials derived from renewable sources, such as plant-based polymers, to avoid risk.

These techniques optimize production processes, mitigate negative impacts, enhance performance & also minimize the environmental footprint of material production. LCA finds ways to cut waste, energy use, and emissions by examining every phase of material extraction to disposal. Also, it facilitates increased strength and durability, extended product lifespans, improved thermal and chemical resistance, lightweight materials optimizing energy usage. Innovative material design approaches can improve sustainability, safety, and material properties.

9.4.6 Renewable Energy

Conventional fossil fuels have drastic environmental consequences, including the release of atmospheric and aquatic pollutants along with climate change. LCA offers a sustainable opportunity in the energy sector that continues to impact production and consumption, through significant reduction of the ecological footprint. Adoption of these principles is limited due to high energy costs of renewables, energy storage, and infrastructure. Despite all of these, green chemistry has been applied in the energy sector successfully. A substantial use of biofuels as sustainable alternatives to fossil fuels, releases less greenhouse gas emissions. Advancements in solar-cell technology, fuel cells have revolutionised the energy sector with low emission and high efficiency [29]. Applying green chemistry principles in evaluations of biofuel or solar, enhances energy efficiency and reduces carbon emissions.

9.4.7 Textiles

Green chemistry sustainability approach has great potential to reduce the impact for textile by adopting eco-friendly recycled materials, energy-efficient processes, thereby balancing the ecological footprint. Additionally, non-toxic dyes, sustainable fibres, bio-based polyesters, would have reduced water use and chemical pollution. Also, encouraging textile recycling, upcycling, biodegradable fabrics and textiles, responsible consumption and manufacturer reduces the concerned risks of the industry.

9.5 Conclusion

LCA and Green Chemistry, while crucial for sustainable practices, encounter a range of obstacles across diverse industries.

1. Data Acquisition: Companies find it difficult to gather accurate data about raw materials, the specifics of manufacturing processes, and product life which is essential for LCA. The complex supply chains and non-sharing of proprietary details often adds to data acquisition.
2. Methodological Complexity: LCA methodologies involve complex calculations, impact assessments, and the interpretation of results. Moreover, Green Chemistry necessitates knowledge of chemical processes, alternative synthetics, and sustainable materials which require thoughtful action.
3. Upfront Investment Concerns: Despite the long-term cost saving potential, the initial costs associated with green chemistry involves modifying existing equipment, new technologies, or revamping processes, which is a significant barrier.
4. Navigating Regulatory Mazes: Environmental regulations, chemical management policies, and reporting requirements differ considerably with changing landscape, making it difficult to implement a uniform LCA and green chemistry approach [30].
5. Standardization Gaps: The absence of standardized LCA procedures makes it difficult to align results with best practices across businesses.
6. Time Constraints in Implementation: Integrating LCA and Green Chemistry principles can impact project timelines, industrial operations and market demands for fast-paced industries.
7. Resistance to Change: Established routines and a natural resistance to change within organizations can impede the adoption of novel approaches like LCA and Green Chemistry.
8. Knowledge and Awareness Deficits: As green chemistry and LCA concepts are new within industries, the knowledge gap affects decision-makers in addressing sustainability issues.
9. Supply Chain Coordination: Extending sustainability principles across entire supply chains introduces complex coordination and logistical hurdles.
10. Managing Environmental Trade-offs: LCA and Green Chemistry often reveal intricate trade-offs between different environmental impacts which requires a sustainability perspective.

These challenges could be mitigated through wider adoption like,

- Develop in-house expertise through comprehensive training programs, fostering a corporate culture that embraces sustainability.
- Partner with universities and research institutions to access cutting-edge knowledge and methodologies in LCA and Green Chemistry, bridging the knowledge gap.
- Create tailored guidance and best practice frameworks for different sectors to streamline implementation and ensure relevance.
- Invest in sophisticated tools to improve the efficiency and accuracy of data collection, analysis, and reporting for LCA.

- Establish teams that integrate expertise from various departments within organizations to embed sustainability into product development and manufacturing processes from the outset.
- Work with policymakers to promote greater consistency in environmental regulations and standards across different industries and regions, simplifying compliance.
- Create platforms for businesses within specific sectors to share expertise, resources, and best practices related to LCA and Green Chemistry implementation.
- Conduct pilot projects to showcase the practical viability and benefits of sustainable practices, building confidence before large-scale adoption.
- Develop incentive programs to encourage the adoption of sustainable practices and facilitate data sharing throughout the supply network.
- Embed LCA and Green Chemistry principles into corporate sustainability strategies to monitor progress using key performance indicators (KPIs) and align business objectives with environmental responsibility.

By actively addressing these challenges and implementing these targeted solutions, industries can break down the barriers to adopting LCA and Green Chemistry. This will pave the way for more sustainable and responsible manufacturing practices, benefiting both businesses and the environment.

References

1. Anastas, P.T.: Green chemistry: an overview. Crit. Rev. Anal. Chem. **29**, 167–175 (1999)
2. Baird, C.: Environmental chemistry, p 43 (2002)
3. Anastas, P.T., Warner, J.: Green Chemistry: theory and Practice, p. 30. Oxford University Press, Oxford, UK (1998)
4. Horvath, I.T., Anastas, P.T.: Innovations and green chemistry. Chem. Rev. **107**, 2169–2173 (2007)
5. Lancaster, M.: Green Chemistry: an Introductory Text, 2nd edn., p. 95. Royal Society of Chemistry (2010)
6. Hauschild, M.Z., Huijbregts, M.A.J.: Introducing life cycle impact assessment. In: Life Cycle Impact Assessment, pp. 1–20 (2015)
7. Linthorst, J.A.: An overview: origins and development of green chemistry. Found. Chem. **12**(1), 55–68 (2010)
8. Llored, J.P., Sarrade, S.: Connecting the philosophy of chemistry, green chemistry, and moral philosophy. Found. Chem. **18**(2), 125–152 (2016)
9. Roschangar, F., Sheldon, R.A., Senanayake, C.H.: Green chemistry and sustainability. Green Chem. **17**(2), 752–768 (2015)
10. Derbenev, I.N., Dowden, J., Twycross, J., Hirst, J.D.: Current trends in green chemistry. Curr. Opin. Green Sustain. Chem. **35**, 100623 (2022)
11. United States Environmental Protection Agency (EPA) (2023) Basics of green chemistry. https://www.epa.gov/greenchemistry/basics-green-chemistry#definition
12. Curzons, A.D., Constable, D.J.C., Mortimer, D.N., Cunningham, V.L.: So you think your process is green, how do you know?—Using principles of sustainability to determine what is green—A corporate perspective. Green Chem. **3**, 1–6 (2001)

13. Hudlicky, T., Frey, D.A., Koroniak, C.D., Claeboe, L., Brammer, L.E.: Toward a reagent-free synthesis. Green Chem. **1**, 57–59 (1999)
14. Sheldon, R.A.: The E factor 25 years on: the rise of green chemistry and sustainability. Green Chem. **19**, 18–43 (2017)
15. Marteel, A.E., Davies, J.A., Olson, W.W., Abraham, M.A.: Green chemistry and engineering: drivers, metrics, and reduction to practice. Annu. Rev. Environ. Resour. **28**, 401–428 (2003)
16. Andraos, J.: Global green chemistry metrics analysis algorithm and spreadsheets: evaluation of the material efficiency performances of synthesis plans for oseltamivir phosphate (Tamiflu) as a test case. Org. Process Res. Dev. **13**, 161–185 (2009)
17. Płotka-Wasylka, J.: A new tool for the evaluation of the analytical procedure: green analytical procedure index. Talanta **181**, 204–209 (2018)
18. Wang, F.Z., Liu, X.L., Zhang, L., Cheng, W.C.: Does digitalisation promote green technology innovation in resource-based companies? Stud. Sci. Sci. **40**, 332–344 (2022)
19. Cherrafi, A., Garza-Reyes, J.A., Belhadi, A., Kamble, S.S., Elbaz, J.: A readiness self-assessment model for implementing green lean initiatives. J. Clean. Prod. **309**, 12740 (2021)
20. Guinée, J.B., (ed.).: Handbook on Life Cycle Assessment: operational Guide to the ISO Standards, pp. 41–62. Springer (2002)
21. Hauschild, M., Rosenbaum, R.K., Olsen, S.I.: Life Cycle Assessment: theory and Practice, pp. 605–631. Springer (2018)
22. Baumann, H.: Introduction and organisation of LCA activities in industry. Int. J. Life Cycle Assess. **5**, 363–368 (2000)
23. Curran, M.A.: Life Cycle Assessment Handbook: a Guide for Environmentally Sustainable Products, pp. 249–266. Wiley (2017)
24. Hellweg, S., Milà i Canals, L.: Life cycle assessment: a review of the methodology and its applications. Chem. Rev. **114**(11), 6016–6063 (2014)
25. Rebitzer, G., et al.: Life cycle assessment: part 1—Framework, goal and scope definition, inventory analysis, and applications. Environ. Int. **30**(5), 701–720 (2004)
26. Ahmad, S., Jaiswal, R., Yadav, R., Verma, S.: Recent advances in green chemistry approaches for pharmaceutical synthesis. Sustain. Chem. One World **4**, 100029 (2024)
27. Byrne, F.P., Jin, S., Paggiola, G., et al.: Tools and techniques for solvent selection: green solvent selection guides. Sustain. Chem. Process. **4**(1), 7 (2016)
28. Wang, S., et al.: Catalysis and LCA in chemical engineering. Green Chem. Lett. Rev. **13**(3), 187–196 (2020)
29. Zah, R., Böni, H., Gauch, M., Hischier, R., Lehmann, M., Wäger, P.: LCA of energy systems: focus on green chemistry. Energy Policy **35**(12), 6391–6400 (2007)
30. Anastas, P.T., Kirchhoff, M.M.: Barriers to the implementation of green chemistry in the United States. Environ. Sci. Technol. **46**(3), 10892–10899 (2012)

Chapter 10
Approaches to Silica Production from Agriculture Waste Biomass

Gurjinder Singh, Atul Kumar, Ashish Kumar, and Arnab Maity

Abstract Silica has many applications in various sectors such as construction, the chemical industry, agriculture, health care, environment remediation, etc. It is currently being synthesized from inorganic precursors that are indirectly prepared from quartz sand, the second most used nonrenewable resource. Due to environmental hazards associated with the current synthesis process and health hazards associated with inorganic precursors, there is an urgent need for alternative sources for the synthesis of silica and silica-based materials. Agriculture waste biomass which is otherwise mostly a menace to the environment can be a much greener alternative in this scenario. This chapter discussed in detail various approaches (viz. thermal, chemical, and biological) used to obtain silica from agriculture waste biomass with several reports from the literature. It includes examples of the synthesis of various silica materials from agriculture waste materials such as rice husk, rice straw, bagasse, bamboo leaves, etc., and their properties. Several cases of applications of bio-derived silica in various fields and recent pilot scale studies are also discussed at the end.

Keywords Silica nanoparticles · Sodium silicate · Agriculture waste · Amorphous silica

---

G. Singh · A. Kumar (✉)
Department of Chemistry, Akal University, Bathinda, Punjab, India
e-mail: atul79prashar@gmail.com

A. Kumar
Department of Chemistry, School of Applied and Life Sciences, Uttaranchal University, Dehradun-248007, Uttarakhand, India

A. Maity
Assistant Professor (Chemistry), RDS College Salmari, Under Purnea University, Purnia, Bihar-855113, India

P. Jain et al. (eds.), *Towards Green Chemical Processes: Strategies and Innovations*, Intelligent Systems Reference Library 277,
https://doi.org/10.1007/978-3-031-94090-3_10

10.1 Introduction

In the earth's crust silicon comes after oxygen in terms of abundance. Its oxide named silica is a white-colored inorganic substance that constitutes about 75% of the earth's crust [1]. In natural silicates, it exists in the form of various crystalline phases such as tridymite, cristobalite, or quartz [2]. High-purity silica (>95%) is a component of various materials and has a wide range of applications. It is used to improve the quality of building materials such as high-performance cement, concrete, pozzolan additives, etc. For example, its use has been found to increase the strength of concrete and make the process of its production more environmentally sustainable [3]. It is commonly used in paper coating, especially in matte papers used in inkjet printing. Studies have shown that the strength of paper can be improved by using calcium carbonate fillers modified with silica [4]. It is used in various chemical industries involved in the manufacture of detergents to improve their quality texture and performance, in plastic and rubber as a reinforcing filler to improve their mechanical properties, carrier of fertilizers to release nutrients in a controlled manner, additive in animal feed to improve the flow of premixes and trace minerals, thickening of paints, to coat the surfaces to provide chemical and physical protection, anti-caking agent, filler, binder, and adhesive [5]. It is used in the synthesis of various types of electronic materials such as batteries, capacitors, semiconductor, etc., and ceramics materials involved in refractory, thermal insulations, and flame retardation [6]. It is an important constituent in many pharmaceuticals, healthcare, and cosmetics products [7]. It is also used in making polymer, glass, solar panels, graphene, zeolite, and catalyst support [8]. Various silica-based materials such as nanoporous silica (MCM-41, MCM-48, SBA-15, etc.), silica gels, silica carbon composites, hybrid organic–inorganic materials [9], etc. are being studied for the wide range of applications such as adsorption, drug delivery, fillers, catalysis, sensing and water purifications.

On an industrial scale, silica is typically produced at a high temperature (approximately 1300 °C) by heating a mixture of sodium carbonate (Na_2CO_3) and quartz sand which generates sodium silicate powder. This reacts with sulfuric acid to produce silica [10]. The reaction of Na_2CO_3 with sand produces tons of CO_2. Given the extensive use of silica across various industries, its production has been increasing over the years, raising significant environmental concerns about the sustainability of this traditional method. Also, quartz sand is a non-renewable resource and second most used material worldwide after water and its shortage will have a huge impact on various production processes [11]. Moreover, alkoxysilane such as tetraethyl orthosilicate (TEOS), tetramethyl orthosilicate (TMOS), etc. which are other silica precursors are produced from rocks and sand through an energy-intensive and less environmentally friendly process. In addition to that, Nakashima et al. reported that acute inhalation of TEOS could result in fatal outcomes within a specific period [12]. Okumura et al. identified TEOS as a potential cause of acute stomach, bladder, and renal failure [13]. Similarly, Kolesar et al. observed that TMOS exposure could lead to edema, conjunctival hyperemia, and epithelial desquamation of the cornea in humans [14].

Considering these environmental issues and health risks, it is essential to replace commercial silica precursors with alternatives that are more environmentally friendly, safer, and cost-effective. So, scientists are searching for alternative safer, cheaper, abundant, and sustainable silica sources for the synthesis of silica-based materials. In this scenario, agriculture waste biomass is considered an alternative silica source [15].

A large number of waste and by-products are generated by agricultural activities. Generally burning is considered a solution to limit the quantities of agriculture. But this harms the environment because burning results in pollution of air, water, and soil. This has made the researchers find an alternative solution to this problem. A lot of attention is being given to the production of value-added chemicals such as biochar, biofuels, organic fertilizers, graphene quantum dots, etc. [16]. from biomass waste. Silica is also an important constituent in many biomasses so its production from biomass waste is being extensively considered which is also facilitated by the drawbacks of current methods of silica production as mentioned in the previous paragraph. So, there is a lot of research going on worldwide to synthesize silica and silica-based materials using agricultural waste biomass. This chapter discusses the various approaches such as thermal, chemical, and biological which are currently being pursued for the production of silica and silica-based materials from agriculture waste biomass. This also includes applications of silica derived from agriculture waste in various fields along with pilot scale studies done to produce silica from these sources. The novelty of the chapter in comparison to recent reviews lies in the fact that it includes the studies done to produce silica from agricultural waste using microorganisms and fungi described in the biological methods section which are lacking in previous reviews. Also, the review of recent pilot scale studies is unique to this chapter.

10.2 Silica as an Important Constituent in Crop Residue

Silica content has an advantageous impact on improving the yield and quality of various crops. Various monocotyledon crops like rice, wheat, ryegrass, maize, barley, banana, etc., and some cyperaceae plants deposit it by the active transport mechanism. Si transporter proteins present in the plasma membranes of the cells of these plants are responsible for this deposition [17]. The silica is present in soil as insoluble aluminosilicate which is withered and dissolved as silicic acid before uptake by the plants. It is absorbed by plant roots, released into the aerenchyma apoplast, and then carried into the stele, an area of vascular tissue, for additional dispersion. Then through stream transpiration, it is translocated in shoot by the xylem. Silicic acid is polymerized and converted to amorphous silica at a concentration greater than 2 M. Depending upon the species and age of the plant silica can be found in various locations such as leaves, seeds, fruits, and inflorescence bracts [18]. The silica content in plant crops varies from 9–93% and it is maximum in mature plants due to its irreversible deposition in cell walls. Some dicots such as mustard, ground nut, and rapeseed also contain silica content but to a lesser extent. It is higher in monocots due to the presence of active transport mechanisms. Silica creates a stiff structure that provides a physical defense to the seeds and it is more concentrated in pericarp cells than stem cells [19].

10.3 Silica Content of Various Crop Residues

Rice straw (76 g silica/Kg crop) and rice husk (40 g/Kg) possess the highest silica yield. Approximately 28,000–55,000 million kilograms of silica per year is extractable from rice husk and straw keeping in mind the global production [20]. If the harvested area is considered sugarcane, the leaf has a better silica yield than rice straw and husk. It can go up to 37,000 million kilograms per year if sugarcane yield is at its best. Based on Kilogram silica per hectare it is best (1415 kg silica/hectare) even better than sugarcane bagasse which is followed by rice straw and husk, and husk from palm oil. Crop residues from barley, coconut, corn, oat, rapeseed, sorghum, and wheat have comparatively lower silica yields [19].

Rice husk is produced by the rice milling industry and constitutes about 20 weight percent of rice. Silica (20%) is its major inorganic constituent. It is mostly disposed of in landfills as waste due to lack of any use. However, it is more acceptable to recycle it to extract its valuable components, given the likelihood of it catching fire and the attendant health and environmental risks. This can also maintain environmental sustainability. Because of its higher silica content, it can be used to produce amorphous silica and various other silica products [21]. Rice straw contains more silica than any other straw. It is also considered waste and has no commercial use due to a lack of awareness [22]. Bagasse and corn cobs are other agricultural wastes that are being extensively studied for the production of silica because of their higher silica content [21, 23].

10.4 Various Strategies for Silica Production from Agricultural Residue

The silica is extracted from biomass in various forms such as amorphous silica, silica gel, mesoporous silica, and zeolites. Currently, silica extraction from agricultural biomass involves three different strategies (1) thermal method, (2) chemical method, and (3) biological method. The main steps or techniques used in these methods along with the type of silica produced by each of these are given in Fig. 10.1. Each method in detail is discussed in the following section.

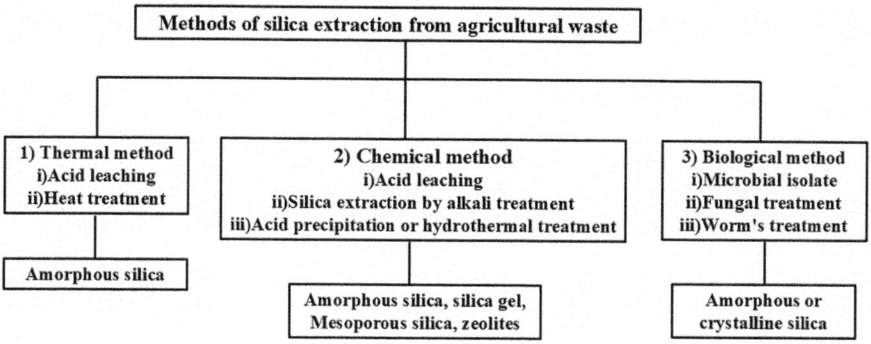

Fig. 10.1 Flow diagram representing various approaches or methods for silica production from agricultural biomass

10.4.1 Thermal Treatment Method

This method involves the heat treatment of biomass. The direct combustion of agricultural waste to produce silica without any chemical treatment is one of the oldest methods [24]. It was the first time in 1984 when amorphous silica was synthesized from the direct combustion of rice husk. However, it was not pure because no purification treatment was done [25].

To improve the quality and purity of silica the researchers later started acid treatments. The acid treatment to remove the metallic impurities can be done either before burning the biomass or after obtaining the ash. Both ways are reported in the literature. Chakravarty et al. studied the effect of acid treatment by three different types of acids (HCl, HNO_3, and H_2SO_4) before burning the rice husk. Among the three acids when compared for the removal of metal ions HCl was found to be most effective. Table 10.1 represents the effect of different acids and their concentrations. They found that completely white-colored ash can be produced by treating rice husk with 1N HCl before heating at 500 °C. Several such studies are done on the effect of acid leaching of rice husk before calcination. Since strong acids are environmentally harmful milder organic acids were also studied for their leaching efficiency. Umeda and Kondoh found that amorphous silica of almost 99% purity can be obtained by combustion of rice husk after its treatment with citric acid. They concluded that various metal ions such as Na, K, Ca, Mg, Fe, and Cu get removed by the chelation reactions of the carboxylic groups [27, 28]. They also found that acid treatment delays the crystallization temperature of silica. The crystalline phases were found at 1073 K when the rice husk was burned without acid treatment while they appeared at 1323 K or higher temperature in the case of treated rice husk. Faizul et al. demonstrated that citric acid treatment of palm ash results in the production of high-purity due to the removal of impurities. They found that under optimum treatment conditions of citric acid, the purity of silica reaches up to 92% compared to 45.5% obtained from untreated palm ash [29]. They achieved this purity under optimum reaction

Table 10.1 Effect of acid leaching on the removal of metallic ingredients from rice husk [26]

Treatment	Metallic ingredients as oxides (p. p. m.)								
	Na	K	Ca	Fe	Mg	Mn	Zn	Cu	Total
Untreated	5125	11,750	7042	946	2562	1192	179	31	28,827
1N HCl treated	46	50	58	362	187	19	19	6	747
3N HCl treated	33	33	46	262	146	7	19	6	552
5N HCl treated	25	HD	25	146	125	3	19	6	349
11.3N HCl treated	12	HD	8	129	125	3	17	6	300
1N H_2SO_4 treated	87	25	2500	233	729	25	102	12	3713
4.5N H_2SO_4 treated	92	25	2417	225	646	25	98	6	3534
9N H_2SO_4 treated	75	25	2500	229	625	19	85	6	3564
18 N H_2SO_4 treated	75	50	2627	217	625	24	83	6	3747
36 N H_2SO_4 treated	117	100	3333	237	421	171	98	6	4483
4.5 HNO_3 treated	17	25	19	500	83	19	17	8	688
9 HNO_3 treated	8	25	11	442	83	12	12	6	599
18 HNO_3 treated	4	25	7	392	62	6	12	HD	508
1NHCl + 4.5N HNO_3(1:3)	4	25	3	375	52	12	19	6	496

conditions which were 70 °C temperature, 60 min of reaction time, and a citric acid concentration of more than 2%.

Chen et. al reported that silica obtained from rice husk burned at 700 °C for 2 h after its treatment with ionic liquids is of almost the same purity as produced after treatment with HCl. The synthesized silica nanoparticles of average 70 nm size and surface area of 241.1m^2g^{-1} from ionic-liquid treated rice husks are almost comparable to HCl-treated rice husks [30]. They claimed that this method is better as both silica and lignocellulose can be obtained from rice husk simultaneously (see Fig. 10.2) and ionic liquids are more environmentally friendly than strong acids. Real et al. compared how the acid treatment affects the silica yield and purity when done before or after heat treatment. They found that acid leaching before heat treatment led to comparatively high surface area silica. They produced amorphous silica of almost 99.5% purity with a surface area of 260 m^2g^{-1} when rice husk was heated to 600 °C after treatment with HCl. But surface area decreased to 1 m^2g^{-1} when the acid treatment was performed after heat treatment at 600 °C [31]. These decreases in surface area were predicted due to the interaction of K^+ with the silica, which had not been removed before heating.

10.4.2 *Chemical Treatment Method*

The thermal method suffers from the drawback of poor control of the shape and size of silica particles. This can be addressed by the chemical treatment method. This

Fig. 10.2 Ionic liquid extracted rice husk as a source of both silica and lignin cellulosic biomass. Reproduced with permission Ref. [30].

method generally involves the extraction of silica from biomass by alkali treatment. The biomass is given either acid or heat treatment or both treatments before extracting silica with alkali. Both of these help in the decomposition of organic matter. The acid treatment also helps in the leaching of various inorganic constituents which results in the production of high-purity silica while the sole purpose of thermal treatment is the facilitation of removal of organic content only. Both the economics and the quality of the silica that is extracted are impacted by the order of these two treatments as well as the lack of one of them. The extraction of silica with alkali happens due to the reaction of NaOH with silica present in raw biomass or biomass ash which results in the formation of sodium silicate as given below [32]

$$SiO_2(s) + 2NaOH(l) \rightarrow Na_2SiO_3(l) + H_2O \tag{10.1}$$

The obtained sodium silicate after acid neutralization or hydrothermal treatment results in the formation of silica in various forms.

10.4.2.1 Synthesis of Silica Nanoparticles by Acid Neutralization

By lowering the pH of the extracted silica, the procedure causes silica to precipitate [33]. The following reaction best describes the formation of silica nanoparticles by this method:

$$Na_2SiO_3 + nH_aX \rightarrow SiO_2 + nNa_aX + H_2O \tag{10.2}$$

Kalapathy et al. studied the treatment of rice husk with NaOH for the extraction of silicate solution and then subsequent production of silica by acid treatment. The sodium silicate solution was made by boiling RH ash with 1N NaOH solution for 1 h with constant stirring. This solution was then titrated with 1N HCl solution until

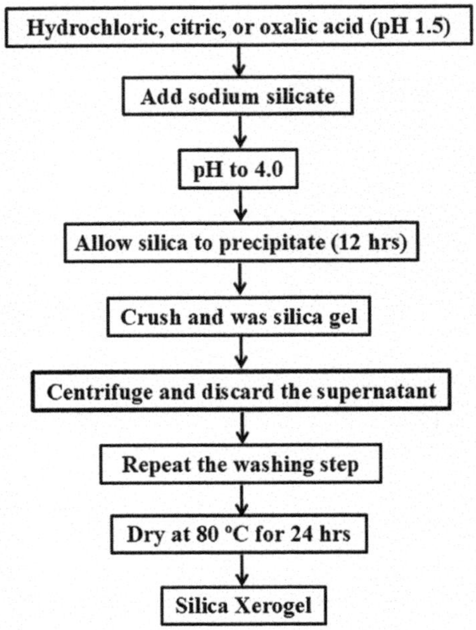

Fig. 10.3 Flow diagram of the procedure used to produce silica xerogel from sodium silicate at pH 4.0. Reproduced with permission Ref. [35]

pH became equal to 7 with constant stirring. The final xerogel was obtained from the obtained gel after aging it for 18 h followed by washing and drying at 80 °C. The total Na, K, and Ca contents of xerogel were found to be less than 0.1% when ash was washed with acid before NaOH treatment as well as when it was finally washed with water [34]. They further improved their method by carrying out the precipitation of sodium silicate by adding it to acid at pH 4. They used three different acids viz. HCl, oxalic acid, and citric acid and found that xerogel produced with oxalic acid and citric acid had a sodium content of 0.22 and 0.52%, respectively lower than one produced with HCl. Figure 10.3 shows the flow diagram of the procedure used to produce silica xerogel from sodium silicate at pH 4.0[35].

Selvakumar et al. found that washing of the rice husk ash with acid results in the higher percentage yield of extracted silica. They treated rice husk ash with three different acids and then extracted it with 1N NaOH as sodium silicate after which it was precipitated with HCl to obtain silica gel [36]. Rungrodnimitchai et al carried-out sodium hydroxide treatment of rice husk ash in a microwave oven at 800 W and they found that even a few minutes of treatment time resulted in a better yield of silica gel than the conventional heating method [37]. Liou et al. produced nano silica from the rice husk using a similar method and they found that the properties of nano-silica are dependent on pH, the concentration of sodium silicate, the type of acid, and time and temperature of aging. According to their finding, the silica produced by using HCl as a precipitating agent has a higher surface area than that produced by sulphuric, oxalic, and citric acids. They also concluded that oxalic acid or citric acid gave better silica yield than mineral acids such as HCl or H_2SO_4. The characteristics of silica particles produced by different acids and a precipitating agent at pH 7 are given in Table 10.2.

They concluded that optimum conditions for producing nano silica particles are obtained at gelation pH 3, silicate concentration of 0.15 M, and aging temperature and time of 50 °C and 12 h respectively [38]. Yuvi Kumar et al. studied the effect of the concentration of NaOH (0.5, 1, 1.5, 2, 2.5 N) used for the extraction of silica from rice husk ash. They found that 2.5 N NaOH gave the best results. The silica nanoparticles produced were found to have a 25 nm average size, 274 m^2g^{-1} surface area, and pores of 1.46 nm average size [39]. There are also reports in which silica is directly extracted from biomass without any heat treatment. In this context, mesoporous silica nanoparticles of spherical morphology were produced by Adam et al. They used a simple sol–gel technique without any template at ambient temperature. They treated rice husk powder with HNO_3 before treating it with 1N NaOH. The obtained sodium silicate solution was titrated with HNO_3 which resulted in the formation of silica nanoparticles of an average size of 50 nm and a surface area of 245 m^2g^{-1}. Although this method shows the formation of small-size silica particles with narrow size distribution the EDX analysis of samples shows the presence of 5.28% carbon [40]. Recently Rakesh et al. studied that type of agriculture waste as well as the calcination temperature used to produce ash also has an impact on the basic properties of silica obtained from their alkali-extracted silicate solution. They extracted silica from ash obtained from various agriculture wastes such as paddy straw, sugarcane bagasse, maize cobs and coconut husk calcined at three different temperatures viz. 500 °C, 600 °C, and 700 °C. They found that paddy straw showed the highest recovery of silica with the highest reactivity but lower pH, EC, and bulk density. Moreover, they found that higher calcination temperature increased reactivity while decreasing pH, EC, bulk density, and solubility in all these cases. These properties are important for the application of silica in the agriculture sector [41].

Table 10.2 Surface features of silica particles produced by various acids as precipitating agents

	S_{BET} (m^2/g)	V_t (cm^3/g)	V_{mic} (cm^3/g)	V_{meso} (cm^3/g)	V_{meso}/V_t (%)	Yields (%)
Hydrochloric acid	350	0.811	0.008	0.803	99.01	92.39
Sulphuric acid	319	0.717	0.009	0.708	98.74	89.33
Oxalic acid	313	0.672	0.004	0.668	99.40	95.05
Citric acid	285	0.567	0.003	0.564	99.47	99.64

10.4.2.2 Production of Silica Particles with Controlled Morphologies

Although the method of precipitating silica by adding acid to sodium silicate is simple and convenient it does not lead to precise control of the morphology of silica nanoparticles. Most of the silica prepared by this method contains small primary particles that are aggregated into large irregularly shaped particles. Some more methods achieve better control of the particle morphology and these methods use organic cosolvents, polymer additives, and water in an oil emulsion system [32].

Zulkifli et al. employed ethanol as a co-solvent to the silicate solution during the precipitation step [42]. They added different amounts of ethanol to the silicate solution which was derived from the rice husk before titrating it with phosphoric acid. In their study, a sodium silicate solution derived from rice husks was combined with varying amounts of ethanol. The resulting mixture was then titrated with phosphoric acid. Without ethanol, irregular and highly aggregated particles were formed, whereas increasing ethanol concentrations led to the synthesis of dispersed spherical particles. They found that when the sodium silicate: H_2O: ethanol ratio was kept at 1:1:0.25 at pH 7 the silica nanoparticles of the size range 75 to 252 nm were obtained with minimum aggregation. BET surface area was found to be approximately 364 m^2g^{-1}. Figure 10.4 gives the SEM and TEM images of silica nanoparticles prepared under optimal conditions. A similar phenomenon was reported by Zulfiqar et al. [43], they added ethanol to the silicate solution during its neutralization with phosphoric acid which resulted in the formation of spherical silica particles. Additionally, they also found that when precipitation is carried out at room temperature the increase in the sodium silicate concentration results in the increase in the average size of silica particles.

They found that silica particles of unimodal size (181 ± 17 nm) are formed when 0.7 wt % white rice husk (WRH) amount is added to the sodium hydroxide solution (Fig. 10.5). However, increasing the WRH concentration to 5.6 wt% resulted in the

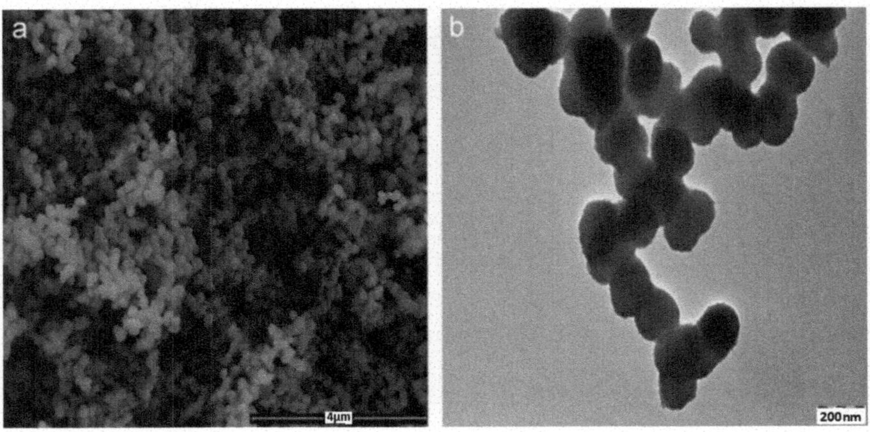

Fig. 10.4 SEM (**a**) and TEM images (**b**) of silica nanoparticles prepared at optimal conditions. Reproduced with permission Ref. [42]

Fig. 10.5 SEM images of uni-modal silica particles produced by sodium silicate solution (SSS) containing 0.7 wt.% white rice husk (WRH). Reproduced with permission Ref. [43]

production of silica particles with three different sizes i.e. trimodal size distribution (352 ± 77 nm, 1.8 ± 0.3 μm, and 7.1 ± 1.3 μm) Fig. 10.5.

They found that the size of particles along with the presence of uni-, bi- and tri modal particles increases with an increase in the concentration of NaOH to form a sodium silicate solution. In contrast, the particle size was found to decrease with an increase in the concentration of silicate solution when the precipitation reaction was carried out at 65 °C. The preparation of spherical silica particles using polymer additives has been reported in several studies. For example, using a polymer with sufficient ethylene oxide chains as an additive leads to spherical-size silica particles [44]. It is considered that the formation of silica particles is the result of interaction between EO groups and silicate species during solidification. In studies by Li et al. [45, 46], polyethylene glycol (PEG) with a molecular weight of 20,000 was dissolved in a sodium silicate solution. They successfully synthesized spherical silica nanoparticles when the silicate solution was titrated with phosphoric acid. They found that increasing the PEG dosage from 100 to 176 mg leads to an increase in the specific surface area of porous silica from 709 to 936 m^2g^{-1} and materials can be prepared within 10 h. N_2 adsorption isotherm of various prepared silica materials shows that silica prepared with minimum PEG content (S100) shows type I isotherm with an abundance of micropores while increasing PEG content from 100 to 176 mg increases the mesopores.

Similarly, Le et al. [47] reported that using PEG with a molecular weight of 10,000 as an additive enabled the formation of spherical silica nanoparticles from rice husk-derived silicate solutions. Hasan et al. synthesized spherical silica particles from a silicate solution obtained from rice husk using a water-in-oil emulsion system [48]. In this emulsion system of water and toluene cetyltrimethylammonium bromide (CTAB) and n-butanol were added as the surfactant and cosurfactant, respectively. Then silica was hydrolyzed by urea at micellar interphase after the formation of reverse micelle. The resulting silica particles displayed a well-defined spherical shape with 227 m^2g^{-1} surface area and 1.24 cm^3g^{-1} pore volume.

10.4.2.3 Synthesis of Ordered Porous Silica Particles

Mobil Oil Corporation synthesized the first mesoporous silica material (MCM) using CTAB as a template and since then several ordered porous materials have been synthesized [49]. These include the SBA, KIT, FDU, and MSU series after the MCM series has been developed. Agriculture waste-derived silicate has attracted significant attention as a replacement for the costly and hazardous silicon alkoxides used to synthesize these materials. Various MCM-type materials have been synthesized using agriculture waste materials during the last decade. Ramalingam et al. synthesize MCM-41 silica nanoparticles of hexagonal porous structures with 2.3 nm size monomodal pores and a surface area of 1115 m^2g^{-1} using rice husk as a silica source [50]. They added sodium silicate solution extracted from rice husk into a certain amount of CTAB solution while maintaining a specific molar composition of SiO_2, NaOH, CTAB, and H_2O. pH of the solution was adjusted at 10 with nitric acid and then stirred for 24 h at 80 °C. A gel was obtained after aging for 48 h at the same temperature which upon calcination at 600 °C after washing resulted in the formation of mesoporous material. Recently, Sennoun et al. synthesized MCM-41 and organofunctionalized MCM-41 from silica obtained from the calcination of rice husk after its acid treatment [51]. They added RH silica to NaOH solution containing octadecyltrimethylammonium surfactant with continuous stirring while keeping the molar composition to 1 SiO_2:0.1 C_{18}TAB:0.25 NaOH:50 H_2O at 55 ºC. Autoclaving of gel at 115 ºC for 24 h resulted in the formation of MCM-41. The NH_2-functionalized MCM-41 and amino acid-functionalized MCM-41 were synthesized by the same procedure but before pouring the gel into an autoclave APTES or each amide derivative of amino acid triethoxysilane was added into it. This resulted in the formation of highly ordered MCM-41 and organo-functionalized MCM-41 of 800 m^2g^{-1} and 510–720 m^2g^{-1} surface area respectively. The scheme for the synthesis organo(R)-functionalized MCM-41 is represented in Fig. 10.6.

Tuan Vu et al. also prepared amorphous and porous silica materials from sodium silicate extracted from rice husk ash. They added sodium silicate to the CTAB solution

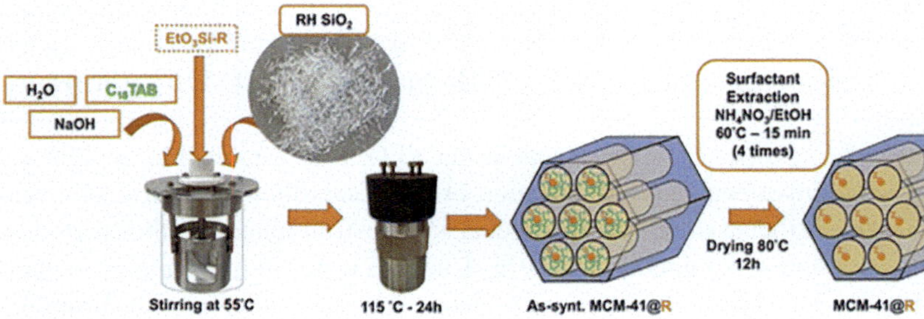

Fig. 10.6 Schematic representation of the synthesis of organo(R)-functionalized MCM-41 and purely siliceous MCM-41 by pseudomorphic transformation by adding or not the corresponding triethoxy organosilane (EtO_3Si–R), respectively. R = aminopropyl, amide derivatives of amino acids (Leu, Ser, Tyr). Reproduced with permission Ref. [51]

and autoclaved the gel for 48 h at 100 °C after aging at 50 °C for 24 h. This after calcination at 600 °C for 6 h gave porous silica material of 10–20 nm average particle size, pore size 2–5 nm, and specific surface area of 544.5 m^2g^{-1} they further modified the material with ethylenediaminetetraacetic acid with binder polyvinyl alcohol to enhance its adsorption capacity to remove Pb^{2+} ions from water [52].

Another MCM-type material known as MCM-48 has a three-dimensional porous structure and cubic symmetry. It is synthesized either by changing the CTAB concentration or mixing it with a neutral surfactant. Looking at a specific example MCM-48 was synthesized by Bhagiyalakshmi et al. in which they used CTAB mixed with polyoxyethylene lauryl ether as a template [53]. They added sodium silicate extracted from rice husk into a surfactant mixture at 80 °C and then aged it for 48 h at 100 °C after its pH was adjusted to 10 with the use of acetic acid. After filtration and drying, the product was calcined at 550 °C. Table 10.3 describes the surface area, volume, and size of pores of MCM-41 and MCM-48 type materials synthesized from silicate solutions derived from the rice husk Table 10.3.

SBA-15 which was first developed by the University of California has two-dimensional hexagonal pores and is another common type of silica [67]. Its pore size ranges from 4 to 12 nm and using organic additives it can be increased by 30 nm. It has more hydrothermal stability than MCM-41 due to more thicker walls. A lot of studies have been done using silicate extracted from the rice husk and

Table 10.3 Morphological properties of rice husk derived MCM type materials [32]

Silica type	Pore diameter(nm)	Surface area (m^2g^{-1})	Pore volume (cm^3g^{-1})	Refs.
MCM-41	2.9	800	0.93	[54]
	2.86	943	-	[55]
	3.54	1101	0.96	[54]
	3.51	1099	0.96	[56]
	3.28	903	-	[57]
	2.3	1115	0.92	[58]
	3.6	602	0.49	[59]
	2.8–3.1	545–1210	0.36–1.00	[60]
	2.92	797	0.57	[61]
	3.16	1347	0.906	[62]
	3.8	500.5	0.45	[63]
	3.0–3.4	552–769	1.025–1.167	[64]
	2.71	972.5	0.87	[65]
	2.3	1115	0.92	[51]
MCM-48	4.02	1024	2.58	[66]
	3.89	1124	0.98	[54]
	2.6	1059	0.68	[57]
	2.5	815	0.75	[61]

other agricultural wastes for the synthesis of SBA-15. It is synthesized by a mechanism represented as the $N^0H^+X^-I^+$ mechanism which involves nonionic surfactant (N), hydrogen ion (H^+), halide ion (X^-), and silica. Triblock non-ionic copolymer Pluronic P-123 acts as structure structure-directing agent in acidic conditions in this case. Henao et al. synthesized SBA-15, SBA-11, and SiO_2 using rice husk ash as a silica source (Fig. 10.7). They compared them for the CO_2 adsorption studies after functionalizing them with polyethyleneimine [68].

Chareonpanich et al. used an ultrasonic technique to synthesize SBA-15 from rice husk [69], which had ordered hexagonal pores and a pore size was 9.5 nm. Various properties of materials of the SBA family such as size and volume of pores and surface area as reported in the literature are given in Table 10.4.

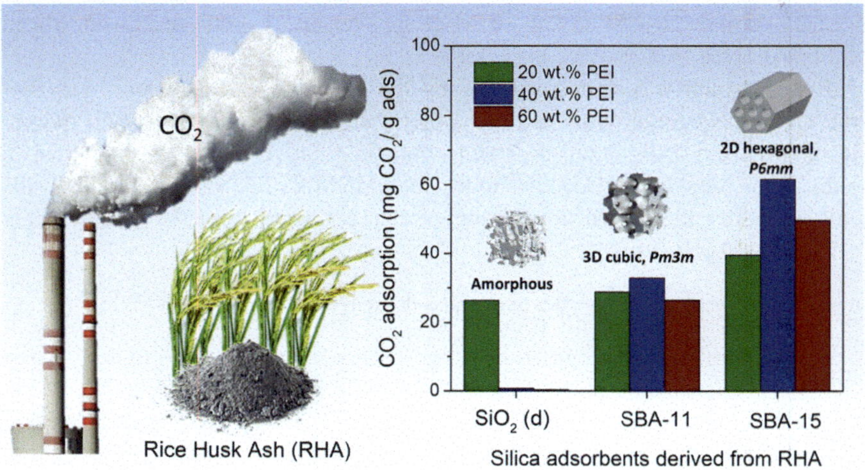

Fig. 10.7 Graphical representation of the production of various types of silica materials along with their CO_2 adsorption capacity. Reproduced with permission Ref. [68]

Table 10.4 Morphological properties of SBA-type materials synthesized from rice husk [32]

Silica type	Pore diameter (nm)	Surface area (m^2g^{-1})	Pore Volume (cm^3g^{-1})	Ref
SBA-15	3.8/7.7/12.9	356	0.792	[72]
	9.5	530–860	0.96–1.27	[69]
	4.8	790	0.82	[73]
	5.8	712	0.68	[61]
	7.4	780	0.95	[56]
	7.7	1095	1.705	[67]
	7.6	604	1.192	[72]
SBA-16	5.8–8.2	775–840	-	[70]

SBA-16 with three-dimensional cubic pores (8.00 nm) was also synthesized from rice husk using Pluronic F-127 polymer as a template [70]. Chun et. al. used Pluronic P-123 and TMB, as templates and pore expanders respectively [71]. These were added to sodium silicate solution in the neutral environment which was maintained by the addition of acetic acid. Amorphous silica with mesocellular foam structure was synthesized with a pore size of 30 nm and 1.77 cm^3g^{-1} pore volume. Rice husk has also been used for the synthesis of various types of zeolites. Since zeolites are microporous aluminosilicates, an additional alumina source is needed for their synthesis. Zeolite A, NaY[74], ZSM-5, ZSM-12, ZSM-48[75], and Linde type J are among the many which have been synthesized from silica derived from rice husk and using alumina source in the form of sodium aluminate.

10.4.3 Biological Method

The biological method of silica extraction from agricultural waste includes microbe, worm, and fungus treatment. This can help to avoid excessive heat treatment and corrosive chemicals. Some specific examples are discussed here. Rice straw was used to synthesize amorphous silica nanoparticles of 50 to 80 nm by Wattanasiri Welch et al. They hydrolyzed rice straw residue with a microbial isolate (T. reesei) and after that, it was heated for 8 h at 500 °C resulting in the formation of silica nanoparticles. However, the obtained silica showed the impurities of manganese and phosphate compounds due to the microbial activities [76]. Silica nanoparticles of spherical shape and narrow size distribution were synthesized from the excreta which is known as humus of the Californian-red worms [77]. The authors collected the humus of the worms fed with rice husk and after drying and neutralization at room temperature, the organic content of the collection was removed by calcining it at different temperatures (500, 600, and 700 °C). After the digestion in acid to remove the traces of inorganic compounds, silica particles in the range of 50–250 nm were obtained depending on the calcination temperature. They could also produce silica nanoparticles using coffee and cane husks but the yield was comparatively low and the size distribution of particles was wide. Shastry et. al. used the fungus *Fusarium oxysporum* to transform the naturally occurring amorphous silica present in rice husk into crystalline silica nanoparticles of 2–6 nm quasi-spherical form at room temperature. After calcination, it leads to the loss of occluded proteins and the formation of porous structures with cubic morphology (Fig. 10.8) [78].

Rice husk was subjected to microbial fermentation for 60 days by Rohatgi et al. They used *Cyathus* which is white rot fungus species grown locally. The fermented husk contained about 49% silica as compared to 23.7% in untreated husk. The fermented rice was subjected to calcination at 400 °C after drying and sieving to remove the fibrous organic material formed after the fermentation. This led to the formation of white silica [20].

Recently Zaki et al. synthesized silica nanoparticles from bagasse as starting material by solid state fermentation (SSD) using a fungal isolate *Curvularias picifera.*

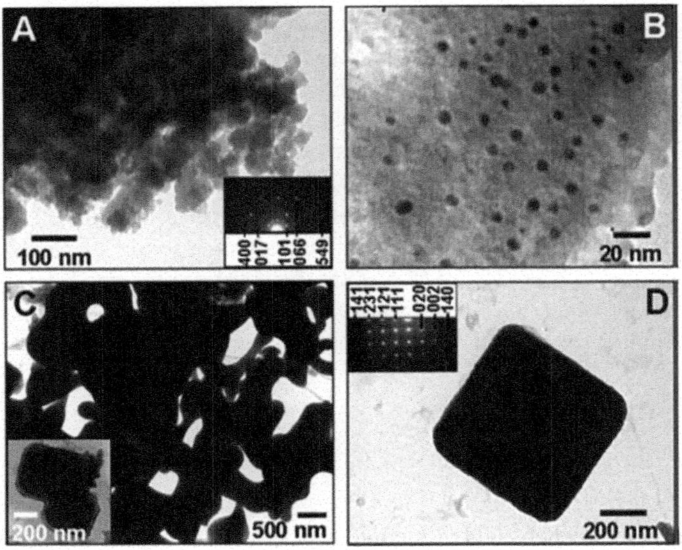

Fig. 10.8 TEM micrographs at different magnifications of silica nanoparticles synthesized by the exposure of rice husk to the fungus *Fusarium oxysporum* before (**a** and **b**) and after calcination at 400 °C for 2 h (**c** and **d**). The inset in (**c**) corresponds to a high-magnification image of silica particles shown in the main figure. The insets in (**a**) and (**d**) are SAED patterns recorded from representative silica nanoparticles in the corresponding images. Reproduced with permission Ref. [78]

They added fungal spore suspension to the bagasse and Czapex's mineral salt mixture and incubated it for 8 days at 28 °C. Then further SiO_2 was leached from the fermented bagasse and silica nanoparticles of size ranging from 30.6 to 130.4 were produced depending upon the biotransformation conditions. SEM and TEM images showed the presence of spherical bio-fabricated silica nanoparticles [79] (Fig. 10.9).

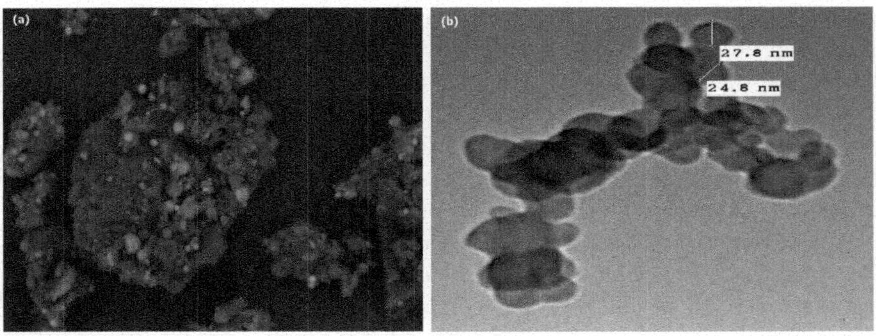

Fig. 10.9 Electron microscope imaging of the biofabricated SNPs under SSF by *Curvularia spicifera* **a** Scanning electron microscope, and **b** Transmission electron microscope [79]

10.5 Applications of Silica Derived from Agriculture Waste

Silica derived from agricultural waste by various approaches has been put into several applications, including CO_2 capturing, catalysis, environmental remediation, drug delivery, bioimaging, and biosensing, to mention a few. A few specific examples from different fields are discussed here. Various mesoporous materials such as MCM-41, MCM-48, and SBA-15 synthesized by Bhagiyalakshmi et al. were grafted with amine compounds, Tris(2-aminoethyl)amine (TREN) and Tetraethylenepentamine (TEPA) and used for CO_2 capture. They found that TREN-modified silica showed better adsorption capacity than the TEPA-modified. Among the various TREN-modified silicas the order of CO_2 adsorption capacity was MCM-48/TREN > MCM-41/TREN > SBA-15/TREN [53]. Jang et al. synthesized MCM-48 and 3-aminopropyltriethoxysilane (APTS) functionalized MCM-48 and found that APTS-MCM-48 showed better CO_2 adsorption capacity than bare MCM-48. The breakthrough curves of both these materials confirmed that these rice husk-derived silica materials can be abundant sources of CO_2 capturing [66]. Appaturi et al. used MCM-41-Imi/Br in the solventless synthesis of styrene carbonate (SC) from CO_2 and styrene oxide (SO) under ambient conditions leading to a maximum selectivity of 99.1% [58]. It was prepared by immobilizing MCM-41 with imidazole using 3-chloropropyltriethoxysilane (CPTES) as an anchoring agent followed by alkylation with 1,2-dibromoethane at 110 °C. Sn-SBA-16 was synthesized by isomorphic substitution of Sn in SBA-16. The mesoporous Sn-SBA-16 was very active for the benzylation of aromatics with 100% selectivity for monoalkylated products [70]. Cu-Ni/SBA-15 synthesized by Pimprom et al. from rice husk ash and tetraethyl orthosilicate as silica source. It was used for the synthesis of dimethyl carbonate from methanol and carbon dioxide [80]. They also found that the addition of Mo to the Cu-Ni/SBA-15 catalyst enhanced its acidity and activity. Silica derived from agricultural waste has been used in several environmental remediation applications. Areerob et al. modified MCM-41 with various silane agents like trimethylchlorosilane (TMS), triisopropyl chlorosilane (TIPS), and phenyldimethylchlorosilane (PDMS). They used these as a selective adsorbent of benzene, toluene, ethylbenzene, and xylenes (BTEX). They found TMS-modified MCM-41 to be better than the other two. They also compared it with commercial graphitized carbon black [61]. MCM-41 synthesized by Nguyen et al. from rice husk had a surface area of 1347 m^2g^{-1} and showed 99% removal efficiency of methylene blue dye [62].

Heavy metal ions such as Cr (IV), Cd (II), Pb (II), Cu (II), Ni (II), and Fe (II) are among the most common contaminants in water. Silica nanoparticles and functionalized silica nanoparticles derived from agricultural waste are efficient adsorbents for removing these metal ions from water. Recently Barani et al. reviewed various agro-waste-derived silica adsorbents used for the removal of heavy metal ions from water [81]. They found that the mechanism of removal of heavy metal ions from water involves eight different processes as mentioned in Fig. 10.10. Prabha et al. reviewed the applications of plant-derived silica nanoparticles and composites

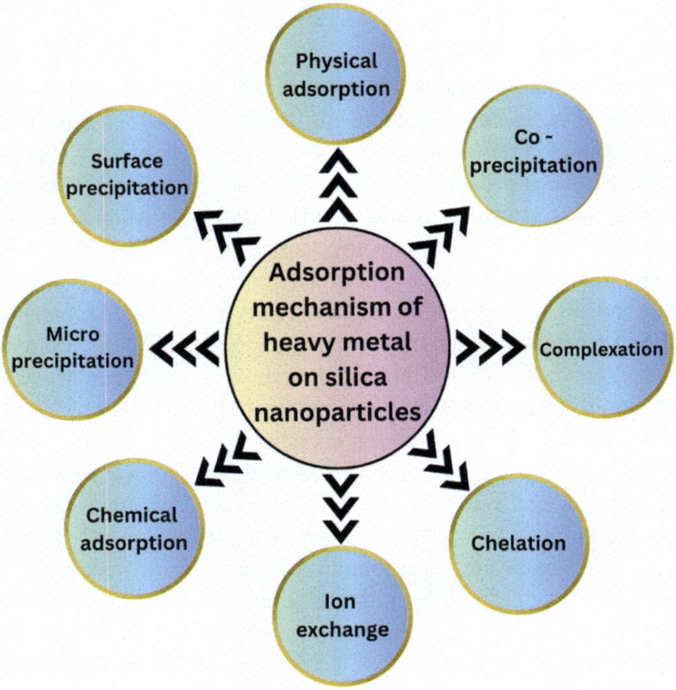

Fig. 10.10 Heavy metal adsorption mechanism in silica nanoparticles (SiNPs). Reproduced with permission Ref. [81]

for biosensors, bioimaging, drug delivery, and supercapacitors [82]. These applications of plant-derived silica nanoparticles are possible due to their biocompatibility, biodegradability, low toxicity, and low cost in comparison to SiNPs synthesized from other precursors such as TEOS.

Rodriguez-Otero took the process of extraction of amorphous silica from ice husk to one step ahead by combining it with the concurrent generation of bio-oil and syngas [83]. The acid-treated rice husk was subjected to pyrolysis which resulted in the production of CO-rich syngas and bio-oil with increased phenolic content. The remaining biochar was subjected to calcination at 700 °C to produce silica particles having a specific surface are 300 m^2g^{-1} which showed Cd adsorption capacity of 407 mgg^{-1}. Thus, they suggested a simultaneous solution to the dual challenges of waste reduction and water contamination.

Errington et al. did a life cycle assessment study based on 8 environmental impact indicators and compared the existing wet and dry industrial methods with the thermal method in which silica is produced from rice husk as alternative biomass feedstock [84]. They concluded that method involving silica production from rice husk as alternative feedstock will be more beneficial as it makes more possibility of reducing the global warming potential of existing silica industry if there is co-recovery of bioenergy. However, this assessment is heavily dependent on the amount of bioenergy produced and sensitive to the GWP of regional electricity grids.

There have also been some pilot-scale studies for the production of silica from agriculture waste. Hidayat et al. started with 3 kg of rice ash husk for the production of silica by sol–gel method on a pilot scale [85]. They mixed rice husk ash(3 kg) with 5% NaOH solution with a ratio of 1:5(w/v) in a 30 L extractor. This results in the formation of sodium silicate in 30 min after hydrothermal treatment at 1.1–1.2 bar pressure at 110–115 °C. It was neutralized with 3% HCl to pH7 which resulted in the formation of 95.41% purity silica with 70.07% yield. Park et al. produced a two-stage continuous process that consisted of attrition ball milling and alkaline leaching for the extraction of silica from rice husk. The process was operated continuously for 80 h and resulted in 89% silica yield [86]. Later they modified it to a one-pot continuous process. They carried it out for 150 h and produced spherical amorphous silica particles of 99.1% purity.4.09 kg silica was produced per day at a resident time of 50 min and energy consumption for 1 kg was 51.6 MJ Kg^{-1}. The equipment used for the process which consists of hydrothermal treatment and ball milling is shown in Fig. 10.11 [87].

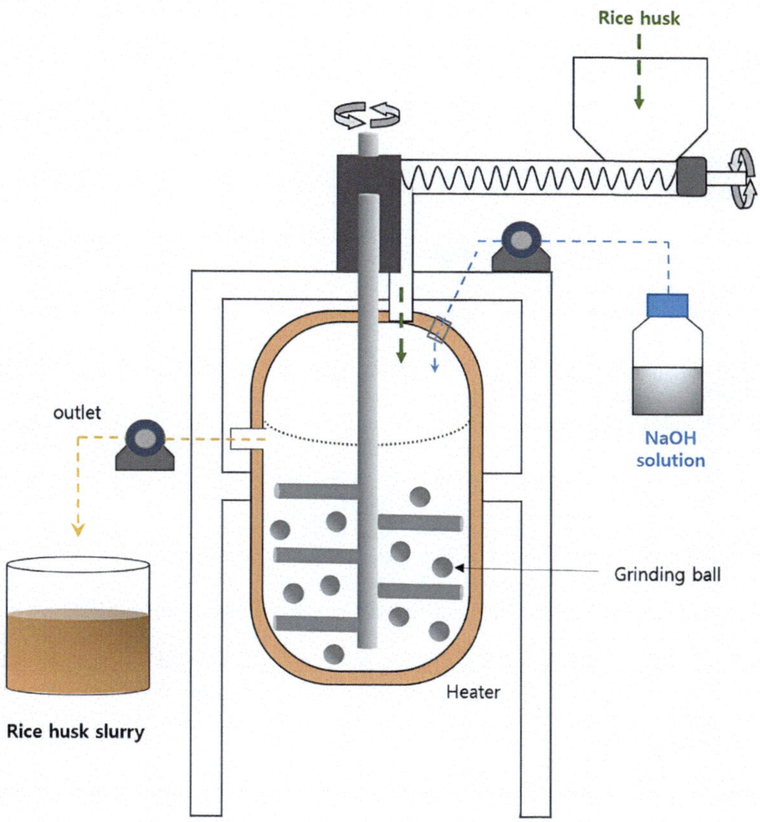

Fig. 10.11 Schematic of the one-pot alkali hydrothermal and ball milling continuous silica extraction process. Reproduced with permission Ref. [87]

10.6 Summary

The application of silica and silica-based materials as mentioned in the introduction makes it one of the valuable materials used by contemporary human society. Despite its immense applications, its current synthesis method needs rethinking. The discussion made in this chapter indicates that agricultural waste biomass can be a better alternative material from which silica and silica-based materials can be synthesized. This will also solve the problems associated with the disposal of agricultural waste. From the discussion in this chapter, it can be concluded that silica synthesized from biomass is of the same quality as that synthesized from current inorganic precursors. However, there is still a need to improvise the three different approaches that are currently used to synthesize silica from agriculture waste biomass. Especially more studies need to be done to replace strong mineral acids with milder organic acids to make thermal and chemical methods more environmentally friendly. Biological methods, although less hazardous, are still in their infancy. Moreover, the large-scale production of silica by these methods should be studied in more detail to see the viability of these approaches against the current synthesis method. It is very clear that to solve the environmental issues created by industrialization such greener approaches will be at the forefront of future research.

References

1. Eddy, D.R., Puri, F.N., Noviyanti, A.R.: Synthesis and photocatalytic activity of silica-based sand quartz as the supporting TiO_2 photocatalyst. Procedia Chem. **17**, 55–58 (2015)
2. Unger, K.K.: Chapter 1 General chemistry of silica. J. Chromatogr. Libr. **16**, 1–14. Elsevier. (Ed.). (1979)
3. Lazaro, A., Brouwers, H.J.H.: Nano-silica production by a sustainable process; application in building materials. In: 8th Fib International-PhD Symposium in Civil Engineering. Kgs. Lyngby, Denmark (2010)
4. Lourenço, A.F., Gamelas, J.A., Sequeira, J., Ferreira, P.J., Velho, J.L.: Improving paper mechanical properties using silica-modified ground calcium carbonate as filler. BioResources **10**(4), 8312–8324 (2015)
5. Wang, Z., Gu, Z., Hong, Y., Cheng, L., Li, Z.: Bonding strength and water resistance of starch-based wood adhesive improved by silica nanoparticles. Carbohydr. Polym. **86**(1), 72–76 (2011)
6. Zharmenov, A., Yefremova, S., Satbaev, B., Shalabaev, N., Satbaev, S., Yermishin, S., Kablanbekov, A.: Production of refractory materials using a renewable source of silicon dioxide. Minerals. **12**(8), 1010 (2022)
7. Costa, J.R., Neto, T., Pedrosa, S.S., Sousa, S.C., Azevedo-Silva, J., Tavares-Valente, D., Madureira, A.R.: Biogenic silica microparticles as a new and sustainable cosmetic ingredient: assessment of performance and quality parameters. Colloids Surf. B BiointerfacesBiointerfaces **226**(113305), 20 (2023)
8. Shinde, P.S., Suryawanshi, P.S., Patil, K.K., Belekar, V.M., Sankpal, S.A., Delekar, S.D., Jadhav, S.A.: A brief overview of recent progress in porous silica as catalyst supports. J. Compos. Sci. **5**(3), 75 (2021)
9. Polarz, S., Smarsly, B.: Nanoporous materials. J. Nanosci. Nanotechnol.Nanosci. Nanotechnol. **2**(6), 581–612 (2002)

10. Pa, F.C., Chik, A., Bari, M.F.: Palm ash as an alternative source for silica production. In: MATEC Web Conferences, vol. 78, p. 01062. EDP Sciences (2016)
11. Bazzi, L., Hesemann, S., Laassiri, S., El Hankari, S.: Alternative approaches for the synthesis of silica nanoparticles and their hybrid composites: synthesis, properties and applications. Int. J. Environ. Sci. Technol. **20**(10), 11575–11614 (2023)32
12. Nakashima, H., Omae, K., Sakai, T., Yamazaki, K., Sakurai, H.: Acute and subchronic inhalation toxicity of tetraethoxy silane (TEOS) in mice. Arch. Toxicol.Toxicol. **68**(277–283), 33 (1994)
13. Okamura, T., Garland, E.M., Johnson, L.S., Cano, M., Johansson, S.L., Cohen, S.M.: Acute urinary tract toxicity of tetraethyl orthosilicate in rats. Fundam. Appl. Toxicol. **18**(3), 425–441 (1992)
14. Kolesar, G.B., Siddiqui, W.H., Crofoot, S.D., Evans, M.G., Meeks, R.G.: Subchronic inhalation toxicity of 1, 1, 1, 3-tetrachloropropane in rats. Fundam. Appl. Toxicol.. Appl. Toxicol. **25**(1), 52–59 (1995)
15. Yadav, M., Dwibedi, V., Sharma, S., George, N.: Biogenic silica nanoparticles from agro-waste: properties, mechanism of extraction and applications in environmental sustainability. J. Environ. Chem. Eng. **10**(6), 108550 (2022)
16. Cho, E.J., Trinh, L.T.P., Song, Y., Lee, Y.G., Bae, H.J.: Bioconversion of biomass waste into high value chemicals. Bioresour. Technol.. Technol. **298**, 122386 (2020)
17. Ma, J.F., Yamaji, N.: A cooperative system of silicon transport in plants. Trends Plant Sci. **20**(7), 435–442 (2015)
18. Shakoor, S., Bhat, M., Mir, S.: Phytoliths in plants: a review. Res. Rev. J. Bot. Sci. **3**(3), 10–24 (2014)
19. Sow, S., Ranjan, S., Ghosh, M., Kumar, S., Choudhury, S.R.: Crop residues as potential sustainable source of silica in the era of climate change: a review **11**, 1629–1641 (2020)
20. Rohatgi, K., Prasad, S.V., Rohatgi, P.K.: Release of silica-rich particles from rice husk by microbial fermentation. J. Mater. Sci. Lett. **6**(7), 829–831 (1987)
21. Vaibhav, V., Vijayalakshmi, U., Roopan, S.M.: Agricultural waste as a source for the production of silica nanoparticles. Spectrochim Acta A Mol Biomol Spectrosc. **139**, 515–520 (2015)
22. Hessien, M.M., Rashad, M.M., Zaky, R.R., Abdel-Aal, E.A., El-Barawy, K.A.: Controlling the synthesis conditions for silica nanosphere from semi-burned rice straw. Mater. Sci. Eng. B **162**(1), 14–21 (2009)
23. Adesanya, D.A., Raheem, A.A.: A study of the workability and compressive strength characteristics of corn cob ash blended cement concrete. Constr. Build. Mater. **23**(1), 311–317 (2009)
24. James, J., Rao, M.S.: Silica from rice husk through thermal decomposition. Thermochim. Acta. Acta **97**, 329–336 (1986)
25. Kapur, P.C.: Production of reactive bio-silica from the combustion of rice husk in a tube-in-basket (TiB) burner. Powder Technol. **44**(1), 63–67 (1985)
26. Chakraverty, A., Mishra, P., Banerjee, H.D.: Investigation of combustion of raw and acid-leached rice husk for production of pure amorphous white silica. J. Mater. Sci. **23**, 21–24 (1988)
27. Umeda, J., Kondoh, K.: High-purity amorphous silica originated in rice husks via carboxylic acid leaching process. J. Mater. Sci. **43**, 7084–7090 (2008)
28. Umeda, J., Kondoh, K.: High-purification of amorphous silica originated from rice husks by combination of polysaccharide hydrolysis and metallic impurities removal. Ind. Crops Prod. **32**(3), 539–544 (2010)
29. Faizul, C.P., Abdullah, C., Fazlul, B.: Extraction of silica from palm ash via citric acid leaching treatment. Adv. Environ. Biol. **7**(12), 3690–3695 (2013)
30. Chen, H., Wang, W., Martin, J.C., Oliphant, A.J., Doerr, P.A., Xu, J.F., DeBorn, K.M., Chen, C., Sun, L.: Extraction of lignocellulose and synthesis of porous silica nanoparticles from rice husks: a comprehensive utilization of rice husk biomass. ACS Sustain. Chem. Eng. **1**(2), 254–259 (2013)

31. Real, C., Alcala, M.D., Criado, J.M.: Preparation of silica from rice husks. J. Am. Ceram. Soc. **79**(8), 2012–2016 (1996)
32. Chun, J., Lee, J.H.: Recent progress on the development of engineered silica particles derived from rice husk. Sustainability **12**(24), 10683 (2020)
33. Soltani, N., Bahrami, A., Pech-Canul, M.I., González, L.A.: Review on the physicochemical treatments of rice husk for production of advanced materials. Chem. Eng. J. **264**, 899–935 (2015)
34. Kalapathy, U., Proctor, A., Shultz, J.: A simple method for production of pure silica from rice hull ash. Bioresour. Technol.. Technol. **73**(3), 257–262 (2000)
35. Kalapathy, U., Proctor, A., Shultz, J.: An improved method for production of silica from rice hull ash. Bioresour. Technol.. Technol. **85**(3), 285–289 (2002)
36. Selvakumar, K.V., Umesh, A., Ezhilkumar, P., Gayatri, S., Vinith, P., Vignesh, V.: Extraction of silica from burnt paddy husk. Int. J. Chemtech Res. **6**(9), 4455–4459 (2014)
37. Rungrodnimitchai, S., Phokhanusai, W., Sungkhaho, N.: Preparation of silica gel from rice husk ash using microwave heating. J. Met. Mater. Miner. **19**(2) (2009)
38. Liou, T.-H., Yang, C.-C.: Synthesis and surface characteristics of nanosilica produced from alkali-extracted rice husk ash. Mater. Sci. Eng. B **176**(7), 521–529 (2011)
39. Yuvakkumar, R., Elango, V., Rajendran, V., Kannan, N.: High-purity nano silica powder from rice husk using a simple chemical method. J. Exp. Nanosci.Nanosci. **9**(3), 272–281 (2014)
40. Adam, F., Chew, T.S., Andas, J.: A simple template-free sol–gel synthesis of spherical nano silica from agricultural biomass. J. Sol-Gel Sci. Technol. **59**, 580–583 (2011)
41. Rakesh, B., Chitdeshwari, T., Maragatham, Sharmila, D.J.S., Semnthil, A.: Silica-derived materials from agricultural wastes: characterization and comparative studies. Digest J. Nanomater. Biostruct. **19**(2), 605–618 (2024)
42. Zulkifli, N.S.C., Ab Rahman, I., Mohamad, D., Husein, A.: A green sol–gel route for the synthesis of structurally controlled silica particles from rice husk for dental composite filler. Ceram. Int. **39**(4), 4559–4567 (2013)
43. Zulfiqar, U., Subhani, T., Husain, S.W.: Towards tenable size of silica particles from rice husk. J. Non-Cryst. SolidsCryst. Solids **429**, 61–69 (2015)
44. Hwang, J., Lee, J.H., Chun, J.: Facile approach for the synthesis of spherical mesoporous silica nanoparticles from sodium silicate. Mater. Lett. **283**, 128765 (2021)
45. Li, D., Chen, D., Zhu, X.: Reduction in time required for synthesis of high specific surface area silica from pyrolyzed rice husk by precipitation at low pH. Bioresour. Technol.. Technol. **102**(13), 7001–7003 (2011)
46. Li, D., Zhu, X.: Short-period synthesis of high specific surface area silica from rice husk char. Mater. Lett. **65**(11), 1528–1530 (2011)
47. Le, V.H., Thuc, C.N.H., Thuc, H.H.: Synthesis of silica nanoparticles from Vietnamese rice husk by sol–gel method. Nanoscale Res. Lett. **8**, 1–10 (2013)
48. Hasan, R., Chong, C.C., Bukhari, S.N., Jusoh, R., Setiabudi, H.D.: Effective removal of Pb (II) by low-cost fibrous silica KCC-1 synthesized from silica-rich rice husk ash. J. Ind. Eng. Chem. **75**, 262–270 (2019)
49. Kresge, A.C., Leonowicz, M.E., Roth, W.J., Vartuli, J.C., Beck, J.S.: Ordered mesoporous molecular sieves synthesized by a liquid-crystal template mechanism. Nature **359**(6397), 710–712 (1992)
50. Ramalingam, R.J., Appaturi, J.N., Pulingam, T., Al-Lohedan, H.A., Al-dhayan, D.M.: In-situ incorporation of ruthenium/copper nanoparticles in mesoporous silica derived from rice husk ash for catalytic acetylation of glycerol. Renew. Energy **160**, 564–574 (2020)
51. Sennoun, L., Lee, C., Fretel, Y., Clavié, M., Subra, G., Ladner, Y., Napoli, A., Galarneau, A., Hesemann, P., Mehdi, A.: Direct pseudomorphic transformation of silica from rice husk into organo-functionalized MCM-41. New J. Chem. **48**, 7052 (2024)
52. Vu, A.T., Nguyen, M.V., Naguyen, T.H.: Fabrication of ethylenediaminetetraacetic modified porous silica composite from rice husk for enhancing the remove of Pb^{2+} from aqueous solution. Results Mater. **21**, 10052 (2024)

53. Bhagiyalakshmi, M., Yun, L.J., Anuradha, R., Jang, H.T.: Utilization of rice husk ash as silica source for the synthesis of mesoporous silicas and their application to CO_2 adsorption through TREN/TEPA grafting. J. Hazard. Mater. **175**(1–3), 928–938 (2010)
54. Chiarakorn, S., Areerob, T., Grisdanurak, N.: Influence of functional silanes on hydrophobicity of MCM-41 synthesized from rice husk. Sci. Technol. Adv. Mater. **8**(1–2), 110 (2007)
55. Artkla, S., Kim, W., Choi, W., Wittayakun, J.: Highly enhanced photocatalytic degradation of tetramethylammonium on the hybrid catalyst of titania and MCM-41 obtained from rice husk silica. Appl. Catal. B. **91**(1–2), 157–164 (2009)
56. Bhagiyalakshmi, M., Yun, L.J., Anuradha, R., Jang, H.T.: Synthesis of chloropropylamine grafted mesoporous MCM-41, MCM-48 and SBA-15 from rice husk ash: their application to CO_2 chemisorption. J. Porous Mater. **17**, 475–484 (2010)
57. Suyanta, S., Kuncaka, A.: Utilization of rice husk as raw material in synthesis of mesoporous silicates MCM-41. Indones. J. Chem. **11**(3), 279–284 (2011)
58. Appaturi, J.N., Adam, F.: A facile and efficient synthesis of styrene carbonate via cycloaddition of CO_2 to styrene oxide over ordered mesoporous MCM-41-Imi/Br catalyst. Appl. Catal. B. **136**, 150–159 (2013)
59. Renuka, N.K., Praveen, A.K., Anas, K.: Influence of CTAB molar ratio in tuning the texture of rice husk silica into MCM-41 and SBA-16. Mater. Lett. **109**, 70–73 (2013)
60. Ahmad-Alyosef, H., Uhlig, H., Münster, T., Kloess, G., Einicke, W.D., Gläser, R., Enke, D.: Biogenic silica from rice husk ash–Sustainable sources for the synthesis of value-added silica. Chem. Eng. **37**, 667–672 (2014)
61. Areerob, T., Grisdanurak, N., Chiarakorn, S.: Utilization of rice husk silica as adsorbent for BTEX passive air sampler under high humidity condition. Environ. Sci. Pollut. Res.Pollut. Res. **23**, 5538–5548 (2016)
62. Nguyen, N.T., Chen, S.S., Nguyen, N.C., Nguyen, H.T., Tsai, H.H., Chang, C.T.: Adsorption of methyl blue on mesoporous materials using rice husk ash as silica source. J. Nanosci. Nanotechnol.Nanosci. Nanotechnol. **16**(4), 4108–4114 (2016)
63. Costa, J.A.S., Sarmento, V.H., Romão, L.P., Paranhos, C.M.: Adsorption of organic compounds on mesoporous material from rice husk ash (RHA). Biomass Convers. Biorefin. **10**, 1105–1120 (2020)
64. Purnawira, B., Purwaningsih, H., Ervianto, Y., Pratiwi, V.M., Susanti, D., Rochiem, R., Purniawan, A.: Synthesis and characterization of mesoporous silica nanoparticles (MSNp) MCM 41 from natural waste rice husk. IOP Conf. Ser.: Mater. Sci. Eng. **541**(1), 012018 (2019). IOP Publishing
65. Kamari, S., Ghorbani, F.: Extraction of highly pure silica from rice husk as an agricultural by-product and its application in the production of magnetic mesoporous silica MCM–41. Biomass Convers. Biorefin. **11**, 3001–3009 (2021)
66. Jang, H.T., Park, Y., Ko, Y.S., Lee, J.Y., Margandan, B.: Highly siliceous MCM-48 from rice husk ash for CO_2 adsorption. Int. J. Greenhouse Gas Control **3**(5), 545–549 (2009)
67. Zhao, D., Feng, J., Huo, Q., Melosh, N., Fredrickson, G.H., Chmelka, B.F., Stucky, G.D.: Triblock copolymer syntheses of mesoporous silica with periodic 50 to 300 angstrom pores. Science **279**(5350), 548–552 (1998)
68. Henao, W., Jaramillo, L.Y., Lopez, D., Romero-Sáez, M., Buitrago-Sierra, R.: Insights into the CO_2 capture over amine-functionalized mesoporous silica adsorbents derived from rice husk ash. J. Environ. Chem. Eng. **8**(5), 104362 (2020)
69. Chareonpanich, M., Nanta-Ngern, A., Limtrakul, J.: Short-period synthesis of ordered mesoporous silica SBA-15 using ultrasonic technique. Mater. Lett. **61**(29), 5153–5156 (2007)
70. Ho, S.T., Dinh, Q.K., Tran, T.H., Nguyen, H.P., Nguyen, T.D.: One-step synthesis of ordered Sn-substituted SBA-16 mesoporous materials using prepared silica source of rice husk and their selectively catalytic activity. Can. J. Chem. Eng. **91**(1), 34–46 (2013)
71. Chun, J., Gu, Y.M., Hwang, J., Oh, K.K., Lee, J.H.: Synthesis of ordered mesoporous silica with various pore structures using high-purity silica extracted from rice husk. J. Ind. Eng. Chem. **81**, 135–143 (2020)

72. Henao, W., Jaramillo, L.Y., López, D., Romero-Sáez, M., Buitrago-Sierra, R.: Insights into the CO_2 capture over amine-functionalized mesoporous silica adsorbents derived from rice husk ash. J. Environ. Chem. Eng. **8**(5), 104362 (2020)
73. Bhagiyalakshmi, M., Do Park, S., Cha, W.S., Jang, H.T.: Development of TREN dendrimers over mesoporous SBA-15 for CO_2 adsorption. Appl. Surf. Sci. **256**(22), 6660–6666 (2010)
74. Wittayakun, J., Khemthong, P., Prayoonpokarach, S.: Synthesis and characterization of zeolite NaY from rice husk silica. Korean J. Chem. Eng. **25**, 861–864 (2008)
75. Wang, H.P., Lin, K.S., Huang, Y.J., Li, M.C., Tsaur, L.: Synthesis of zeolite ZSM-48 from rice husk ash. J. Hazard. Mater. **58**(1–3), 147–152 (1998)
76. Wattanasiriwech, S., Wattanasiriwech, D., Svasti, J.: Production of amorphous silica nanoparticles from rice straw with microbial hydrolysis pretreatment. J. Non-Cryst. SolidsCryst. Solids **356**(25–27), 1228–1232 (2010)
77. Estevez, M., Vargas, S., Castano, V.M., Rodriguez, R.: Silica nano-particles produced by worms through a bio-digestion process of rice husk. J. Non-Cryst. SolidsCryst. Solids **355**(14–15), 844–850 (2009)
78. Bansal, V., Ahmad, A., Sastry, M.: Fungus-mediated biotransformation of amorphous silica in rice husk to nanocrystalline silica. J. Am. Chem. Soc. **128**(43), 14059–14066 (2006)
79. Zaki, A.G., Yousef, S.A., Hasanien, Y.A.: Bioharvesting and improvement of nano-silica yield from bagasse by irradiated Curvularia spicifera. BMC Microbiol. **25**(66) (2025)
80. Pimprom, S., Sriboonkham, K., Dittanet, P., Föttinger, K., Rupprechter, G., Kongkachuichay, P.: Synthesis of copper–nickel/SBA-15 from rice husk ash catalyst for dimethyl carbonate production from methanol and carbon dioxide. J. Ind. Eng. Chem. **31**, 156–166 (2015)
81. Barani, S., Sebastian, S.P., Dhevagi, P., Prasanthrajan, M., Suganthi, A.: Synthesis of silica nanoparticles (SiNPs) from agro-wastes for removal of heavy metals from an aqueous medium—a mini review. Green Chem. Lett. Rev. **17**(1), 2422416 (2024)
82. Prabha, S., Durgalakshmi, D., Rajendran, S., Lichtfouse, E.: Plant-derived silica nanoparticles and composites for biosensors, bioimaging, drug delivery, and supercapacitors: a review. Environ. Chem. Lett. **19**, 1667–1691 (2021)
83. Rodriguez-Otero, A., Galarneau, A., Drané, M., Vargas, V., Sebastian, V., Wilson, A., Grégoire, D., Radji, S., Marias, F., Christensen, J.H., Bouyssiereet, B.: Towards achieving circular economy in the production of silica from rice husk as a sustainable adsorbent. Processes **12**, 2420 (2024)
84. Errington, E., Guo, M., Heng, J.Y.Y.: Synthetic amorphous silica: environmental impacts of current industry and the benefit of biomass-derived silica. Green Chem. **25**, 4244 (2023)
85. Hidayat, T., Setyawan, N., Hoerudin: Semi pilot scale study on production of silica from rice husk ash. IOP Conf. Ser. Earth Environ. Sci. **1172**, 012049 (2023)
86. Park, J.Y., Gu, Y.M., Park, S.Y., Hwang, E.T., Sang, B.I., Chun, J., Lee, J.H.: Two-stage continuous process for the extraction of silica from rice husk using attrition ball milling and alkaline leaching methods. Sustainability **13**, 7350 (2021)
87. Park, J.Y., Gu, Y.M., Chun, J., Sang, B.I., Lee, J.H.: Pilot-scale continuous biogenic silica extraction from rice husk by one-pot alkali hydrothermal treatment and ball milling. Chem. Biol. Technol. Agric. **10**(102) (2023)

Chapter 11
Green Chemical Approaches for Sustainable Organic Synthesis

Dhruvi S. Chaudhari, Kushan Parikh, and Shanta Raj Lakshmi

Abstract The use of green chemistry techniques in the synthesis of organic compounds has generally drawn the attention of numerous researchers recently. For synthetic chemists, using green synthetic procedures and methodologies is crucial since they tend to use less hazardous and environmentally harmful materials or generate less of them. The twelve principles of green chemistry serve as a guide for chemists as they create safe chemical procedures and methods. The traditional method of synthesizing frequently involves the use of hazardous or poisonous organic solvents, catalysts, or other substances that are declared trash when the reaction is finished. New green chemistry synthesis techniques were created to lessen the use of such hazardous materials. One of these techniques is the use of microwaves to speed up the synthesis of desired molecules, produce higher-quality products with higher yields, and work in softer environments. Organic synthesis frequently uses solvent-free synthesis because it is simpler, uses fewer hazardous materials (organic solvents), and is less expensive. Solvent-free synthesis is commonly carried out in conjunction with heating, mixing, and ball milling, as well as using ultrasound and microwaves.

Keyword Green Chemistry · Sustainable synthesis · Eco-friendly reagents · Microwave-assisted synthesis · Sono-chemistry

11.1 Introduction

The definitions of "organic" and "synthesis" have changed considerably throughout time. The term "organic chemistry" came from organic or organized bodies, or living things, as opposed to non-organic bodies, which were disorganized. We are greatly impacted by organic chemistry in our daily lives. During the 1980s and

---

D. S. Chaudhari · K. Parikh · S. R. Lakshmi (✉)
Department of Chemical Science, Faculty of Applied Sciences, Parul University, Vadodara, India
e-mail: rajlakshmi.shanta0@gmail.com

P. Jain et al. (eds.), *Towards Green Chemical Processes: Strategies and Innovations*,
Intelligent Systems Reference Library 277,
https://doi.org/10.1007/978-3-031-94090-3_11

1990s, there was an increase in environmentally sensitive terminology in the chemical industry. Since Anastas and Warner's contributions in 1998, the usage of green chemistry has grown fast and linearly [1]. The basic definition of "green chemistry" is "the design of chemical products and processes to reduce or eliminate the use and generation of hazardous substances"[2]. Green chemistry is a concept and style of thinking that ought to be used in the creation, production, and use of chemical goods. Since the majority of chemicals are not benign, creating a completely "environmentally-friendly" process is nearly difficult. This is by no means a simple undertaking. Consequently, twelve guiding principles were created to help chemists create more sustainable chemical processes in order to lessen the environmental effect that chemical processes produce, as opposed to only aiming for the ideal ecologically benign synthesis [3]. Since the late twentieth century, when Rachel Carson's book Silent Spring brought attention to the harmful consequences of pesticide releases into the environment, there has been a growing awareness of the issues of environmental conservation and sustainable development [4]. In industrial organic synthesis, it became evident that a paradigm shift was required from the conventional ideas of reaction efficiency and selectivity, which primarily concentrate on chemical yield, to one that places importance on the use of raw materials, waste reduction, and avoiding the use of toxic and/or hazardous substances. The 2030 Agenda for Sustainable Development, which was established by the UN, outlines 17 Sustainable Development Goals, reflecting the shift in focus towards the creation of a more sustainable future more than 50 years later. Specifically, Goal 9 calls for the following: "Create Inclusive and Sustainable Industrialization, Build Resilient Infrastructure, and Encourage Innovation" [5]. Sustainability should be defined as an inherent characteristic of a material, chemical, process, reaction, or technology. A different definition of sustainability was recently put forth, one that is unaffected by social or economic factors: waste cannot be produced more quickly than it can be cleaned up, and resources, including energy, should be used at a rate that allows for their natural replacement [6]. The first two principles, sometimes known as system conditions, of the four sustainability principles are comparable to this concept [7]. According to Table 11.1 Green chemistry focuses on synthesis processes that adhere to its twelve main principles, minimizing or eliminating adverse effects on the environment and human health [8] (Fig. 11.1).

The following are some specific benefits of using greener chemical processes:

- Preventing needless waste.
- most solvents, catalysts, and reagents may be recycled, which helps to prevent needless waste in organic synthesis.
- Minimize the loss of precursors and intermediate compounds during the synthesis of the final product.
- Utilize less hazardous chemical reactions that employ safe and low-toxicity substances.
- Use safe and low-toxicity separation agents and solvents, such as water and natural compounds like plant extracts.

Table 11.1 The Effect of green chemistry to improving economic efficiency, environmental preservation, and health

Area	Benefits
Human health	Improved environmental conditions lead to a reduction in the discharge of toxic chemicals, resulting in fewer respiratory health issues. Moreover, enhanced water quality is achieved by minimizing the release of these harmful substances, providing cleaner water for drinking and recreational purposes. Additionally, food safety is bolstered by eliminating persistent toxic chemicals that can contaminate the food chain. Applying pesticides that specifically target certain pests and degrade rapidly after use further contributes to safer food production Enhanced safety in consumer products will pave the way for introducing innovative and safer options in the marketplace. For example, certain pharmaceuticals will be produced with less waste, while cleaning solutions and pesticides will offer safer alternatives to current choices Increased safety for chemical industry employees, less use of dangerous materials, fewer protective outfits needed, and reduced risk of incidents such as explosion and fire
Environment	Fewer poisonous chemicals are in the environment, reducing exposure to plants and animals and reliance on landfills, particularly hazardous waste landfills This reduces climate change risk, smog development, and ozone layer depletion Chemical substances enter the environment through deliberate application (pesticides), unintended emissions, or disposal Green chemicals are designed to be transformed into more environmentally friendly products or recycled for further use
Economy and business	Enhanced efficiency of chemical reactions results in the consumption of less feedstock to yield the same quantity of product Waste generation is reduced, along with the need for hazardous waste disposal, which helps avoid costly remediation and end-of-pipe treatments More efficient processes lead to decreased energy and water usage, increased plant capacity, and faster manufacturing times The replacement of commercial feedstock with waste is facilitated, improving efficiency and requiring less product to achieve the same performance Improved marketability and sales are achieved by promoting a safer product label There is increased competition within the chemical industry and among customers A minimized reliance on petroleum products reduces consumption, helping to avoid risks and fluctuations in costs

- Reduce energy consumption by adhering to ideal conditions, such as room temperature and normal pressure.
- Employ renewable resources as raw materials.
- Minimize the number of reagents to avoid unnecessary reactions, thereby reducing waste from excess chemicals and response steps.
- Utilize highly selective catalysts to achieve higher yields of reaction products.
- Ensure that reaction products are degradable and do not persist in the environment.
- Prevent contamination through continuous (real-time) monitoring and analysis.

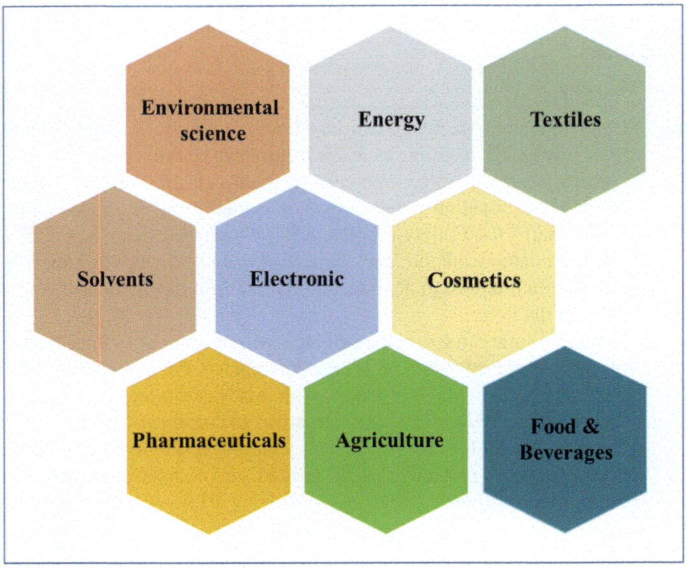

Fig. 11.1 Representative image of usage of green chemistry in a range of domains

11.2 Green Chemical Metrics

11.2.1 E-Factor

Numerous methods have been developed for quantifying chemical waste to enhance waste prevention and reduction efforts. One of the oldest parameters used to assess a process's waste is the environmental factor, often avoid risks and cost fluctuations to as the E-factor or mass efficiency. This metric is defined as the ratio of the mass (kg) of waste generated during a process to the mass (kg) of the desired product produced.

$$E - Factor = \frac{Mass\ of\ Waste\ Generated}{Mass\ of\ Desired\ Product}$$

Any substance produced during the process that is not the desired product, such as inorganic salts, is classified as waste. Typically, the greater the value generated, the more waste is produced. Conversely, lower E-factor values indicate that the process generates fewer undesired by-products, making an ideal value of zero preferable [9].

11.2.2 Atom Economy

Synthetic methods should be designed to maximize the incorporation of all materials used in the process into the final product. The second principle of green chemistry evaluates various metrics that measure the effective incorporation of materials into the final product, aiming to reduce waste and optimize processes. These metrics assess the overall performance of synthetic processes, paving the way for more sustainable and eco-friendly practices. To gain global acceptance, these sustainability metrics should be straightforward to quantify. The concept of atom economy has been developed in recent years as an advancement over traditional metrics such as product yield, which is defined as the ratio of the molecular weight of the final product to the total molecular weight of all reactants [10].

$$Atom\,Economy = \frac{Mol.\,Wt.\,of\;the\,final\,product}{Total\,Mol.\,Wt.\,of\;all\;reagents} \times 100$$

11.3 Green Solvents

The produced environmentally friendly solvents that are a substitute for petrochemical solvents are known as "green solvents" (Fig. 11.2). In addition to the inherent properties of the solvent, green chemistry's initial emphasis on finding purportedly "green" reaction solvents was gradually replaced by a more technology/application-driven approach that acknowledges the significance of the solvent's sourcing, use, management, and disposal [11].

11.3.1 Water as a Green Solvent

Since solvents make up a large portion of the overall mass handled in a chemical process, their selection is crucial. Water is the greenest solvent with the least influence on the environment out of all the potential solvents. In the beginning, the scientific community was hesitant to employ water as a medium in organic processes because of the inherent insolubility of organic compounds in water [12]. However, as water is the most environmentally friendly solvent available and the preferred solvent in biological systems, natures need for cleaner chemistry forces us to take it into account when using it as a solvent in chemical reactions. Furthermore, water is the most readily available liquid and has excellent thermal stabilities and characteristics like non-flammability, moderate viscosity, pH fluctuation, etc. [13]. Reactions occurring in water create a homogeneous mixture, while reactions that happen in water result in

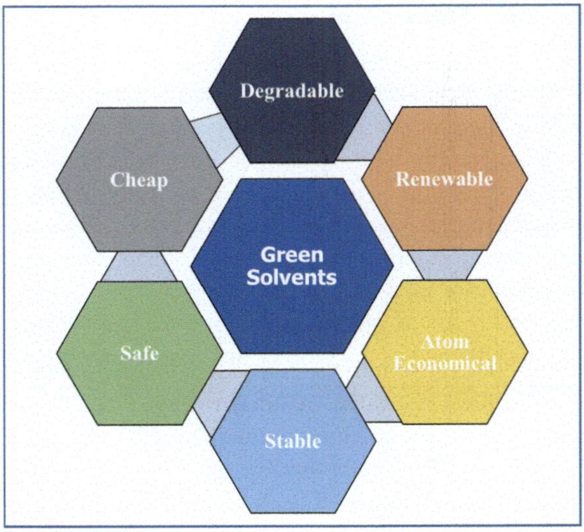

Fig. 11.2 Representing the important characteristics of green solvents

a heterogeneous mixture [14]. The chemistry of reactions involving water has historically been poorly understood until Breslow and his colleagues revealed that water can significantly enhance both the speed and selectivity of the Diels–Alder reaction (Fig. 11.3). This discovery challenged the prevailing notion that a high solubility of organic reactants in a solvent is essential for obtaining better yields [15]. These days, the on-water method is a widely utilized technique for a variety of catalysed and non-catalysed organic processes, particularly the aldol reaction, Claisen rearrangement, allylation, olefin functionalization, multicomponent reactions, cross-coupling, C–H functionalization, etc. [16].

The term "catalyst-free reactions" refers to reactions that occur without a catalyst while maintaining their speed and yield. They can occasionally overcome catalytic processes in speed. Through hydrogen bonding (Fig. 11.4), water stabilizes the transition state more effectively than the reactants, which improves the process [18].

A synthetic protocol for forming various pyran-based heterocyclic compounds utilizing UV365 to initiate eco-friendly multicomponent reactions (Fig. 11.5). This method entails irradiating a mixture of C–H activated cyclic ketones (such as pyrazolone or 1,3-diketone), aromatic aldehydes or isatins, and malononitrile with ultraviolet light (i.e., UV365) in a water–ethanol (1:1) solvent system [19].

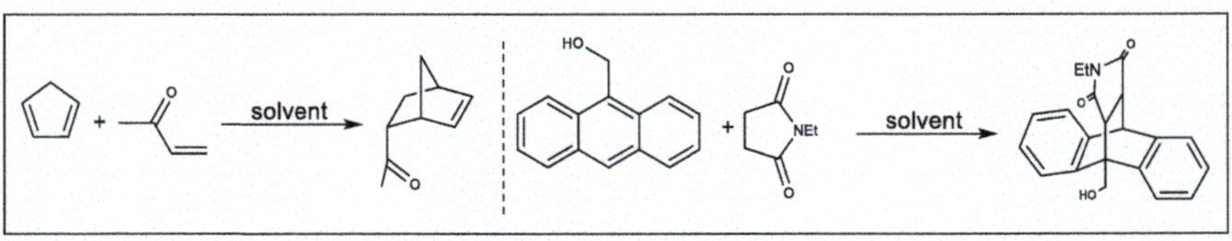

Fig. 11.3 Diels–Alder reactions accelerated by "on water" conditions (Breslow) [17]

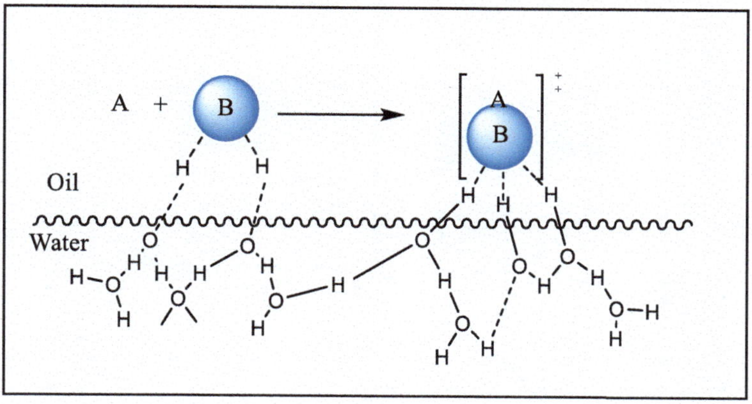

Fig. 11.4 Stabilisation of transition state via hydrogen bonding interactions

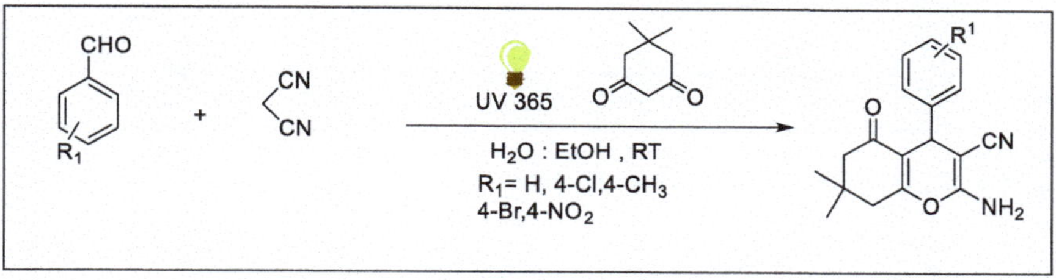

Fig. 11.5 Synthesis of pyran-based heterocyclic scaffolds

11.3.2 Supercritical Fluids (SCFs)

Substances in noncondensing, single-phase fluids that are beyond their critical temperature (Tc) and critical pressure (Pc) are referred to as "supercritical fluids" [20]. The maximum temperature and pressure at which SCF may coexist in equilibrium as a vapour and a liquid are Tc and P_C (Fig. 11.6). SCF simultaneously displays both liquid and gas characteristics. Because SCF has a high solvent power, it can be used as an extracting agent [21]. CO_2, C_2H_6, NH_3, H_2O, and other gases are frequently utilized as SCFs. Due to their remarkable versatility, supercritical fluids (SCFs) have been used in a wide variety of new technologies and are still being used today. The creation of eco-friendly, sustainable, safer, and inventive solutions for a range of geological, industrial, and environmental applications [22].

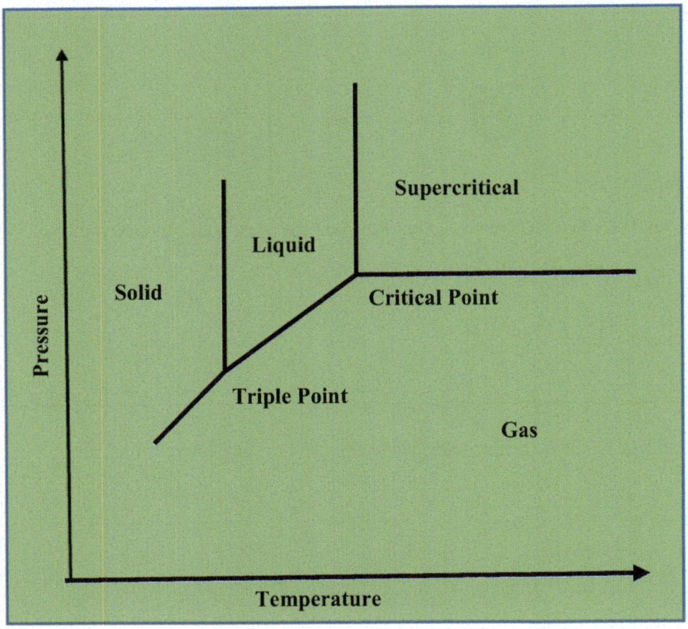

Fig. 11.6 Supercritical state for a pure component

11.3.3 Supercritical Carbon Dioxide

Carbon dioxide (CO_2) 's critical properties, such as its critical temperature (T_c) of 31 °C, critical pressure (Pc) of 7.38 MPa, and $\rangle = 0.468\ g/cm^3$, are known to be relatively low. Supercritical carbon dioxide ($scCO_2$) is a readily available, affordable, non-toxic, non-explosive, and non-flammable energy source and is most frequently utilized as a reaction medium, an organic reaction catalyst, an organic synthesis solvent, and for the extraction of natural raw materials from plants and animals [23]. Applications for supercritical carbon dioxide as a processing solvent include polymerization, particle generation, microcellular foaming, polymer modification, and the creation of polymer composites [24].

$$R\text{-}NH_2 + CO_2 \circledast RNHCOOH \oplus RNH_3 + RNHCOO\text{-}RNH_2 + 2H_2C(O)CH_2RN$$

$$(CH_2CH_2OH)_2 \circledast$$

$$H_2C\text{-}O\text{-}CH_2 + CO^{2-} \circledast [COO\text{-}CR\text{-}C\text{-}O] \text{ (polycarbonate)}$$

Catalysis, healthcare, energy, and environmental disciplines can all benefit from the production, stimulation, and characterization of inorganic nanomaterials using

$scCO_2$-assisted technology. When compared to their bulk equivalents, these synthesized $scCO_2$-assisted nanomaterials often showed improved physicochemical characteristics (such as specific area, particle size, amorphous, and distribution), which usually resulted in better performance in their respective domains. The metal oxide was prepared as the reactants and reaction precursors using SAS-based techniques. Furthermore, despite extensive research into the special benefits of $scCO_2$-based procedures, they are far less common in producing inorganic nanomaterials than other traditional techniques, including chemical precipitation, micro-emulsion, and hydrothermal processes. Due to its widespread usage in chemical processes, $scCO_2$ is a compelling alternative solvent for green chemistry and will likely remain so in the future. The next ten years may see the development of more efficient, cost-effective, and environmentally friendly chemical manufacturing methods thanks to chemical reactions in supercritical fluids, which may continue to be a developing field of study and technology.

11.4 Catalysis as a Green Strategy

A process known as catalysis occurs when a material, referred to as a catalyst, speeds up a chemical reaction without being consumed by the reaction. By offering a different reaction route with a lower activation energy, the catalyst transforms reactants into products. Catalysis boosts reaction selectivity, improves overall efficiency, and permits reactions to take place under more benign circumstances. A key component of green chemistry is catalysis, which enables the development of more ecologically friendly chemical processes, Numerous catalysts may be recovered and used again with little loss of activity or selectivity, This makes catalytic processes more economically viable, reduces waste, and requires fewer catalysts [25].

11.4.1 Homogeneous Catalysis

Recent advancements in homogeneous catalysis (Fig. 11.7) provide exciting opportunities for organic synthesis, allowing for precise control over chemo-, regio-, and enantioselectivities in reactions. In addition to enabling the processes to take place in relatively moderate environments, homogeneous catalysis improves our knowledge of molecular-level reaction mechanisms, opening up amazing possibilities for catalytic process enhancement [26].

When EtONa was combined with [Ir(cod)(acac)] (Ir8), where acac refers to acetylacetonate and cod denotes 1,5-cyclooctadiene, along with the ligand 1,3-bis-(diphenylphosphino)propane (dppp) and 1,7-octadiene as an additive, n-butanol was produced with a selectivity of up to 61%, though the conversion was reduced to 12% (Fig. 11.8) [27].

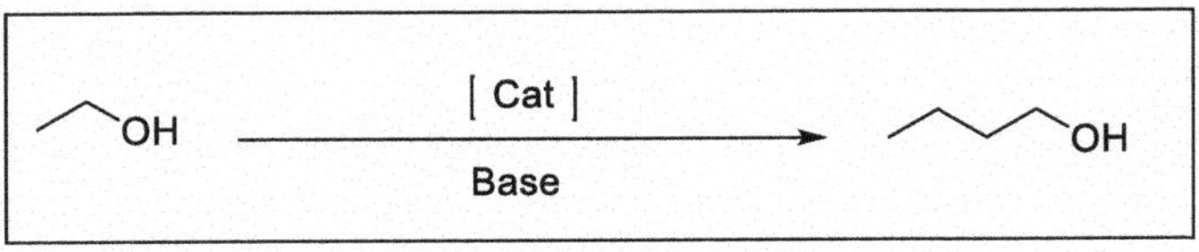

Fig. 11.7 Guerbet reaction catalyzed by homogeneous catalysts

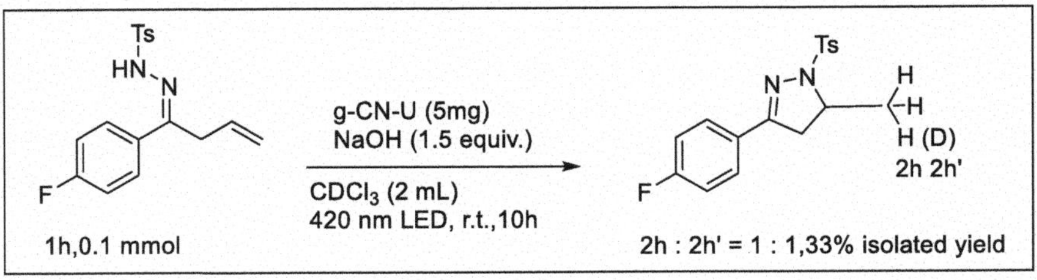

Fig. 11.8 Reaction conducted in the deuterated solvent

The reaction was performed in $CDCl_3$, producing a 1:1 mixture of 2 h and 2 h', with a total yield of 33%, which indicates that the partial hydrogen source came from $CHCl_3$ [28].

Ligand-free catalysis utilizing copper salts presents notable challenges, primarily due to the risk of reductive deposition of copper at the cathode (Fig. 11.9). The use of ligands in copper catalysis is often favored to address this issue. In electrochemical homogeneous Chan–Lam coupling reactions involving amines (5–1) and aryl boronic acids (5–2), $Cu(OAc)_2$ serves as the catalyst, with Fe^+ acting as the anodic mediator. This approach has yielded moderate to excellent results in the synthesis of anilines [29].

Electrochemical aryl C–H oxygenation utilizing Cp*Rh + catalysis in an undivided cell at room temperature showed that the anodic oxidation of a rhodacyclic carboxylate intermediate promoted reductive elimination (Fig. 11.10) [30].

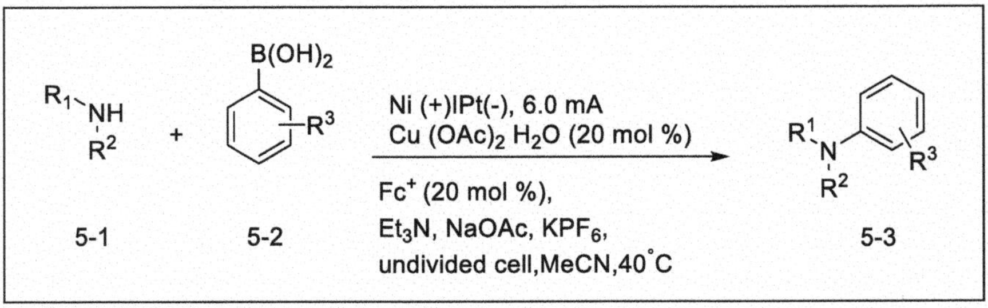

Fig. 11.9 Ligand-free Cu-catalyzed electrochemical homogeneous Chan–Lam coupling reactions

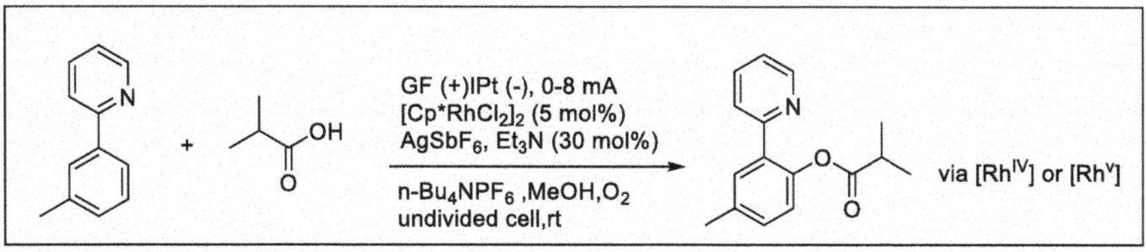

Fig. 11.10 Stabilization of Rh catalysts by Cp* ligands in electrochemical reactions

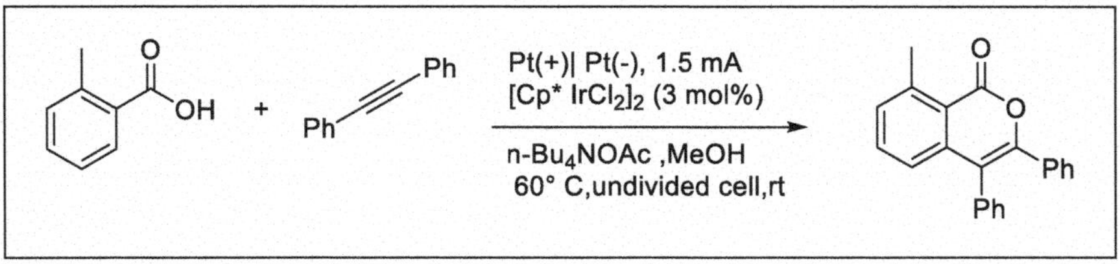

Fig. 11.11 Electrochemical reactions catalyzed by Cp*-stabilized cationic Ir

The reduction of iridium cations (Ir^{3+}) to iridium metal (Ir^0) is more favorable than the reduction of protons in aqueous solutions (Fig. 11.11), with a reduction potential (E^{red}) that ranges from 0.77 to 1.16 V versus the normal hydrogen electrode (NHE). In organic solvents, however, the coordination of iridium species with suitable compounds can stabilize the metal against cathodic reduction. This stabilization facilitates the regioselective annulation of benzoic acids with internal alkynes [31].

11.4.2 Heterogeneous Catalysis

In heterogeneous catalysis, the catalyst and reactants are in separate phases. Typical instances include solid catalysts utilized in liquid-phase or gas-phase processes. Heterogeneous catalysis usually works by surface reactions, in which the reactant's molecules adsorb onto the catalyst's surface, change chemically, and then desorb as products. Compared to homogeneous catalysis, heterogeneous catalysis may have lesser catalytic activity, but it does have benefits such easier catalyst recovery and reuse. In the field of heterogeneous catalysis, immobilization of nanoparticles (NPs) on solid supports—such as porous silica [32], zeolites [33], polymers [34], and carbon-based supports—such as carbon nanotubes (CNTs) [35], graphene [36], graphene oxide (GO) [37], and graphene nano-sheets (GNSs) is a technique that is well suited to the principles of green chemistry and has gained popularity recently [38]. From the perspective of catalysis, these materials are also noteworthy for their porosity, limited interaction with supporting materials, and strong mechanical resistance [39]. One advantage of these mixed catalytic systems is that they can be readily separated from the reaction media, which makes them stand out in the quest for more environmentally friendly synthetic processes [40].

A model reaction involving benzaldehyde, dimedone, ammonium acetate, and ethyl acetoacetate was selected to evaluate the effectiveness of Cu@IRMOF-3 as a porous catalyst (Fig. 11.12) [41].

A novel in-situ doping and carbonization method has been developed to create highly porous nitrogen and phosphorus co-doped carbon materials embedded with cobalt nanoparticles (Co@NCP) through the thermolysis of bimetallic zeolitic imidazolate frameworks (BMZIFs). Ammonium phosphate nanospheres act as sacrificial templates for the porous structures and provide phosphorus. The resulting Co@NCP demonstrates excellent performance in the base-free hydrogen auto-transfer synthesis of quinoxalines from diols and o-nitroanilines (Fig. 11.13) [42].

The LnNi-PHI catalysts, containing up to 6 wt % site-isolated bipyridyl-Ni species, present several advantages and exhibit remarkable metallaphotocatalytic activity, chemical stability, and recyclability (Fig. 11.14). These catalysts are highly effective in a variety of C–P, C–S, C–O, and C–N cross-coupling reactions, demonstrating a wide substrate scope and a strong tolerance for diverse functional groups [43].

Heterogeneous photocatalysis can selectively oxidize toluene to benzaldehyde or benzoic acid under mild conditions (Fig. 11.15) [44].

Primary amines and diamines can react with various 1,3-dicarbonyl compounds to produce β-enaminones, β-enaminoesters, bis(β-enaminones), and bis(β-enaminoesters) with excellent yields and in a relatively short time (Fig. 11.16) [45].

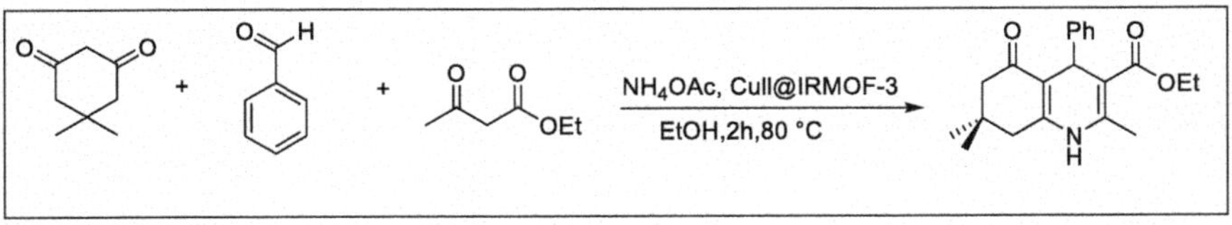

Fig. 11.12 The model reaction for preparation of polyhydroquinolines

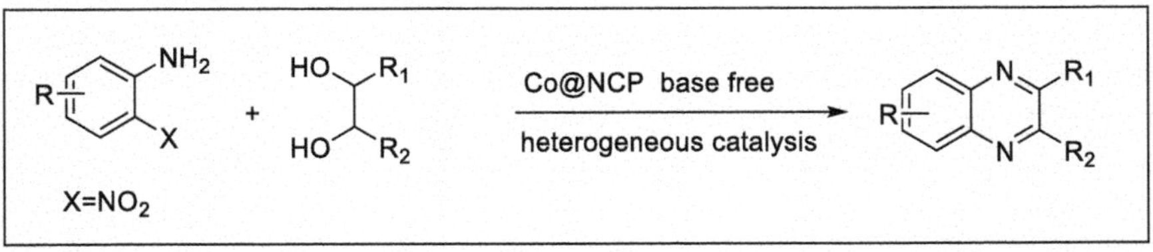

Fig. 11.13 Heterogeneous catalysis as a pathway for the synthesis of quinoxalines

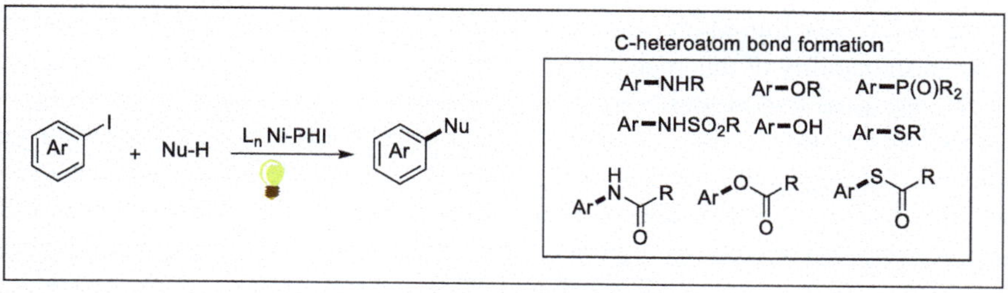

Fig. 11.14 Approach for heteroatom ligand exchange and its applications

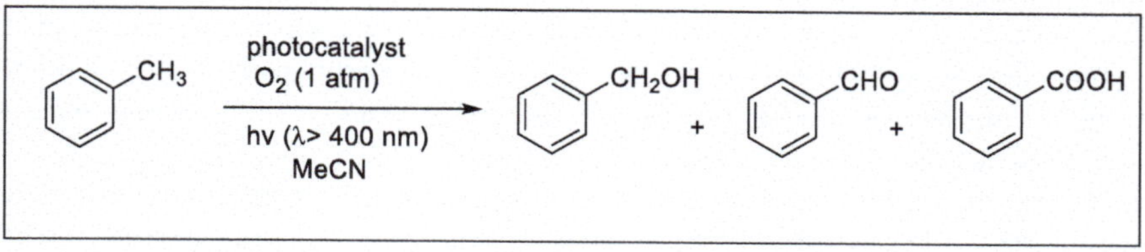

Fig. 11.15 Typical reaction performance in toluene oxidation via p-BWO photocatalyst

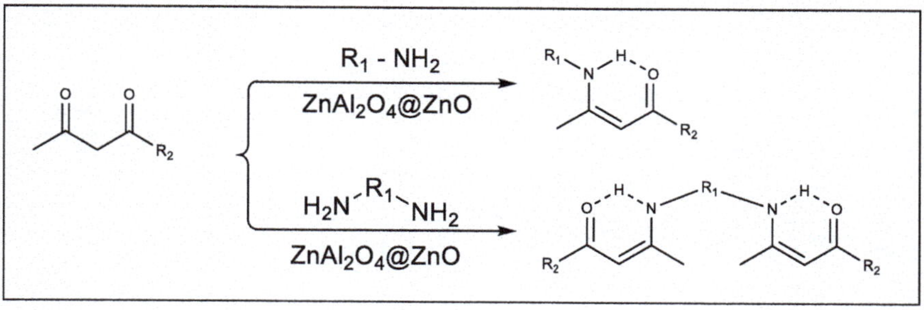

Fig. 11.16 General scheme for the synthesis of β-enaminones, β-enaminoesters, bis(β-enaminones), and bis(β-enaminoesters)

11.4.3 Bio-catalysis

Utilizing microbes and enzymes as catalysts for organic synthesis processes, or bio-catalysis, is a practical method for producing a wide range of compounds [46]. Wild-type enzymes have been employed on an industrial basis for many years [47]. Larger and more reproducible enzymes may now be created because to more recent advancements in molecular biology methods Furthermore, biocatalysts are now being used on an industrial scale due to the potential to customize them to certain reactions [48]. Large amounts of wastewater must be handled as a result of biocatalysis in aqueous conditions [49]. The creation of a simple and clean method to extract the glycerol by-product, the synthesized biodiesel, and the complete recovery of the entire

Sponge-Like Ionic Liquid/biocatalytic system for additional usage served as evidence of the sustainability of this biocatalytic reaction system [50]. At room temperature, the Sponge-Like Ionic Liquids are solid and immiscible with water; thus, adding a little amount of water causes the Sponge-Like Ionic Liquids to precipitate, creating a semisolid heterogeneous mixture. A high level of biocatalytic activity was achieved with this method. The Sponge-Like Ionic Liquids /biocatalyst system showed full recovery and was effectively repurposed; over the course of 12 operating cycles, the activity remained essentially unaltered. This outcome shows that the procedure might be easily scaled up. When using waste cooking oils, even "worsened" by the addition of FFAs (up to 30% w/w), and/or methanol as nucleophile acceptors in the transesterification/esterification reactions in 6 h at 60 °C, the exceptional suitability of this biocatalytic sustainable approach based on SLILs was also demonstrated [51].

The enzyme cytochrome KtnC is expressed in Pichia pastoris to facilitate the homocoupling of its native substrate and cross-coupling with various phenol derivatives, leading to the formation of non-natural biaryl products (Fig. 11.17) [52].

The enzymatic kinetic resolution of the atropisomeric intermediate during the asymmetric synthesis of JNJ-4355, which is an inhibitor of a protein associated with myeloid leukaemia (Fig. 11.18) [53].

The initial synthesis route for Molnupiravir began with uridine and involved the use of organic solvents, various derivatizations, and protective steps, resulting in a low final yield (Fig. 11.19). However, by rearranging the reaction steps in what is referred

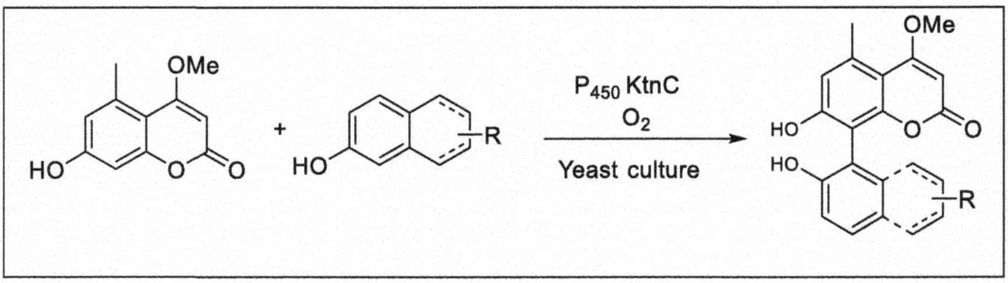

Fig. 11.17 Cytochrome P450 enzyme-mediated atroposelective oxidative homocoupling

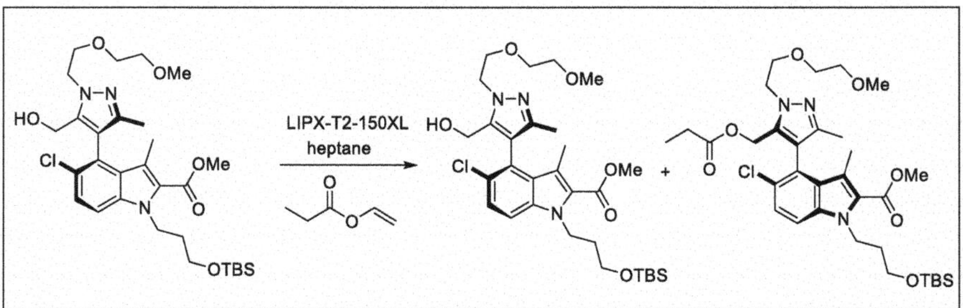

Fig. 11.18 Enantioenrichment of synthetic intermediate

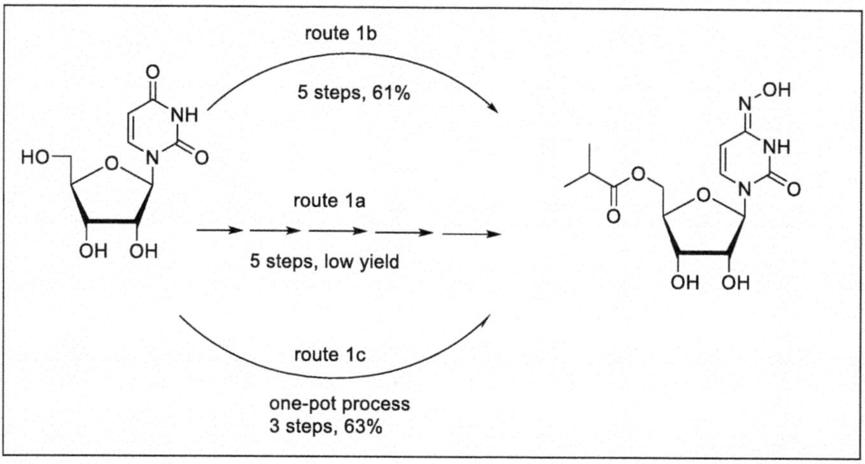

Fig. 11.19 Chemical routes to Molnupiravir synthesis

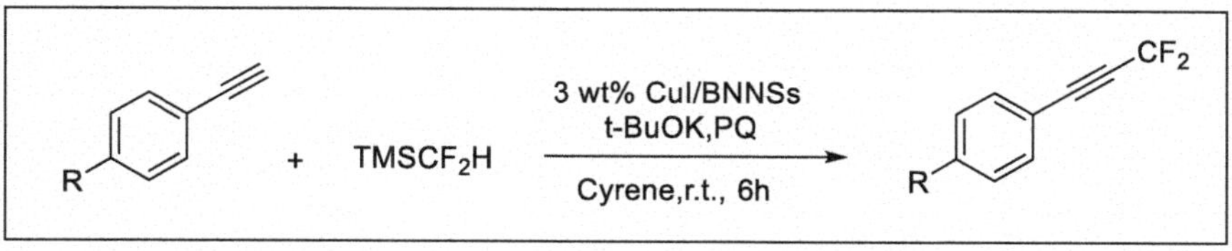

Fig. 11.20 Terminal alkynes using Cyrene as the solvent

to as route 1b, Steiner and colleagues enhanced the previously mentioned synthetic protocol. This revised approach included a continuous flow procedure for the final acetonide deprotection, which improved both the selectivity and reproducibility of the process [54].

CuI nanoparticles acted as a heterogeneous catalyst for reactions involving terminal alkynes. They began their investigation with benzothiazole and $TMSCF_2H$, using t-BuOK as the base and Cyrene as the solvent (Fig. 11.20) [55].

Screening was conducted for non-natural activity using a small set of three fungal laccases and two bacterial laccases (Fig. 11.21). Among the fungal blue-copper enzymes, the laccases from Myceliophthora thermophila and Moniliophthora roreri did not exhibit significant nitroso-ene reactivity [56].

11.4.4 Application in Eco-friendly Evolution

The impact of various conventional and renewable solvents on biomass conversion processes—such as pre-treatment, hydrolysis, and dehydration reactions—and

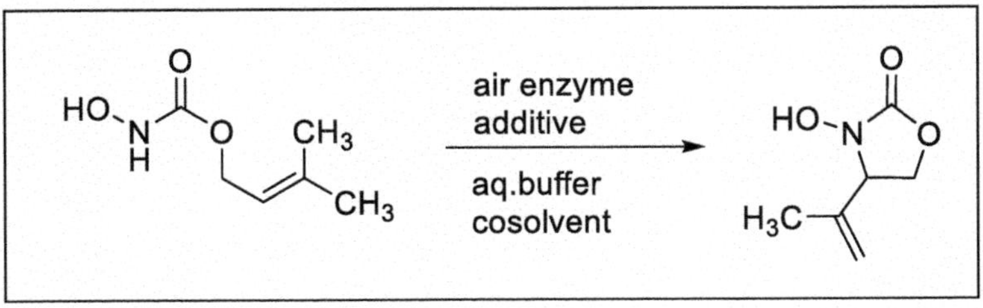

Fig. 11.21 Enzyme screening for the oxidative cyclization of N-hydroxycarbamate

the extraction of products or intermediates has been investigated. For instance, methyl tetrahydrofuran (MTHF), produced from biomass-derived furfural (FUR), shows promise as a viable alternative to tetrahydrofuran (THF) for industrial solvent applications (Fig. 11.22) [57].

The combined effects of TEAB and DMC as cosolvents enabled the complete dissolution of D-fructose and facilitated the extraction of HMF from the reaction mixture (Fig. 11.23) [58].

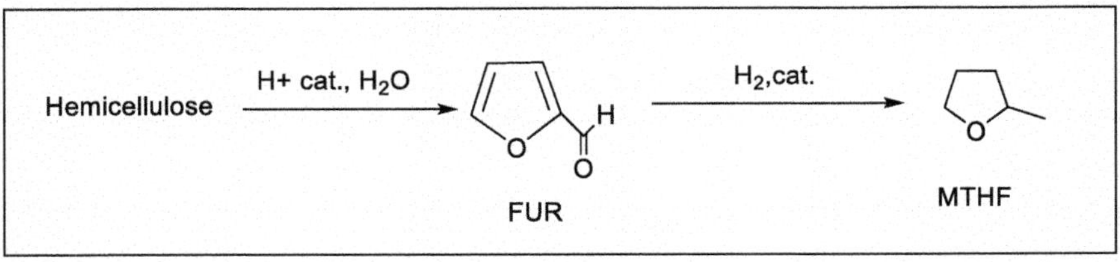

Fig. 11.22 Preparation pathways for key carbohydrate-derived renewable solvents

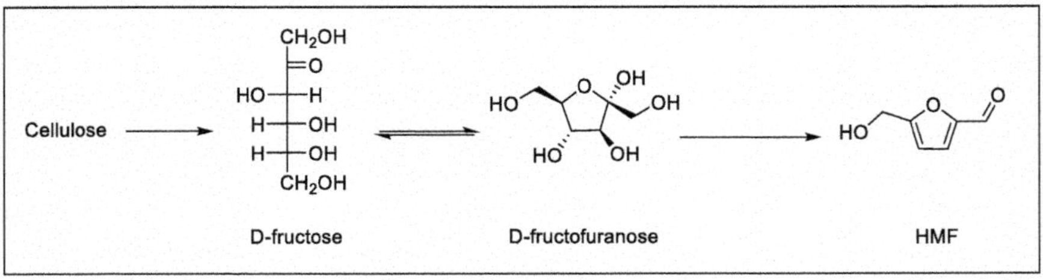

Fig. 11.23 Production of HMF crystals from D-Fructose

11.5 Energy-Sustainable Methods in Green Chemistry

The processes and materials are especially examined from the standpoints of sustainability, renewability, economical and clean energy consumption, and general adherence to sustainable and green chemistry principles [59].

11.5.1 Microwave-Assisted Synthesis

Microwave synthesis is a powerful and versatile technique that has transformed the field of chemical synthesis. Its ability to achieve rapid reaction rates, high yields, and precise control over reaction parameters positions it as an ideal method for synthesizing various materials, including organic compounds and inorganic nanomaterials. Furthermore, the alignment of microwave synthesis with the principles of green chemistry highlights its potential as a sustainable and environmentally friendly alternative to traditional synthesis methods [60].

Using microwave irradiation, 1-(2-hydroxyphenyl)-3-(p-tolyl) prop-2-en-1-one reacts with various methylene compounds to produce the derivatives cyanopyridine, cyclohexanone, and chromen-4-one (Fig. 11.24) [61].

Under MW irradiation, a porous MOF was formed using nickel(II) dihydroxyterephthalate (Ni-DHTP), which was then working as a catalyst to oxidize cyclohexene in the presence of H_2O_2 (Fig. 11.25) [62].

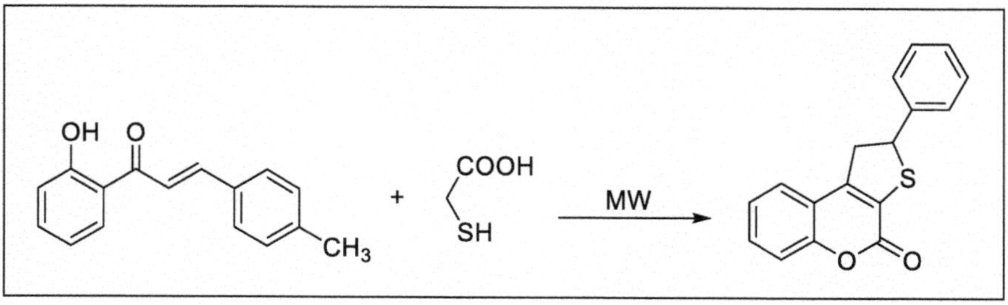

Fig. 11.24 Synthetic route used for 1-(2-hydroxyphenyl)-3-(p-tolyl) prop-2-en-1-on

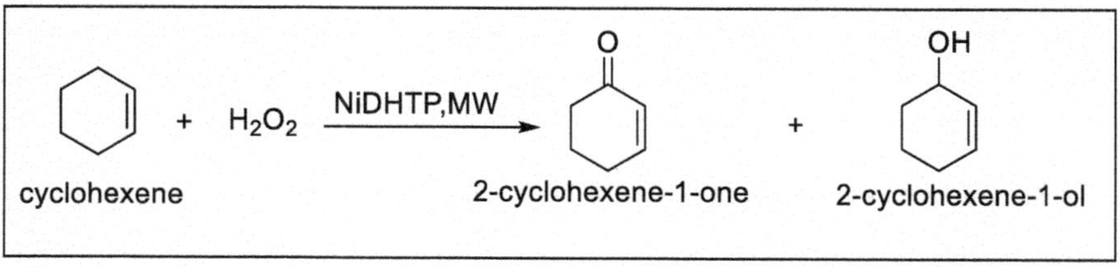

Fig. 11.25 Oxidation of cyclohexene

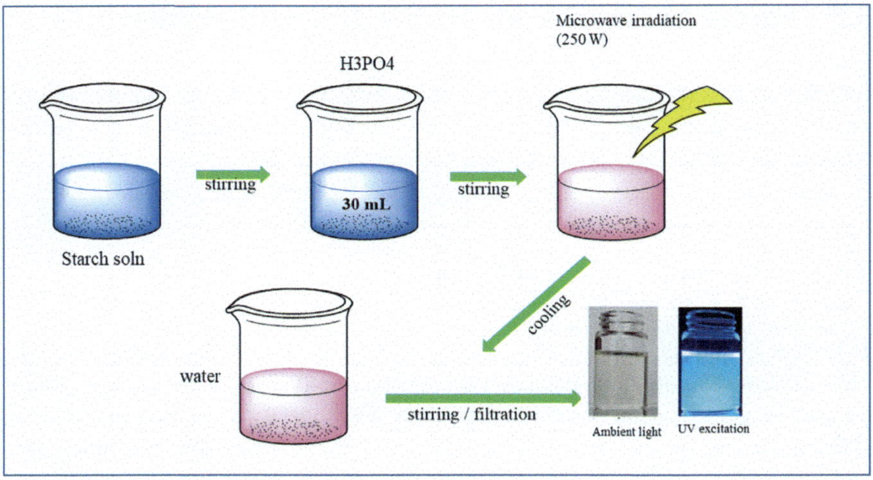

Fig. 11.26 Schematic of the synthetic method of CDs from starch

The absorption and fluorescence characteristics (Fig. 11.26) of the CDs prepared using different phosphoric acid concentrations were compared. MW irradiation was performed for 12 min [63].

The efficient synthesis of ten diverse functionalized scaffolds of the target molecule in a single pot highlights the effectiveness of this methodology in aligning with the rigorous principles of green chemistry. The successful synthesis of pyrazole [3, 4-b] pyridine using a multi-component reaction (MCR) approach has been effectively accomplished (Fig. 11.27) [64].

The chalcones were synthesized through a well-established Claisen-Schmidt condensation reaction using conventional and microwave-assisted methods. An equimolar mixture of 4-morpholinoacetophenone and various substituted benzaldehydes was reacted under basic conditions to produce the target compound (Fig. 11.28). Additionally, a non-conventional microwave irradiation methodology was employed to compare reaction times and product yields [65].

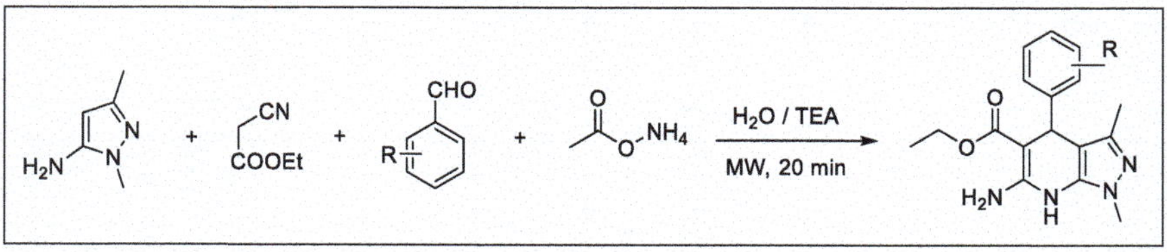

Fig. 11.27 Synthesis of pyrazolo [3, 4-b] pyridines via MCR approach

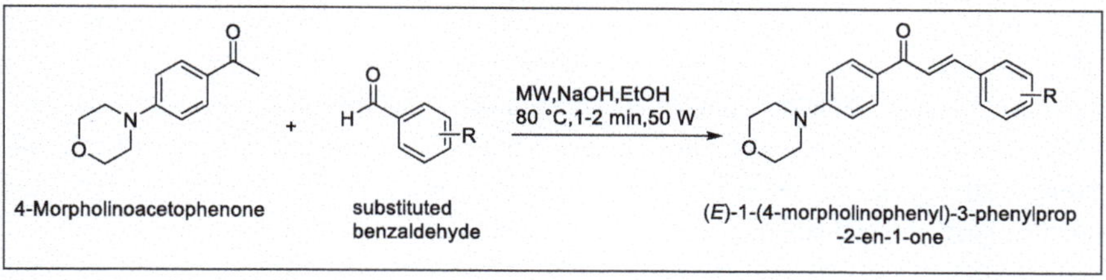

Fig. 11.28 Synthetic route to obtain (E)-1-(4-morpholinophenyl)-3-phenylprop-2-en-1-one

11.5.2 Ultrasound-Assisted Reactions

Ultrasonic sound waves' impact on chemical processes is referred to as sonochemistry. This nomenclature is consistent with more established methods that employ electricity (electrochemistry) and light (photochemistry) to activate chemicals [67]. Green chemistry is becoming more interested in sono chemical synthesis since it is economical and can reduce the need for extra reagents during synthesis (Fig. 11.29). The nanostructure is not subjected to any additional heat treatment because it is a one-step process [68].

Fig. 11.29 ZnO–Cu–GO particles were prepared by the ultra-rapid sono-chemical approach [66]

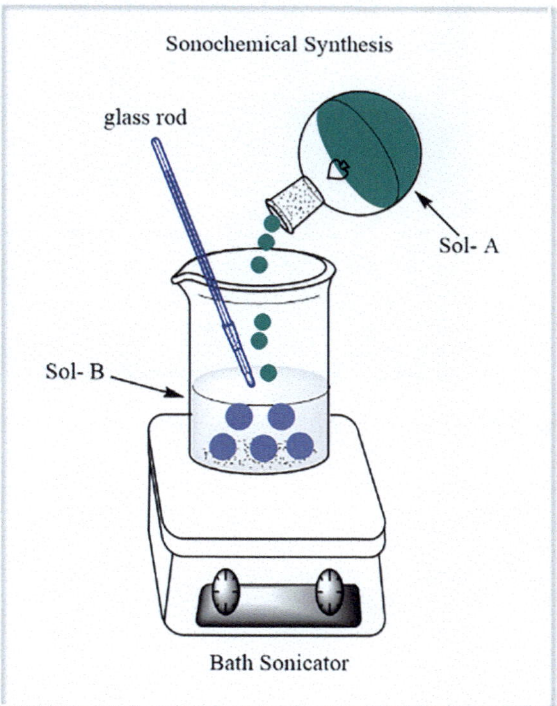

Given the current green chemistry movement, which stresses switching from traditional energy sources to more ecologically friendly ones like microwaves and ultrasounds, the sonochemical method yields a variety of nanoporous materials, such as Metal oxides, carbon materials, metal–organic frameworks (MOFs), covalent-organic frameworks (COFs), silicas, organosilica, and various combinations of other materials (Fig. 11.30) [69].

The synthesis of iso-thioureas by multi-component reactions is required when stoichiometric oxidants or heavy metal catalysts are required for responses and the sulphur supply is constrained (Fig. 11.31). The electro-catalytic three-component process uses thiols as the sulphur source and passes through a radical route [70].

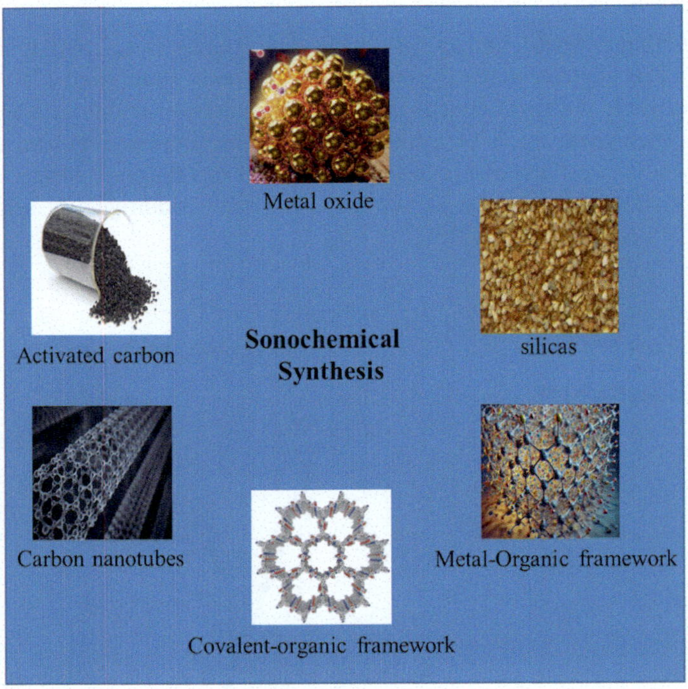

Fig. 11.30 Representative applications of sono-chemical method

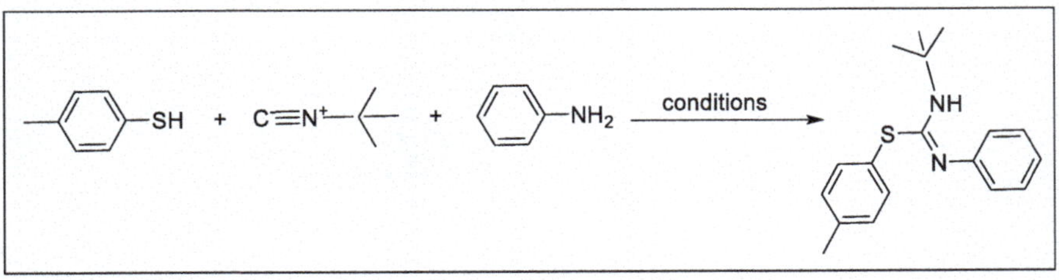

Fig. 11.31 Synthesis of iso-thioureas

11.5.3 Photochemical Reactions

The area of chemistry that studies the chemical reactions of light is called photochemistry. Photochemical excitation significantly alters chemical compounds' reactivity, accounting for the molecular variety or chemical structure space [71].

The effectiveness of this design (Fig. 11.32), the maleimide reaction partner is photo excited by a medium-pressure mercury lamp (400 W) during [2 + 2] photo cyclo addition [72].

High yield and high enantioselectivity were produced by converting the quinoxalinone derivative into the correlation azetidine (Fig. 11.33) [73].

The polycyclic valerolactam derivatives' enantioselective synthesis, including the complexation of the chiral borane derivative with the dihydropyridinone derivative (Fig. 11.34) [74].

Ultrasound sonication is essential for establishing a new C–C bond during the Barbier-type reaction (Fig. 11.35). In this process, a known starting aldehyde reacts with prenyl bromide in freshly activated zinc powder under ultrasound sonication, resulting in a separable mixture (1.7:1) with an impressive yield of 92% [75].

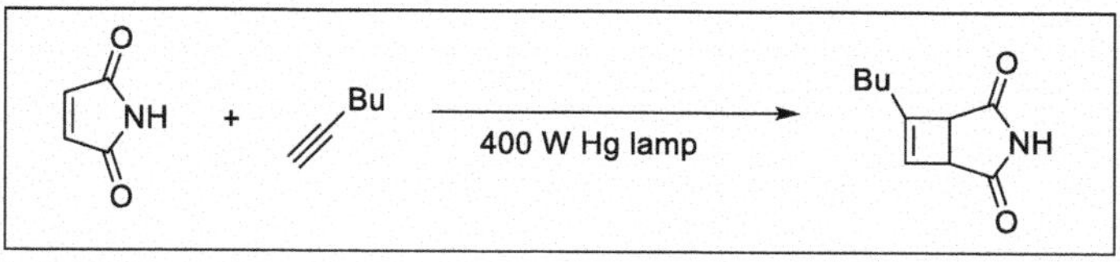

Fig. 11.32 Intermolecular [2 + 2] photo cyclo addition

Fig. 11.33 Enantioselective intramolecular synthesis of an azetidine

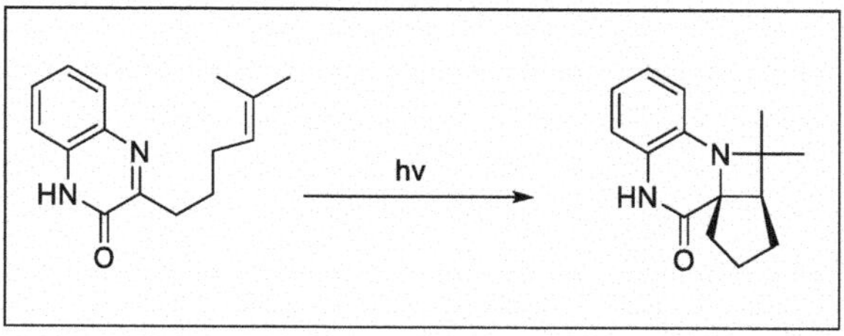

Fig. 11.34 Enantioselective acid catalyzed construction of a polycyclic valerolactam derivative via intramolecular [2+2] photocycloaddition

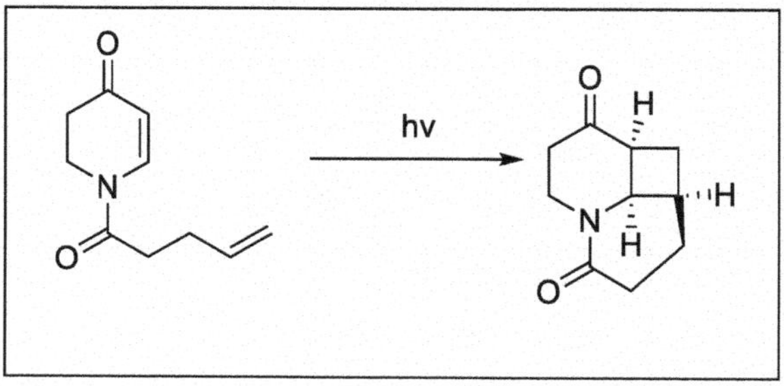

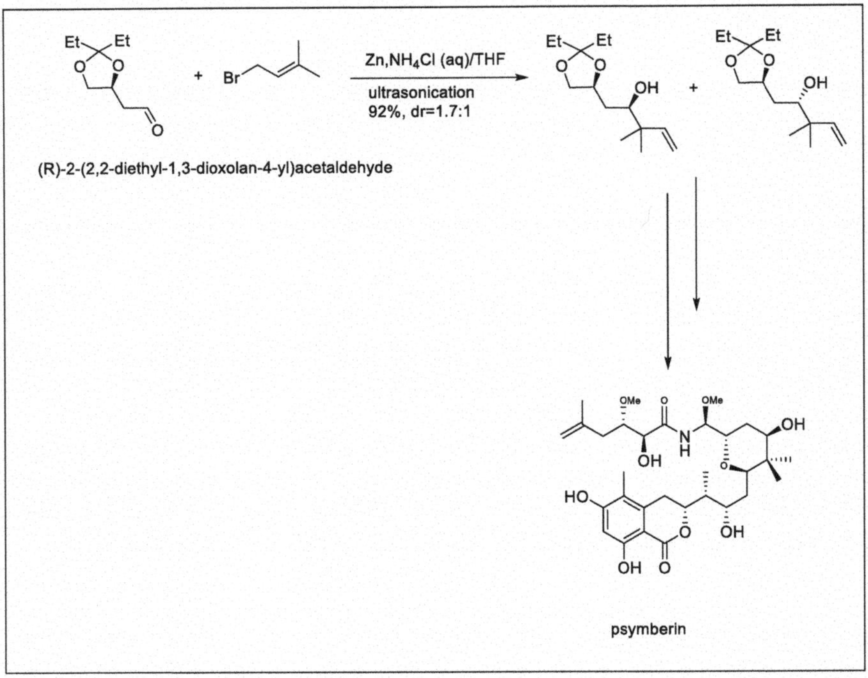

Fig. 11.35 Ultrasound-assisted total synthesis of psymberin

11.5.4 Flow Chemistry

Flow chemistry enhances mass and heat transfer processes by preventing temperature gradients, heat accumulation, and hotspots, which can lead to a loss of reaction selectivity. This approach is particularly effective for highly exothermic reactions [76].

A reaction was conducted using tributyl methylammonium methyl carbonate alongside a supercritical liquid as the catalytic base, marking an intriguing connection to safer auxiliary materials (Fig. 11.36). The reaction conditions were transformed into high-temperature and high-pressure flow processes, which were further optimized. Ultimately, they successfully produced 250–370 mg of N-methylated indole [77].

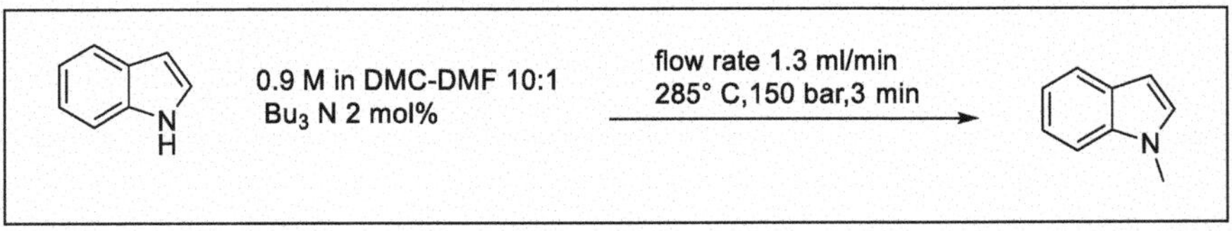

Fig. 11.36 Flow-chemistry-based implementation of indole synthesis

11.6 Towards Industrial Applications of Green Chemistry

Green synthetic chemistry is an important emerging field focusing on understanding various chemical reactions to develop more environmentally friendly synthetic pathways for starting materials and active pharmaceutical ingredients (APIs) [78]. This field encompasses several classical multi-component reactions (MCRs) that align with the core Principles of Green Chemistry, including the Mannich and Hantzsch reactions.

A more environmentally friendly Mannich reaction was conducted under solvent-free conditions at room temperature. In this reaction, benzaldehyde, acetophenone, and aniline were combined in a single pot, resulting in the formation of the desired amide derivative, 1,3-diphenyl-3-(phenylamino)propane-1-one (Fig. 11.37). The catalyst used in this process was $(C_4H_{12}N_2)_2[BiCl]Cl{\cdot}HO$, which can be reused without loss of efficiency and is recyclable. This method significantly reduces hazards and solvent waste while enhancing atom economy [79].

The Hantzsch Multi-Component Reaction (MCR) enables the synthesis of dihydropyridine derivatives (DHPs) by condensing an aldehyde with two equivalents of a β-ketoester using an ammonia donor. A more efficient and cost-effective one-pot Hantzsch reaction for producing 1,4-DHPs has been developed (Fig. 11.38). This improved reaction employs an aromatic aldehyde, ethyl acetoacetate, and ammonium acetate, with phenylboronic acid as the catalyst [80].

Starting with the compound 2-methylpropylbenzene, a Friedel–Crafts acetylation was performed using anhydrous hydrogen fluoride as the catalyst instead of the conventional aluminum chloride ($AlCl_3$). This reaction yielded 1-(4-isobutylphenyl) ethenone. The next step involved hydrogenation with RANEY® nickel, resulting in the formation of the alcohol 1-(4-isobutylphenyl) ethanol. Following this, carbonylation using cobalt and palladium catalysts produced ibuprofen. This process effectively recycled and recovered the waste by-products generated during the reactions

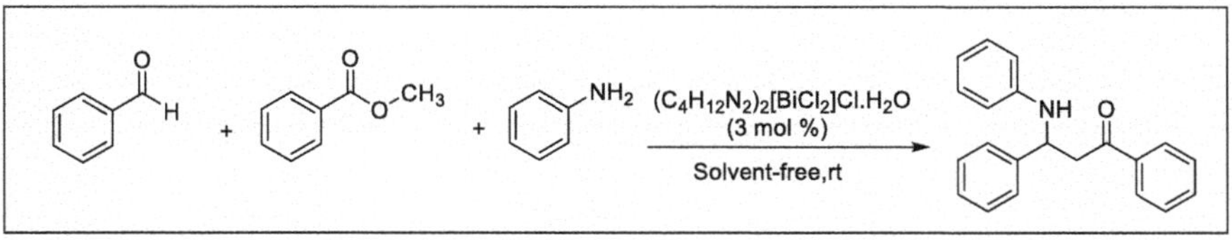

Fig. 11.37 Utilizing the Mannich reaction in the synthesis of amide derivatives

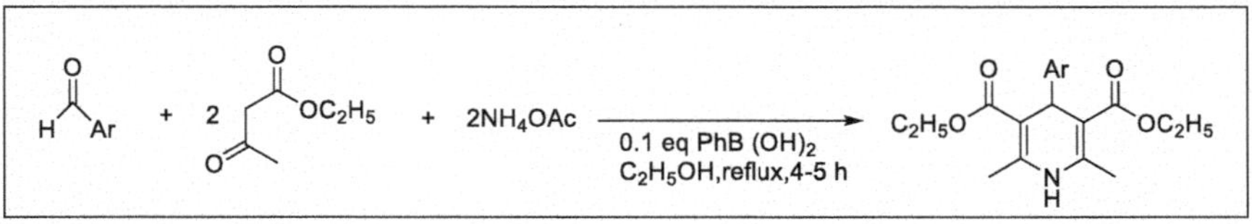

Fig. 11.38 Synthesis of 1,4-Dihydropyridines Using Phenylboronic Acid as a Catalyst

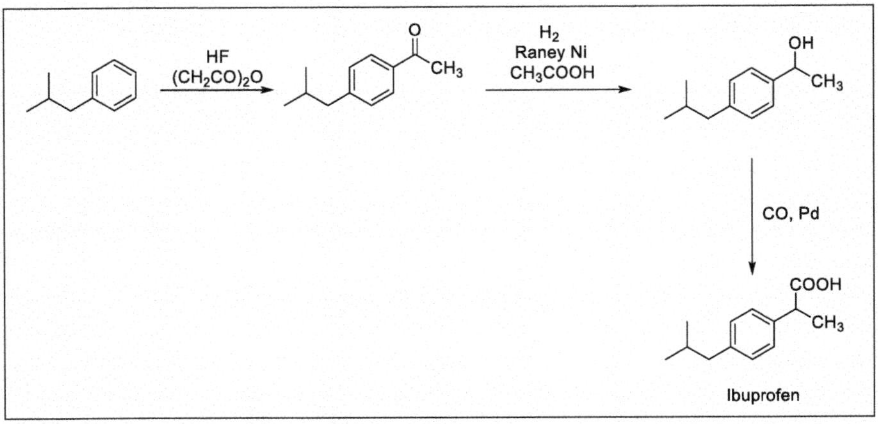

Fig. 11.39 New commercial green synthesis of ibuprofen

(Fig. 11.39), achieving a significant atom utilization rate of 77%, compared to just 40% from the previous six-step method [81].

11.7 Conclusion

The chapter on "Green Chemical Approaches for Sustainable Organic Synthesis" highlights how important green chemistry concepts are to changing organic synthesis in the direction of a more ecologically friendly and sustainable future. Green chemistry is a means of lowering the environmental impact of chemical manufacture through the integration of non-toxic reagents, renewable feed stocks, energy-efficient procedures, and waste reduction techniques. Novel approaches like microwave-assisted synthesis, bio-catalysis, and the utilization of ionic and supercritical fluids show how cutting-edge processes may improve reaction efficiency while upholding environmentally favourable standards. Furthermore, the problems of solvent waste and toxicity may be effectively addressed by the use of solvent-free systems and green solvents. Collaboration between academics, business, and regulatory agencies is necessary for the effective use of these strategies in order to promote innovation, standardize procedures, and deal with practical and financial obstacles. The chemical community can satisfy the needs of contemporary business and promote a healthy world by putting sustainability first. In order to achieve a genuinely sustainable future in organic synthesis, this chapter emphasizes the significance of ongoing research and development in green chemical techniques.

References

1. Lint, H.J.A.: An overview: origins and development of green chemistry. Found. Chem. **12**(1), 55–68 (2010)
2. Mammino, L.: Green chemistry and computational chemistry: a wealth of promising synergies. Sustain. Chem. Pharm. **34**, 101151 (2023)
3. Anas, P.T., Eghbali, N.: Green chemistry: principles and practice. Chem. Soc. Rev. **39**, 301–312 (2010)
4. Parra-Arroyo, L., González-González, R.B., Castillo-Zacarías, C., Martínez, E.M.M., Sosa-Hernández, J.E., Bilal, M., Parra-Saldívar, R.: Highly hazardous pesticides and related pollutants: toxicological, regulatory, and analytical aspects. Sci. Total. Environ. **807**, 151879 (2022)
5. Halkos, G., Gkampoura, E.C.: Where do we stand on the 17 sustainable development goals? An overview on progress. Econ. Anal. Policy **70**, 94–122 (2021)
6. Csefalvay, E., Akien, G.R., Qi, L., Horvath, I.T.: Definition and´ application of ethanol equivalent: sustainability performance metrics for biomass conversion to carbon-based fuels and chemicals. Catal. Today. Today **239**, 50–55 (2015)
7. Ruggerio, C.A.: Sustainability and sustainable development: a review of principles and definitions. Sci. Total. Environ. **786**, 147481 (2021)
8. Alberto, C., Ade, M.: Environmental sustainability: implications and limitations to green chemistry. Found. Chem. **16**, 125–147 (2013)
9. Ahsan, H., Islam, S.U., Ahmed, M.B., Lee, Y.S., Sonn, J.K.: Significance of green synthetic chemistry from a pharmaceutical perspective. Curr. Pharm. Des.. Pharm. Des. **26**(45), 5767–5782 (2020)
10. Jimenez-Gonzalez, C., Ponder, C.S., Broxterman, Q.B., Manley, J.: Using the right green yardstick: why process mass intensity is used in the pharmaceutical industry to drive more sustainable processes. Org. Process Res. Dev. **15**(4), 912–917 (2011)
11. Winterton, N.: The green solvent: a critical perspective. Clean Technol. Environ. Policy **23**(9), 2499–2522 (2021)
12. Li, C.J., Chan, T.H.: Comprehensive Organic Reactions in Aqueous Media. Wiley, Hoboken (2007)
13. Harry, N.A., Radhika, S., Neetha, M., Anilkumar, G.: Recent advances and prospects of organic reactions "on water." Chem. Select. **4**, 12337–12355 (2019)
14. Breslow, R.: Determining the geometries of transition states by use of antihydrophobic additives in water. Acc. Chem. Res. **37**, 471–478 (2004)
15. Cebular, K., Stavber, S.: Metal-free and acid-free activation of carbonyl moiety using molecular halogens or N-halamines. Adv. Org. Synth. **15**, 1–97 (2021)
16. (a) Li, C.J., Chen, L.: Organic chemistry in water. Chem. Soc. Rev. **35**, 68–82 (2006) (b) Engberts, J.B.F.N., Blandamer, M.J.: Understanding organic reactions in water: from hydrophobic encounters to surfactant aggregates. Chem. Comm. **18**, 1701–1708 (2001) (c) Sinou, D.: Asymmetric organometallic-catalyzed reactions in aqueous media. Adv. Synth. Catal. **344**, 221–237 (2002) (d) Rideout, D.C., Breslow, R.: Hydrophobic acceleration of Diels-Alder reactions, J. Am. Chem. Soc. **102**, 7816–7817 (1980)
17. Harry, N.A., Cherian, R.M., Radhika, S., Anilkumar, G.: A novel catalyst-free, eco-friendly, on water protocol for the synthesis of 2,3-dihydro-1H-perimidines. Tetrahedron Lett. **60**, 150946–150949 (2019)
18. Schüth, F., Weitkamp, J.: Handbook of Heterogeneous Catalysis, pp. 1–8. Wiley-VCH (2008)
19. Nishanth, R.R., Jena, S., Mukherjee, M., Maiti, B., Chanda, K.: Green synthesis of biologically active heterocycles of medicinal importance: a Review. Environ. Chem. Lett. **19**, 3315–3358 (2021)
20. Ahmad, T., Masoodi, F.A., Rather, S.A., Wani, S.M., Gull, A.: Supercritical fluid extraction: a review. J. Biol. Chem. Chron. **5**(1), 114–122 (2019)
21. Topuz, B., Günal, G., Guler, S., Aydin, H.M.: Use of supercritical CO_2 in soft tissue decellularization. In: Methods in Cell Biology, vol. 157, pp. 49–79. Academic Press (2020)

22. Razzak, F.S.A.: Review on green solvent supercritical carbon dioxide and its chemical reactions. J. Sci. Res. **65**(8), 17–22 (2021)
23. Chanshetti, U.: Concepts of green chemistry. Curr. Perspect. Chem. Sci. **7**, 144–150 (2021)
24. Dong, M., Wang, G., Zhang, X., Tan, D., Ren, J., Colorado, H., Hou, H., Toktarbay, Z., Guo, Z.: An overview of polymer foaming assisted by supercritical fluid. Adv. Compos. Hybrid Mater. **6**(6), 207 (2023)
25. Crabtree, R.H.: Green Catalysis: Homogeneous Catalysis, vol. 1. Wiley-VCH, Hoboken NJ (2013)
26. Crisci, A.J., Tucker, M.H., Lee, M.Y., Jang, S.G., Dumesic, J.A., Scott, S.L.: Acid functionalized SBA-15-type silica catalysts for carbohydrate dehydration. ACS Catal.Catal. **1**(7), 719–728 (2011)
27. Motamedi, R., Bardajee, G.R., Shakeri, S.: Facile one-pot synthesis of chromeno [4,3-b] quinoline derivatives catalyzed by Cu(II)-Schiff base/SBA-15. Heterocyclic Comm. **20**, 181–184 (2014)
28. Yang, M., Lian, R., Zhang, X., Wang, C., Cheng, J., Wang, X.: Photocatalytic cyclization of nitrogen-centered radicals with carbon nitride through promoting substrate/catalyst interaction. Nature Comm. **13**(1), 4900 (2022)
29. Walker, B.R., Manabe, S., Brusoe, A.T., Sevov, C.S.: Mediator-enabled electrocatalysis with ligandless copper for anaerobic Chan-Lam coupling reactions. J. Am. Chem. Soc. **143**, 6257–6265 (2021)
30. Jin, S., Kim, J., Kim, D., Park, J.-W., Chang, S.: Electrolytic C–H oxygenation via oxidatively induced reductive elimination in Rh catalysis. ACS Catal.Catal. **11**, 6590–6595 (2021)
31. Yang, Q.L., Jia, H.W., Liu, Y., Xing, Y.K., Ma, R.C., Wang, M.M., Qu, G.R., Mei, T.S., Guo, H.M.: Electrooxidative iridium-catalyzed regioselective annulation of benzoic acids with internal alkynes. Org. Lett. **23**(4), 1209–1215 (2021)
32. Kosinov, N., Liu, C., Hensen, E.J.M., Pidko, E.A.: Engineering of transition metal catalysts confined in zeolites. Chem. Mater. **30**(10), 3177–3198 (2018)
33. Dzhardimalieva, G.I., Zharmagambetova, A.K., Kudaibergenov, S.E., Uflyand, I.E.: Polymer-immobilized clusters and metal nanoparticles in catalysis. Kinet. Catal.. Catal. **61**(2), 198–223 (2020)
34. Yue, Y.N., Meng, W.J., Liu, L., Hu, Q.L., Wang, H., Lu, J.X.: Amino acid-functionalized multi-walled carbon nanotubes: a metal-free chiral catalyst for the asymmetric electroreduction of aromatic ketones. Electrochim. Acta. Acta **260**, 606–613 (2018)
35. Rasal, S.A., Tamore, M.S., Shimpi, N.G.: Ultrasound-mediated synthesis of novel α-aminophosphonates using graphene nanosheets-silver nanoparticles (GNS- AgNPs) as a recyclable heterogeneous catalyst. Chem. Select. **4**(8), 2293–2300 (2019)
36. Achary, L.S.K., Nayak, P.S., Barik, B., Kumar, A., Dash, P.: Ultrasonic-assisted green synthesis of β-amino carbonyl compounds by copper oxide nanoparticles decorated phosphate functionalized graphene oxide via Mannich reaction. Catal. Today. Today **348**, 137–147 (2020)
37. Gopiraman, M., Deng, D., Ganesh, B.S., Hayashi, T., Karvembu, R., Kim, I.S.: Sustainable and versatile CuO/GNS nanocatalyst for highly efficient base free coupling reactions. ACS Sustain. Chem. Eng. **3**(10), 2478–2488 (2015)
38. Mo, Y., Zhao, X., Yuan, T., He, B.: Factors affecting the separation performance of graphene oxide membranes: mechanical support, properties of graphene oxide, and exotic species. J. Chem. Tech. Biotechnol. **93**(5), 1388–1393 (2018)
39. Naeimi, H., Ansarian, Z.: Functionalized polytriazoles on graphene oxide supported copper (I) complex as an effective reusable catalyst for sonochemical click synthesis of triazoles in aqueous media. Inorg. Chim. Acta. Chim. Acta **466**, 417–425 (2017)
40. Alcántara, A.R., Domínguez, D.M.P., Littlechild, J.A., Schürmann, M., Sheldon, R.A., Wohlgemuth, R.: Biocatalysis as Key to sustainable industrial chemistry. ChemSus Chem. **15**, 1–11 (2022)
41. Taghavi, R., Rostamnia, S.: Four-component synthesis of polyhydroquinolines via unsymmetrical hantzsch reaction employing Cu-IRMOF-3 as a robust heterogeneous catalyst. Chem. Methodol. **6**(9), 639–648 (2022)

42. Sun, K., Li, D., Lu, G.P., Cai, C.: Hydrogen auto-transfer synthesis of quinoxalines from o-nitroanilines and biomass-based diols catalyzed by MOF-derived N, P Co-doped cobalt catalysts. Chem. Cat. Chem. **13**(1), 373–381 (2021)
43. Xing, L., Yang, Q., Zhu, C., Bai, Y., Tang, Y., Rueping, M., Cai, Y.: Poly (heptazine imide) ligand exchange enables remarkable low catalyst loadings in heterogeneous metallaphotocatalysis. Nature Comm. **14**(1), 1501 (2023)
44. Dai, Y., Xiong, Y.: Control of selectivity in organic synthesis via heterogeneous photocatalysis under visible light. Nano Res. Energy. **1**(1), e9120006 (2022)
45. Harrad, M.A., Semane, A., Badereddine, M., Tounsi, A.: ZnAl2O4@ ZnO an effective, heterogeneous catalyst for the synthesis of bis-(β-enaminones) and bis-(β-enaminoesters). Chem. J. Moldova. **19**(2), 74–82 (2024)
46. Wu, S., Snajdrova, R., Moore, J.C., Baldenius, K., Bornscheuer, U.T.: Biocatalysis, enzymatic synthesis for industrial applications. Angew. Chem. Int. Ed.. Chem. Int. Ed. **60**, 88–119 (2021)
47. Gonzalo, G., Domínguez, D.M.P.: Biocatalysis: An Industrial Perspective. RSC Publishing, Cambridge, UK (2017)
48. Holtmann, D., Hollmann, F.: Is water the best solvent for biocatalysis? Mol. Catal. **517**, 112035 (2022)
49. Villa, R., Alvarez, E., Porcar, R., Garcia-Verdugo, E., Luis, S.V., Lozano, P.: Ionic liquids as an enabling tool to integrate reaction and separation processes. Green Chem. **21**, 6527–6544 (2019)
50. Lozano, P., Bernal, J.M., Sanchez-Gómez, G., LopezLopez, G., Vaultier, M.: How to produce biodiesel easily using a green biocatalytic approach in sponge-like ionic liquids. Energy Environ. Sci. **6**(4), 1328–1338 (2013)
51. Shuai, L., Luterbacher, J.: Organic solvent effects in biomass conversion reactions. Chemsuschem **9**(2), 133–155 (2016)
52. Zetzsche, L.E., Yazarians, J.A., Chakrabarty, S., Hinze, M.E., Murray, L.A.M., Lukowski, A.L., Joyce, L.A., Narayan, A.R.H.: Biocatalytic oxidative cross-coupling reactions for biaryl bond formation. Nature **603**(7899), 79–85 (2022)
53. Jouffroy, M., Neufeld, K.: Synthesis of atropisomeric biaryls via chiral suzuki–miyaura/ enzymatic kinetic resolution. ACS Catal.Catal. **12**(14), 8380–8385 (2022)
54. Steiner, A., Znidar, D., Ötvös, S.B., Snead, D.R., Dallinger, D., Kappe, C.O.: A high-yielding synthesis of EIDD-2801 from uridine. Eur. J. Org. Chem. **2020**(43), 6736–6739 (2020)
55. Veerabagu, U., Jaikumar, G., Lu, F., Quero, F.: High yield and greener C-H difluoromethylation reactions using copper iodide nanoparticles/boron nitride nanosheets as a versatile and recyclable heterogeneous catalyst. React. Chem. Eng. **6**(10), 1900–1910 (2021)
56. Jäger, C., Haase, M., Koschorreck, K., Urlacher, V.B., Deska, J.: Aerobic C–N bond formation through enzymatic nitroso-ene-type reactions. Angew. Chem. Int. Ed.. Chem. Int. Ed. **62**(7), e202213671 (2023)
57. Musolino, M., Andraos, J., Aricò, F.: An easy scalable approach to HMF employing DMC as reaction media: reaction optimization and comparative environmental assessment. Chem. Select. **3**(8), 2359–2365 (2018)
58. Akpan, E.I., Wetzel, B., Friedrich, K.: Eco-friendly and sustainable processing of wood-based materials. Green Chem. **23**(6), 2198–2232 (2021)
59. Schmidt, R., Prado-Gonjal, J., Morán, E.: Microwave-Assisted Hydrothermal Synthesis of Nanoparticles. CRC Press Taylor & Francis Group, CRC Concise Encyclopedia of Nanotechnology (2022)
60. Shalaby, M.A., Fahim, A.M., Rizk, S.A.: Microwave-assisted synthesis, antioxidant activity, docking simulation, and DFT analysis of different heterocyclic compounds. Sci. Rep. **13**(1), 4999 (2023)
61. Phan, P.T., Hong, J., Tran, N., Le, T.H.: The properties of microwave-assisted synthesis of metal–organic frameworks and their applications. Nanomaterials **13**(2), 352 (2023)
62. Shibata, H., Abe, M., Sato, K., Uwai, K., Tokuraku, K., Iimori, T.: Microwave-assisted synthesis and formation mechanism of fluorescent carbon dots from starch. Carbohydr. Polymer Tech. App. **3**, 100–218 (2022)

63. Meng, A., Cheng, B., Tan, H., Fan, J., Su, C., Yu, J.: TiO_2/polydopamine S-scheme heterojunction photocatalyst with enhanced CO_2-reduction selectivity. App. Catal. B: Environ. **289**, 120039 (2021)
64. Pachipulusu, S., Merugu, K.S., Bhonsle, R.R., Surapaneni, S., Kurnool, A., Kerru, N.: Microwave-assisted, multicomponent synthesis of pyrazolo [3, 4-b] pyridines under green conditions. Chem. Select. **9**(38), e202403569 (2024)
65. Choudhary, D., Kumar, B., Chandrasekaran, B., Singh, T.G., Kaur, R., Aldahish, A., Vasudevan, R., Balaji, P., Balaji, P.: Microwave-assisted synthesis of morpholine-based chalcones as reversible MAO-A inhibitors in the management of mental depression. Pharmaceuticals **18**(3), 309 (2025)
66. Kamali, M., Dewil, R., Appels, L., Aminabhavi, T.M.: Nanostructured materials via green sonochemical routes–sustainability aspects. Chemosphere **276**, 130–146 (2021)
67. Shenoy, S., Ahmed, S., Lo, I.M., Singh, S., Sridharan, K.: Rapid sonochemical synthesis of copper doped ZnO grafted on graphene as a multi-component hierarchically structured visible-light-driven photocatalyst. Mater. Res. Bull. **140**, 111290 (2021)
68. Głowniak, S., Szczęśniak, B., Choma, J., Jaroniec, M.: Recent developments in sonochemical synthesis of nanoporous materials. Molecules **28**(6), 26–39 (2023)
69. He, M., Zhong, P., Liu, H., Ou, C., Pan, Y., Tang, H.: Electrochemically mediated three-component synthesis of isothioureas using thiols as sulfur source. Green Syn. Catal. **4**(1), 41–45 (2023)
70. Balzani, V., Ceroni, P., Juris, A.: Photochemistry and Photophysics: Concepts, Research, Applications. John Wiley & Sons, Hoboken, NJ (2014)
71. (a) Gerry, C.J., Schreiber, S.L.: Chemical probes and drug leads from advances in synthetic planning and methodology. Nat. Rev. Drug Discov. **17**, 333–352 (2018) (b) Schreiber, S.L.:Target-oriented and diversity-oriented organic synthesis in drug discovery. Science **287**, 1964–1969 (2000) (c) Galloway, W.R.J.D., Isidro-Llobet, A., Spring, D.R.: Diversity-oriented synthesis as a tool for the discovery of novel biologically active small molecules. Nature Comm. **1**, 80 (2010) (d) Beletskaya, I.P., Nájera, C., Yus, M.: Chemodivergent reactions. Chem. Soc. Rev. **49**(19), 7101–7166 (2020)
72. Zondag, S.D., Mazzarella, D., Noël, T.: Scale-up of photochemical reactions: transitioning from lab scale to industrial production. Annu. Rev. Chem. Biomol. Eng.. Rev. Chem. Biomol. Eng. **14**(1), 283–300 (2023)
73. Li, X., Großkopf, J., Jandl, C., Bach, T.: Enantioselective, visible light mediated aza paternò-büchi reactions of quinoxalinones. Angew. Chem.. Chem. **60**, 2684–2688 (2021)
74. Hoffmann, N.: Enantioselective synthesis of heterocyclic compounds using photochemical reactions. Photochem. Photobiol. Sci.. Photobiol. Sci. **20**, 1657–1674 (2021)
75. Majhi, S.: Applications of ultrasound in total synthesis of bioactive natural products: a promising green tool. Ultrason. Sonochem.. Sonochem. **77**, 105665 (2021)
76. Jiao, J., Nie, W., Yu, T., Yang, F., Zhang, Q., Aihemaiti, F., Yang, T., Liu, X., Wang, J., Li, P.: Multi-step continuous-flow organic synthesis: opportunities and challenges. Chem. Eur. J. **27**, 4817–4838 (2021)
77. Colella, M., Degennaro, L., Luisi, R.: Continuous flow synthesis of heterocycles: a recent update on the flow synthesis of indoles. Molecules **25**(14), 3242 (2020)
78. Kar, S., Sanderson, H., Roy, K., Benfenati, E., Leszczynski, J.: Green chemistry in the synthesis of pharmaceuticals. Chem. Rev. **122**(3), 3637–3710 (2021)
79. Lu, H., Wu, R., Cheng, H., Nie, S., Tang, Y., Gao, Y., Luo, Z.: An efficient, mild, solvent-free, one-pot three-component mannich reaction catalyzed by $(C_4H_{12}N_2)$ 2 [BiCl6] Cl·H_2O. Synthesis **47**(10), 1447–1454 (2015)
80. Debache, A., Boulcina, R., Belfaitah, A., Rhouati, S., Carboni, B.: One-pot synthesis of 1, 4-dihydropyridines via a phenylboronic acid catalyzed Hantzsch three-component reaction. Synth. Lett. **4**, 509–512 (2008)
81. Dicks, A.P. (ed.): Green Organic Chemistry in Lecture and Laboratory. CRC Press (2011).

Chapter 12
Green Techniques in Drug Discovery: Sustainable Pathways to Pharmaceutical Innovation

Glanish Jude Martis, Santosh L. Gaonkar, and Hiroki Shimizu

Abstract Green techniques have evolved over the years in the field of synthetic organic chemistry, chemical biology and medicinal chemistry. Their impact on these diverse subjects have laid a strong foundation as their advantages reflect more than the intent for which they were first developed or introduced. Pharmaceutical innovations mainly deal with the advancements and breakthroughs in developing newer strategies and bringing out greater outcomes than persisting at present. Sustainability is the major concern for the medicinal chemists, those who work on furnishing potent drug molecules and green techniques stretch them a helping hand in reducing their efforts with the harmful effects of the solvents and other chemicals used in bulk scale. Thus, this chapter deals with the new synthetic protocols developed for producing molecules of medicinal value both in small as well as large scale. The importance and advantages can be drawn as the chapter progresses eventually.

Keywords Green techniques · Drug design · Microwave · Ultrasound · Ionic liquids · Medicinal targets

12.1 Introduction

The twelve principles of green chemistry formulated by Paul Anastas and John Warner boosted the pharmaceutical industries to grow exponentially and strive hard towards developing drug molecules [1]. The sustainable strategies imbibe chemical reactions with low cost and less time to make them more effective towards a green environment [2]. These techniques are beneficial for profit, people and planet, too

G. J. Martis · S. L. Gaonkar (✉)
Department of Chemistry, Manipal Institute of Technology, Manipal Academy of Higher Education, Manipal, Karnataka, India
e-mail: sl.gaonkar@manipal.edu

H. Shimizu
Cellular and Molecular Biotechnology Research Institute, National Institute of Advanced Industrial Science and Technology (AIST), Tsukuba, Ibaraki, Japan

P. Jain et al. (eds.), *Towards Green Chemical Processes: Strategies and Innovations*, Intelligent Systems Reference Library 277,
https://doi.org/10.1007/978-3-031-94090-3_12

[3]. It is really a challenging task for a chemist to balance with the time pressure in delivering a compound of medicinal interests [4]. Various green techniques are now employed which include, ultrasound promoted synthesis [5], microwave synthesis [6], ionic liquids [7], sealed-tube reactions [8] etc. Thus, it is very interesting to know the various techniques involved under green principles in the verge of developing drug molecules. Green principles are not only restricted towards synthesizing molecules rather used to determine various drug constituents in the samples viz., paracetamol [9]. Most of the ionic liquids synthesized in recent years are dealt with in upbringing as efficient catalysts for synthesizing biologically important drug molecules. Later, computational studies reveal their binding affinity and molecular modelling helps to understand the varied dimensional area of organic hybrid molecules [10]. Even though green techniques use promising factors like solvent-free, water-mediated synthesis, one should have in-depth knowledge on the thermal characteristics of such reactions to avoid disasters in the laboratories and pharmaceutical industries [11].

Therefore, this chapter presents various techniques, strategies and methodologies used in the pharmaceutical industries in developing novel medicinal targets.

12.2 Green Techniques Towards Potential Drug Discovery

12.2.1 Microwave-Assisted Organic Synthesis

Microwave-induced organic synthesis has emerged as one of the most significant tools used for the ease of organic reactions. The utility of microwave technique has several advantages over the traditional methods employed for many years. Many multicomponent reactions have been reported with high yields that facilitated in the production of medicinally important and biologically relevant heterocyclic compounds. Some of the well-known multicomponent reactions such as Biginelli, Hantzsch, Ugi etc., have driven the drug discovery processes.

Biginelli Reactions: Dihydropyrimidinones are the classic moieties obtained via Biginelli reaction. One of these approaches made use of urea, benzyl bromide and ethyl acetoacetate as the main substrates in DMSO and irradiated under microwave for 25 min at 80 °C to give dihydropyrimidone with excellent yield of 91% (Scheme 12.1) [12]. Zinc triflate catalyst also facilitates solvent-free Biginelli reactions under microwave conditions using urea, β-bisphosphonates and aldehydes to give dihydropyrimidinones. Aliphatic aldehydes were not suitable for this reaction in giving products with moderate to high yield (Scheme 12.2) [13]. Cu-catalyzed Biginelli reactions involving methyl acetoacetate, aromatic aldehydes and urea are also prominent in showing promising yields. Furthermore, the reaction with ceric ammonium nitrate oxidised hydroxyl group to keto produces regioselectivity (Scheme 12.3) [14].

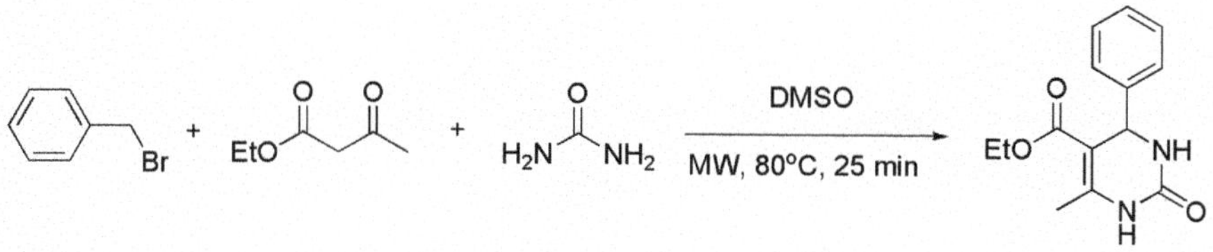

Scheme 12.1 Microwave synthesis of dihydropyrimidone via Biginelli reaction

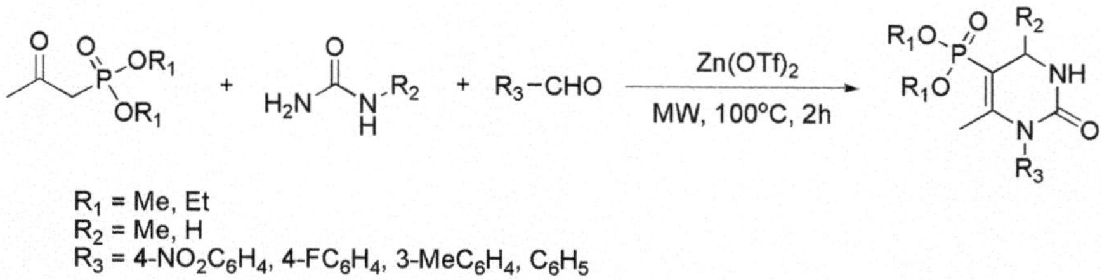

Scheme 12.2 Zn-catalyzed Biginelli reaction under microwave conditions

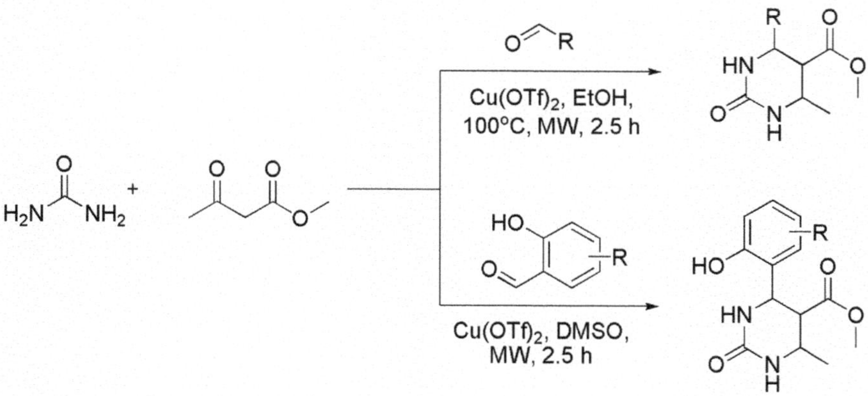

Scheme 12.3 Cu-catalyzed Biginelli reaction under microwave conditions

2-amino-4*H*-chromene derivatives, well-accomplished medicinal targets were synthesized from aromatic aldehydes, malononitrile and naphthols under microwave irradiation for 2–3 min to give desired products in 90–91% yield (Scheme 12.4) [15].

Hantzsch Reactions: The condensation of 2 mol of ethyl acetoacetate, one mole of aldehyde and ammonium acetate using microwave irradiation gave 1,4-dihydropyridines in much better yields with less consumption of the time compared to the conventional synthetic protocols. The use of alumina as solid support also resulted in giving desired products with yield slightly less than the reaction without solid support. This kind of modified Hantzsch reaction is much more prominent as that of Biginelli reactions (Scheme 12.5) [16, 17].

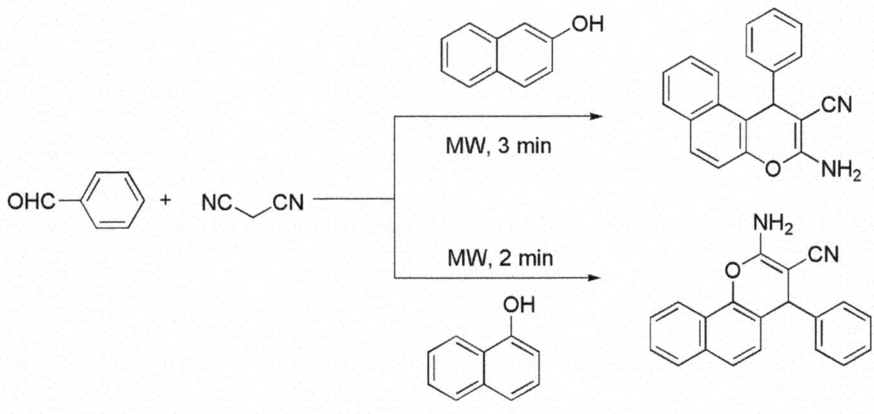

Scheme 12.4 Synthesis of 2-amino chromenes via Biginelli reaction

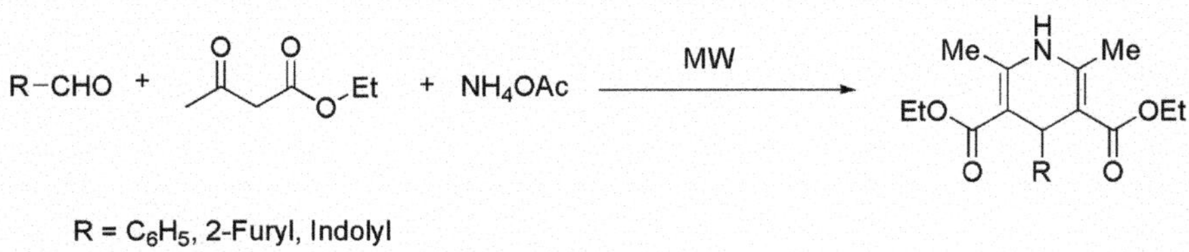

Scheme 12.5 Microwave-assisted Hantzsch reaction

Nifedipine, a calcium channel blocker was produced using a flow-type microwave reactor by charging ammonia, methyl acetoacetate and 2-nitrobenzaldehyde in ethanol at 800W. The main advantage of this reaction is its utility in the health-care sector by lowering the blood pressure. This can be effectively used in the drug discovery wherein the utility of calcium channel blockers plays their pivotal roles (Scheme 12.6) [18].

Ugi reactions: The astonishing work reported by Karl Ugi in 1959 paved a remarkable direction for setting drug discovery targets as it furnishes easy-going reaction between aldehydes or ketones, isocyanides, amines, carboxylic acid in a protic solvent to yield amido amides. However, the four-component Ugi reaction

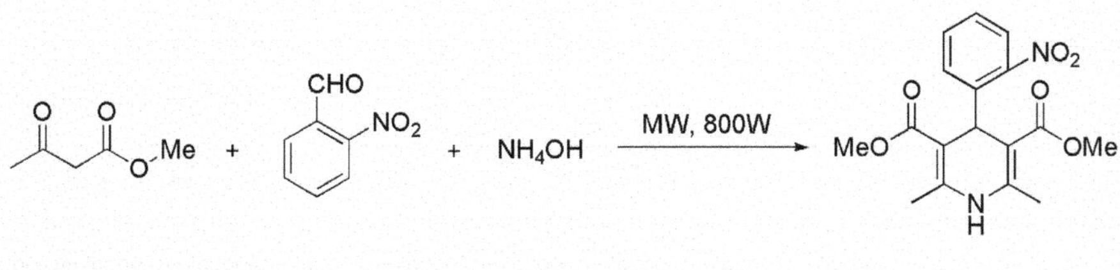

Scheme 12.6 Synthesis of nifedipine using flow-type microwave reactor

has dragged great attention in producing five different centres for variable functionality. Among the prominent medicinal targets, quinoline-based peptides resulted in great biological imperative by making use of microwave techniques in saving energy, time and solvents etc. Lipoic acid was condensed with 2-methoxyquinoline-3-carbaldehyde, cyclohexyl isocyanide and substituted aniline to give quinoline-peptide under microwave conditions. The screening of these products was done for assessing toxicity, antimicrobial and antioxidant evaluation and reported significant results favouring biological potentiality (Scheme 12.7) [19].

Bidirectional functionalization is also possible via Ugi reaction showcasing the versatility of multicomponent reaction in giving macrocyclic compounds. 1,3,5-thiadiazinane-2-thiones (THTTs) were the attractive substrates that attributed bidirectional functionality by reacting with substituted amines, formaldehydes and isocyanide in ethanol at 70 °C under microwave irradiation (100 W) for an hour to give Ugi products (Scheme 12.8) [20].

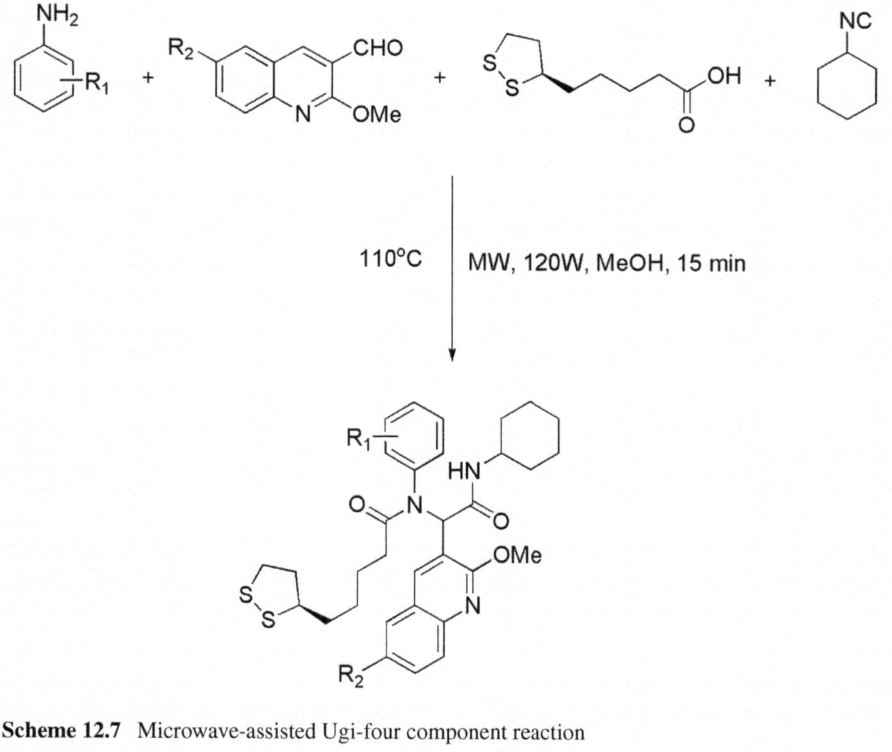

Scheme 12.7 Microwave-assisted Ugi-four component reaction

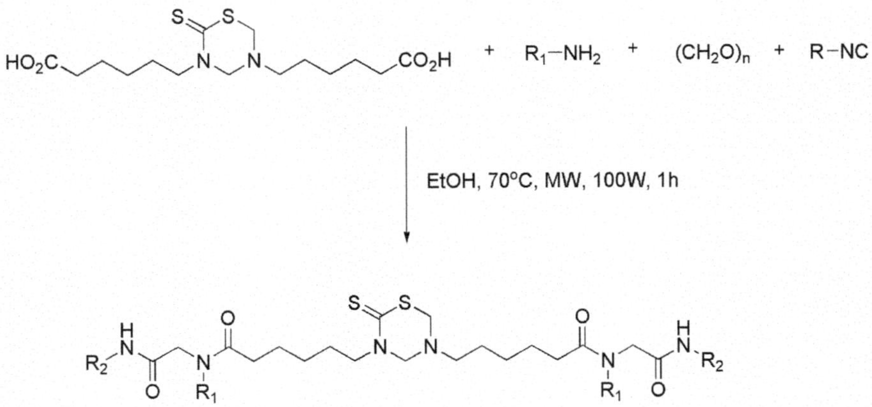

Scheme 12.8 Bidirectional Ugi-reaction

12.2.2 Sonochemistry

Ultrasound in chemistry is also referred to as sonochemistry or ultrasonic chemistry which has left an indelible mark in the field of synthetic organic chemistry. Initially, this kind of technique was developed with an intent to activate metallic or solid reagents. Their wide applicability had paved a way for numerous applications including drug discovery, medicinal chemistry aspects etc. The numerous experiments on ultrasound synthesis furnished several advantages such as less reaction-time, sustainable synthesis, improved yield and selectivity, isolation and extraction of natural products and more. These advantages are attributed from the basic principles involving increased mass transfer, acoustic streaming and cavitation [21–23].

Prominent name reactions such as Diels–Alder, Buchwald-Hartwig, Suzuki–Miyaura, Sonogashira, Heck etc. that afforded potential drug moieties could be efficiently carried out using ultrasonic chemistry principles.

Suzuki-Miyuara coupling: The ultrasonic synthesis of palladium nanoparticles with chitosan-starch functionalized Fe_3O_4 nanocomposite catalyst was very helpful for the Suzuki coupling reaction to link C–C bonds. Similarly, bromobenzene and phenyl boronic acid were coupled using Fe_3O_4@CS-Starch/Pd catalyst. The preparation of the magnetic Fe nanocomposite was done using ultrasound irradiation of Fe_3O_4 NPs in water and sonicated with formerly prepared CS-Starch composite. Glutaraldehyde was used for the cross-linking purpose and Na_2PdCl_4 was introduced and the catalyst was obtained on one hour sonication (Scheme 12.9) [24].

Heck reaction: The Pd catalyzed C–C coupling reactions are very prominent and one among them is Heck reaction. One of the attractive palladium complexes, which was later turned to be a great anticancer moiety was prepared using Pyridine-2-carbaldehyde-oxime with $PdCl_2$ in a solution of acetonitrile to form a yellow needle shaped Pd complex. The same complex was then used for sonicated Heck reactions of 4-bromobenzaldehyde with methyl acrylate at 80 °C (Scheme 12.10) [25].

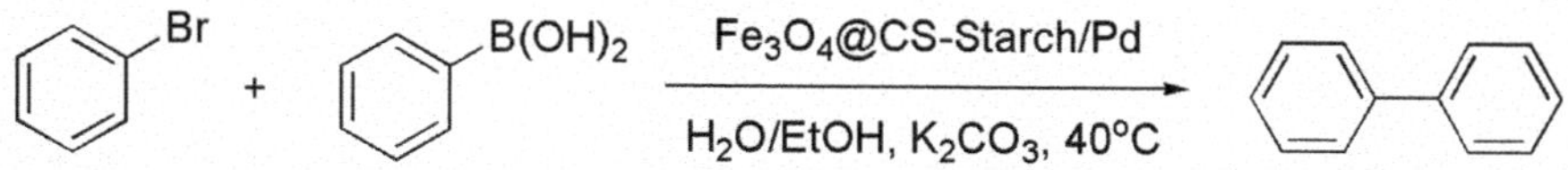

Scheme 12.9 Ultrasonic Suzuki–Miyaura cross coupling reaction using Fe_3O_4@CS-Starch/Pd catalyst

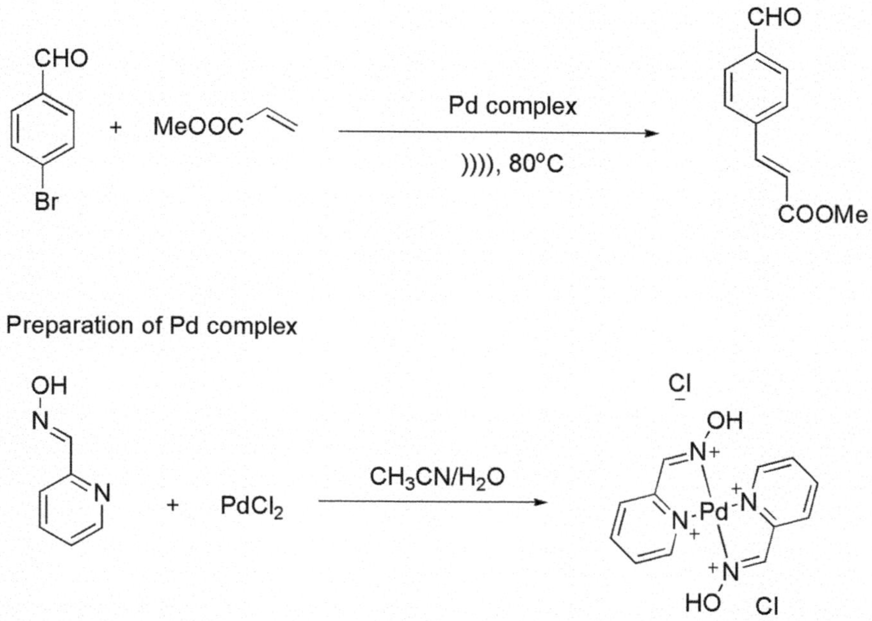

Scheme 12.10 Ultrasound-promoted Heck reaction

Sonogashira cross coupling reaction: As discussed in the Suzuki–Miyaura coupling reaction, Sonogashira reaction is also promoted by ultrasound irradiation by the preparation of Cu/cellulose-modified Fe_3O_4 nanocomposites. Fe_3O_4 NPs were sonicated in water at room temperature and cellulose nanofiber (CNF) was added and continued sonication for about an hour followed by the dropwise addition of $CuCl_2$ in ethanol. Ultimately, Fe_3O_4@CNF@Cu nanocomposite was then obtained by isolation. This nanocomposite was used as a catalyst for the coupling reaction of phenyl iodide and phenylacetylene using ultrasonic conditions at 40 °C (Scheme 12.11) [26].

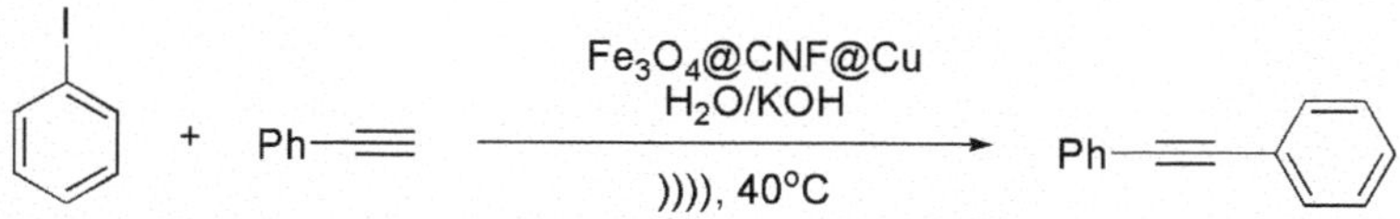

Scheme 12.11 Ultrasonic Sonogashira reaction of phenyl iodide

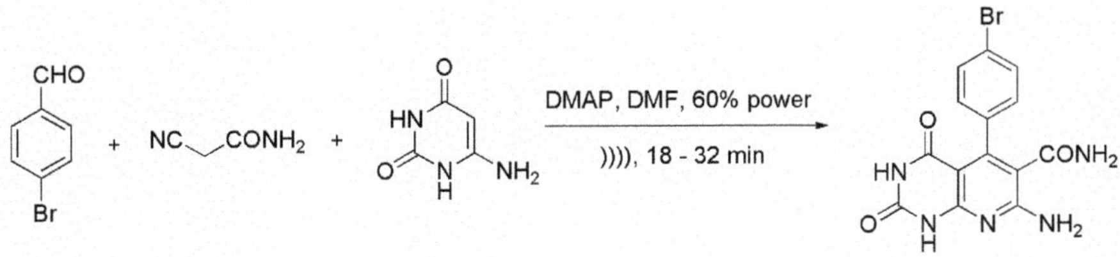

Scheme 12.12 Ultrasound-assisted Knoevenagel-Michael multicomponent reaction

Knoevenagel-Michael reaction: Most of the drug targets involve fused heterocyclic hybrids pyridopyrimidines, quinazolines etc. Similarly, pyridopyrimidines could be easily achieved by driving ultrasonic chemistry involving domino Knoevenagel-Michael type reaction of 4-bromobenzaldehyde, cyanoacetamide, 6-aminouracil and 4-dimethylaminopyridine (DMAP) in DMF at 60% power ultrasonication for an average of 18–32 min. The organocatalyst employed in this reaction is mild in nature and most importantly is one of the inexpensive catalysts available. These products are of high medicinal interests which have created their own impact by their potential and mesmerizing biological activity and SAR studies (Scheme 12.12) [27].

Coumarin-Pyrazole hybrids: Coumarins and pyrazoles have already shown their importance as individual cores. Further, their union has brought promising results and has taken a diversified path in thriving deeper roots in the world of chemical biology and drug discovery. Rather than moving with the traditional methods of synthesizing such linked heterocycles, ultrasonic chemistry has given another track to proceed the reactions with higher yield and less time. Likewise, 4-hydroxycoumarin when condensed with pyrazole aldehydes in the presence of citric acid for accelerating the reaction and water as a green solvent in ultrasound for 15–20 min at room temperature afforded pyrazolyl-bis-coumarin hybrid. These hybrids were then proved to be effective biological scaffolds for the treatment of different ailments threatening human life (Scheme 12.13) [28].

Pyranocoumarins: Using ultrasound technology as an eco-friendly energy source, pyranocoumarins can be obtained from 4-hydroxycoumarin, benzaldehyde and 2-(phenylsulfonyl)-acetonitrile in piperidine as the base and ethanol for 30 min. The yield of the product was 94% which is more than the reported literature with conventional methods (Scheme 12.14) [29].

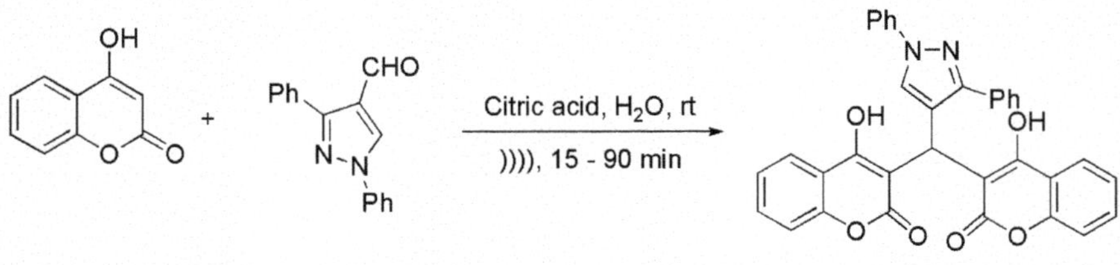

Scheme 12.13 Ultrasound synthesis of pyrazolyl-bis-coumarin hybrids

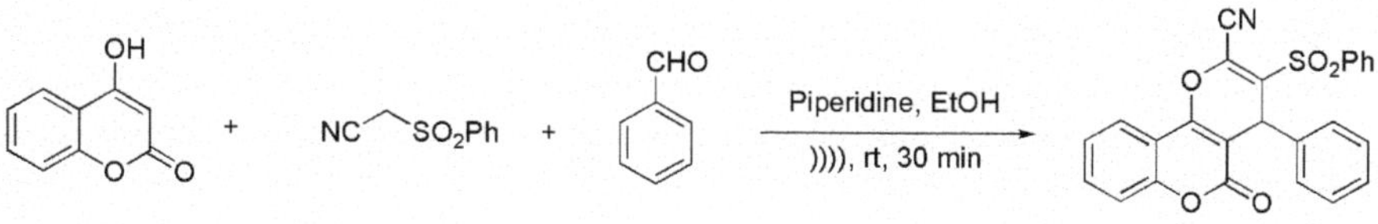

Scheme 12.14 Ultrasound-assisted synthesis of pyrano coumarins

12.2.3 *Solvent Free Green Synthesis*

Synthetic protocols involving no usage of solvents has gained a lot of importance due to its remarkable sustainability. Such solvent free syntheses have enabled easy production of heterocyclic compounds of medicinal significance in short duration. They can be obtained by microwave, ultrasound and mechanochemical methods as discussed below.

Coumarins: Pechmann condensation is the best way to get coumarins from a varied class of phenols with β-ketoesters. One of the breakthroughs was created using a ball milling method to produce coumarins from phloroglucinol and ethyl acetoacetate using methanesulfonic acid as the catalyst (Scheme 12.15) [30].

Bis-indoles: Hamacanthin A, Hamacanthin B and Gelaganimine B are the noted bis-indole derived compounds that have shown outstanding activity against different classes of fungal infections and inflammations. In order to construct basic bis-indole derivatives, microwave technique is noteworthy in giving results in the shortest time duration with enhanced yield and productivity. One of such methods used to produce simple bis-indoles involved the utility of a catalytic amount of acetic acid for reaction acceleration. Indole reacted with salicylaldehyde and 4-methoxybenzaldehyde, two different bis-indoles of medicinal value (Scheme 12.16) [31].

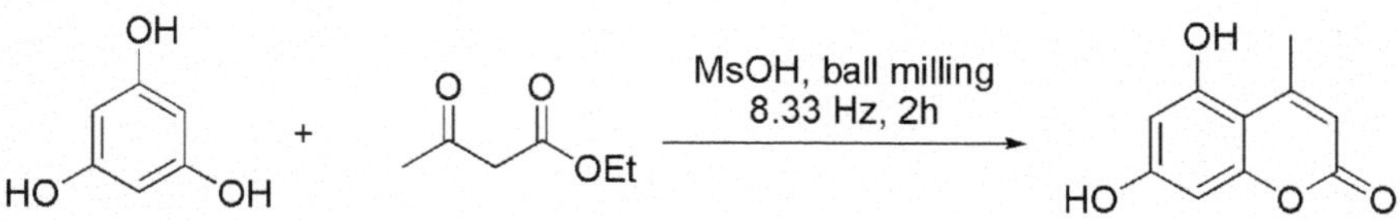

Scheme 12.15 Mechanochemical solvent free synthesis of 5,7-hydroxycoumarin

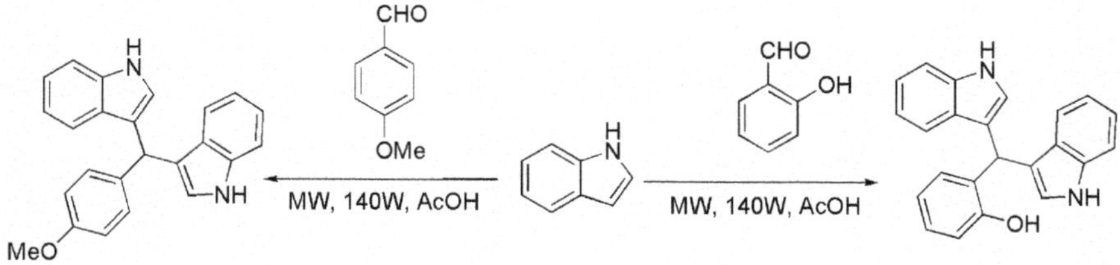

Scheme 12.16 Solvent free microwave synthesis of bis-indoles

4,7-dihydro pyrazolo pyrimidines: Ultrasound-promoted solvent free synthesis of 4,7-dihydropyrazolopyrimidines was achieved efficiently via heterocyclization of 3-aryl-5-amynopyrazole and enone such as 3-aryl-5-aminopyrazole with acetic acid as the catalyst (Scheme 12.17) [32].

Benzil bis-hydrazones: Another important green technique that is simple to perform a reaction just by grinding the substrates. Likewise, Benzil bis-hydrazones are obtained by reacting benzil with 2-cyanoacetohydrazide resulting in the formation of the precursor N′, N″-(1,2-diphenylethane-1,2-ylidene)bis(2-cyanoacetohydrazide). Grinding is done with moist L-Proline which is made to undergo Knoevenagel condensation of the precursor with different aldehydes to give the desired product in excellent yield (Scheme 12.18) [33].

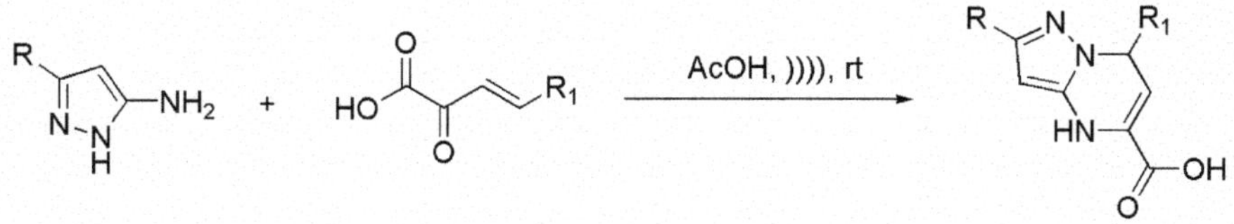

Scheme 12.17 Ultrasound-promoted solvent free synthesis of 4,7-dihydro pyrazolo pyrimidines

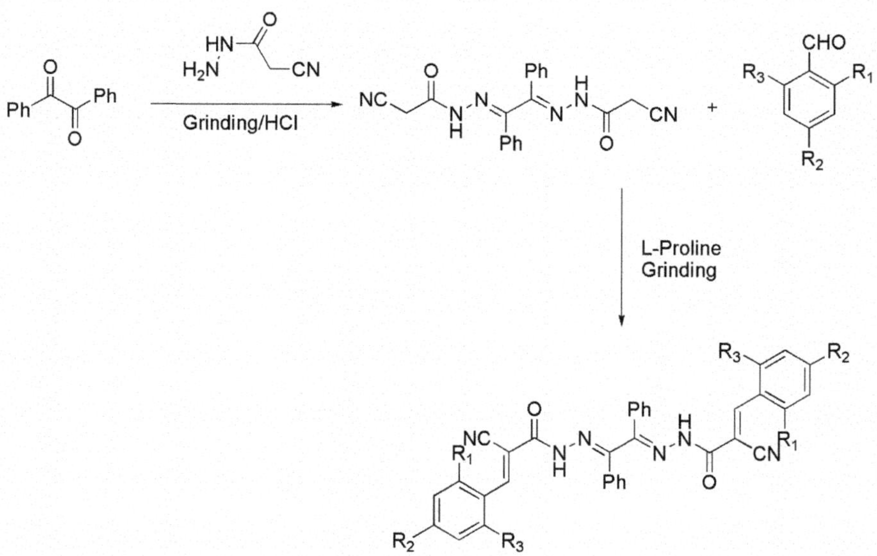

Scheme 12.18 Solvent free synthesis of Benzil bis-hydrazones by grinding method

12.2.4 Ionic Liquids

Several organic chemists employed the method of generating ionic liquids for the replacement of conventional catalysts for the successful reactions enabling good yield and sustainability. Chloroaluminate based catalysts $[HN_{222}][AlCl_4]$ and $[HN_{222}][Al_2Cl_7]$ are used instead of HF or $AlCl_3$ for the acylation of isobutylbenzene using acetic anhydride to yield 4-isobutylacetophenone (intermediate of ibuprofen) along with a side product 2-isobutylacetophenone (Scheme 12.19) [34]. 3-vinyl indoles can also be prepared using sulfonyl containing ionic liquid. The augmented activity of this catalyst is due to the presence of sulfonyl and sulfonic acid groups simultaneously. 2-methyl indole when reacted with acetophenone under solvent free conditions results in the formation of 3-vinyl indole in high yield (Scheme 12.20) [35]. Some of the prominent ionic liquids used for the organic synthesis are shown in Fig. 12.1.

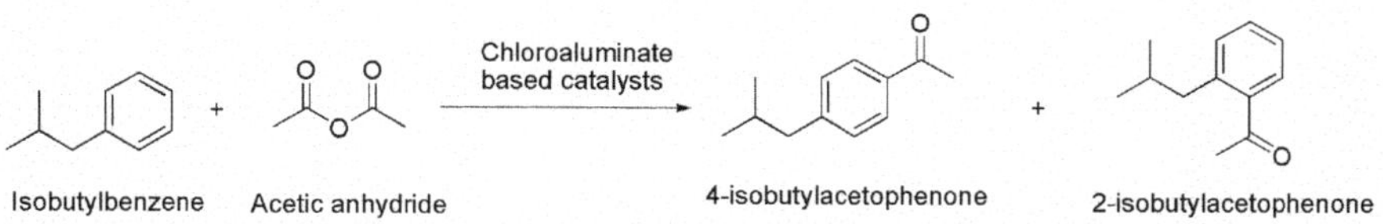

Scheme 12.19 Acylation of isobutylbenzene catalyzed by chloroaluminate based ionic liquids

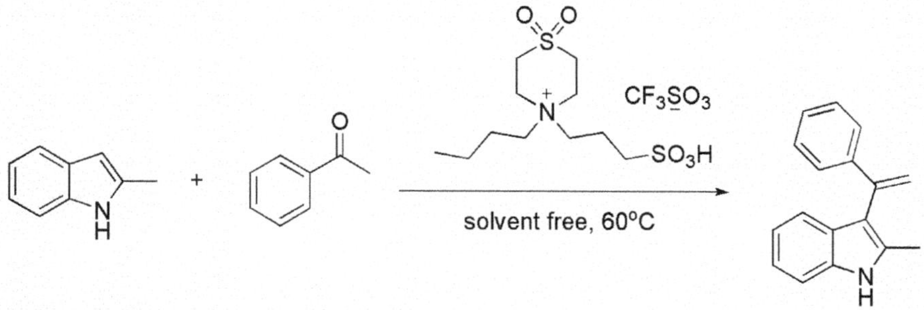

Scheme 12.20 Synthesis of 3-vinyl indole using sulfonyl-containing ionic liquid

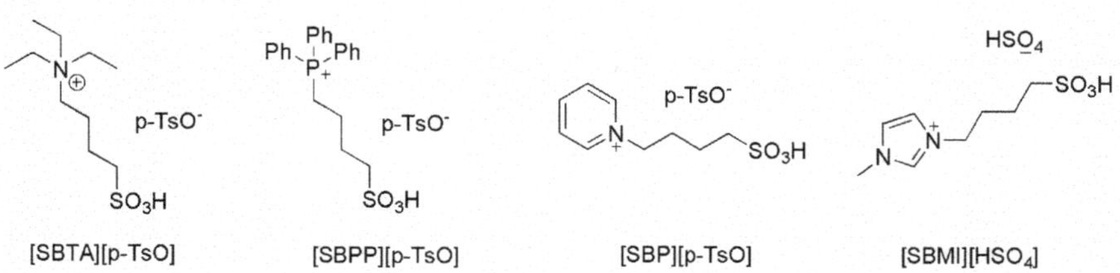

Fig. 12.1 Ionic liquids acting as effective catalysts

12.2.5 Phase Transfer Catalysis

Among various catalysis processes, phase transfer catalysis (PTC) has crossed the borders of applications and is now a widespread area of research in synthetic organic chemistry. The reactions involving PTC take place in two phases namely, organic and aqueous. The major role of such catalysts is to regulate the mass transport over the phase border, allowing the reactants to undergo reaction. Crown ethers, quaternary ammonium salts and cyclodextrins, all having hydrophobic and hydrophilic units. Some of the commonly used phase transfer catalysts are given in Fig. 12.2.

Based on the chiral phosphonium salts, few phase transfer catalysts employed for asymmetric transformations developed by Manabe and Maruoka are depicted in Fig. 12.3.

Pyrazolines with good enantioselectivity are obtained from N-Boc-hydrazine and chalcone in the presence of PTC under freezing conditions for 24 h (Scheme 12.21)

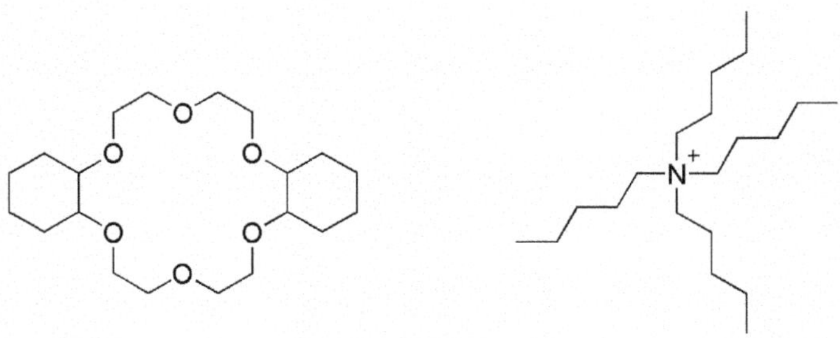

18-crown-6 TBAB

Fig. 12.2 Commonly used phase transfer catalysts

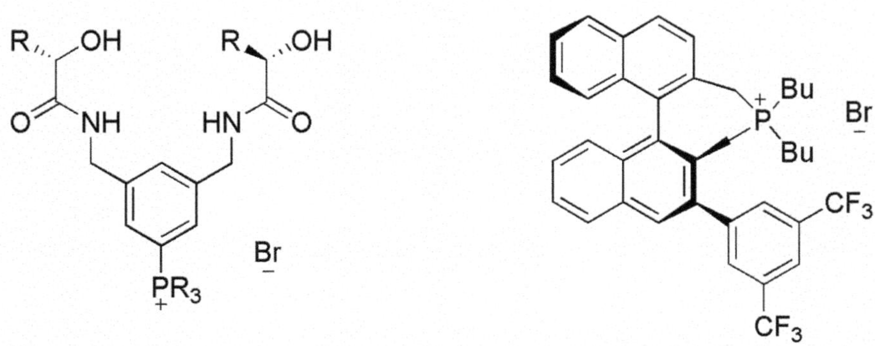

Manabe Maruoka

Fig. 12.3 Chiral Phosphonium salts as PTC

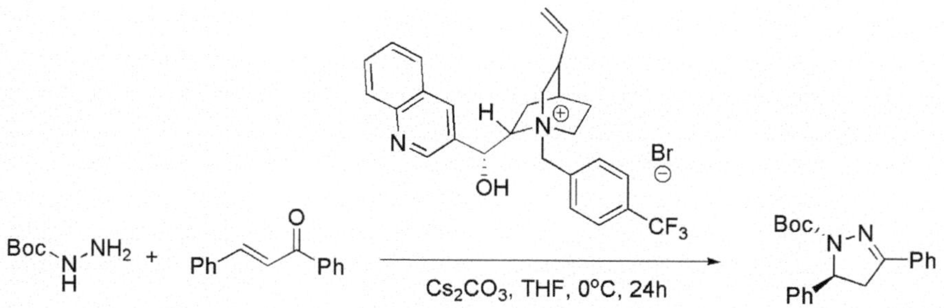

Scheme 12.21 Synthesis of Pyrazolidine derivatives using PTC

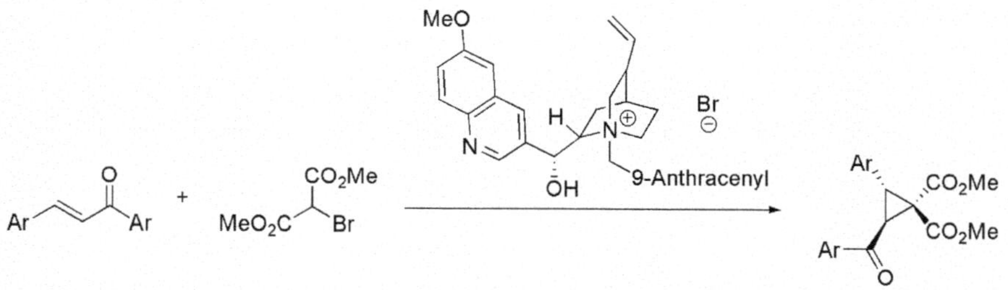

Scheme 12.22 Synthesis of activated chalcones using PTC

[36]. Likewise, carbocyclic compounds are very helpful as building blocks for synthesizing many natural products and biologically significant heterocycles. Cyclopropanation of chalcones with bromomalonate using PTC produced electron deficient activated chalcones with high yield. (Scheme 12.22) [37].

12.3 Conclusion

The advancements of green techniques following the principles of green chemistry have shown their spectacular mark in chemical biology and pharmaceutical chemistry. The various microwave-assisted synthetic works discussed here are the strong evidence for the applicability of green principles in thriving deeper roots under drug discovery. The utility of ultrasound-promoted organic synthesis is itself a boon over the generally practiced traditional mode of synthesis. The yields obtained after employing such techniques were remarkably good which has reduced the efforts of the chemists struggling for the yield for days altogether. Both sonochemistry and microwave technology has opened a new way for accomplishing potent organic heterocyclic molecules which are later screened for various biological activity. The role of ionic liquids as efficient and effective catalysts for organic reactions is well discussed and their significance is noteworthy in accelerating reaction rate above the expected outcome. The use of phase transfer catalysts is also important for the reactions travelling in aqueous and organic phases. The solvent-free methodologies retard the harmful effects of the solvents on human health and their knowledge should be very precise for carrying out such green reactions.

Altogether, green techniques are now being used in the pharmaceutical industries for the rapid synthesis without the toxic effects of the solvents on the reactions. Hence, green techniques are very helpful for the innovation of different pharmaceutically active compounds that could aid in developing as potent drug molecules in upcoming years.

References

1. Anastas, P.T., Warner, J.C.: Green chemistry. Org. Process Res. Dev. **4**, 437–438 (2000)
2. Tucker, J.L., Faul, M.M.: Industrial research: drug companies must adopt green chemistry. Nature **534**, 27–29 (2016)
3. Watson, W.J.W.: How do the fine chemical, pharmaceutical, and related industries approach green chemistry and sustainability? Green Chem. **14**, 251–259 (2012)
4. Lombardino, J.G., Lowe, J.A.: The role of the medicinal chemist in drug discovery—then and now. Nat. Rev. Drug Discov. **3**, 853–862 (2004)
5. Draye, M., Chatel, G., Duwald, R.: Ultrasound for drug synthesis: a green approach. Pharmaceuticals **13**, 23 (2020)
6. Mavandadi, F., Pilotti, Å.: The impact of microwave-assisted organic synthesis in drug discovery. Drug Discov. Today **11**, 165–174 (2006)
7. Shukla, M.K., Tiwari, H., Verma, R., et al.: Role and recent advancements of ionic liquids in drug delivery systems. Pharmaceutics **15**, 702 (2023)
8. Mahurkar, N.D., Gawhale, N.D., Lokhande, M.N., et al.: Benzimidazole: a versatile scaffold for drug discovery and beyond—a comprehensive review of synthetic approaches and recent advancements in medicinal chemistry. Results Chem. **6**, 101139 (2023)
9. Mezaal, E.N., Sadiq, K.A., Jabbar, M.M., et al.: Green methods for determination of paracetamol in drug samples: a comparative study. Green. Anal. Chem. **10**, 100123 (2024)
10. Madgula, K., Dandu, S., Kasula, S., Halady, P.: Microwave synthesized ionic liquids as green catalysts for the synthesis of benzimidazoles: spectral and computational studies for potential anticancer activity. Inorg. Chem. Commun. **138**, 109218 (2022)
11. Kar, S., Sanderson, H., Roy, K., et al.: Green chemistry in the synthesis of pharmaceuticals. Chem. Rev. **122**, 3637–3710 (2022)
12. Malik, A.A., Dangroo, N.A., Ara, T.: Microwave-assisted tandem Kornblum oxidation and Biginelli reaction for the synthesis of dihydropyrimidones. ChemistrySelect **5**, 12965–12970 (2020)
13. Popovics-Tóth, N., Tajti, Á., Bálint, E.: Biginelli reaction of β-ketophosphonates, aromatic or aliphatic aldehydes and urea derivatives. Phosphorus Sulfur Silicon Relat. Elem. **197**, 597–598 (2022)
14. Huseynzada, A.E., Jelch, C., Akhundzada, H.V.N., et al.: Synthesis, crystal structure and antibacterial studies of dihydropyrimidines and their regioselectively oxidized products. RSC Adv. **11**, 6312–6329 (2021)
15. O, S., Martis, G.J., Mugali, P.S.: Multicomponent single step microwave induced organic synthesis. Int. J. Sci. Res. Chem. **8**(6), 1–6 (2023)
16. Kidwai, M., Saxena, S., Mohan, R., Venkataramanan, R.: A novel one pot synthesis of nitrogen containing heterocycles: an alternate methodology to the Biginelli and Hantzsch reactions. J. Chem. Soc. Perkin. Trans. **1**, 1845–1846 (2002)

17. Kidwai, M., Mohan, R., Saxena, S.: Solid-supported Hantzsch—Biginelli reaction for syntheses of pyrimidine derivatives. Russ. Chem. Bull. **52**, 2457–2460 (2003)
18. Khrustalev, D., Yedrissov, A., Khrustaleva, A., et al.: An efficient method for the synthesis of 1,4-dihydropyridine derivatives in a flow microwave reactor. Mater. Today. Proc. **81**, 1204–1208 (2023)
19. Thangaraj, M., Gengan, R.M., Ranjan, B., Muthusamy, R.: Synthesis, molecular docking, antimicrobial, antioxidant and toxicity assessment of quinoline peptides. J. Photochem. Photobiol. B. **178**, 287–295 (2018)
20. Echemendía, R., Rabêlo, W.F., López, E.R., et al.: A bidirectional access to novel thiadiazine hybrid molecules by double multicomponent reactions. Tetrahedron Lett. **59**, 4050–4053 (2018)
21. Chemat, F., Rombaut, N., Sicaire, A.G., et al.: Ultrasound assisted extraction of food and natural products: mechanisms, techniques, combinations, protocols and applications: a review. Ultrason. Sonochem. **34**, 540–560 (2017)
22. El Ghani, N., Miralles, S., Botton, V., et al.: Acoustic streaming enhanced mass transfer at a wall. Int. J. Heat Mass Transf. **172**, 121090 (2021)
23. Shabalala, N.G., Pagadala, R., Jonnalagadda, S.B.: Ultrasonic-accelerated rapid protocol for the improved synthesis of pyrazoles. Ultrason. Sonochem. **27**, 423–429 (2015)
24. Veisi, H., Joshani, Z., Karmakar, B., et al.: Ultrasound assisted synthesis of Pd NPs decorated chitosan-starch functionalized Fe_3O_4 nanocomposite catalyst towards Suzuki-Miyaura coupling and reduction of 4-nitrophenol. Int. J. Biol. Macromol. **172**, 104–113 (2021)
25. El-bendary, M.M., Saleh, T.S., Al-Bogami, A.S.: Synthesis and structural characterization of a palladium complex as an anticancer agent, and a highly efficient and reusable catalyst for the Heck coupling reaction under ultrasound irradiation: a convenient sustainable green protocol. Polyhedron **194**, 114924 (2021)
26. Ghamari Kargar, P., Len, C., Luque, R.: Cu/cellulose-modified magnetite nanocomposites as a highly active and selective catalyst for ultrasound-promoted aqueous O-arylation Ullmann and sp-sp^2 Sonogashira cross-coupling reactions. Sustain. Chem. Pharm. **27**, 100672 (2022)
27. Bhat, A.R., Naikoo, G.A., Hassan, I.U., et al.: Ultrasound assisted one pot expeditious synthesis of new pyrido[2,3- d]pyrimidine analogues using mild and inexpensive 4-dimethylaminopyridine (DMAP) catalyst. Beni-Suef. Univ. J. Basic Appl. Sci. **6**, 238–246 (2017)
28. Nikpassand, M., Zare Fekri, L., Hematinezhad, K.: Ultrasound-promoted regioselective synthesis of pyrazolyl-bis coumarinyl methanes using citric acid as a natural and efficient catalyst. Polycycl. Aromat. Compd. **40**, 456–464 (2020)
29. Al-bogami, A.S.: One-pot, three-component synthesis of novel pyrano[3,2-c]coumarins containing sulfone moiety utilizing ultrasonic irradiation as an eco-friendly energy source. Res. Chem. Intermed. **41**, 93–104 (2015)
30. Sharapov, A.D., Fatykhov, R.F., Khalymbadzha, I.A., et al.: Mechanochemical synthesis of coumarins via Pechmann condensation under solvent-free conditions: an easy access to coumarins and annulated pyrano[2,3-*f*] and [3,2-*f*]indoles. Green Chem. **24**, 2429–2437 (2022)
31. Hegde, D.C., Martis, G.J., Kerikai, A.S., Mugali, P.S.: A simple strategy for obtaining bis-indole derivatives under solvent free conditions. J. Chem. Lett. **5**, 139–143 (2024)
32. Chebanov, V.A., Sakhno, Y.I., Desenko, S.M.: High regioselective ultrasonic-assisted synthesis of 2,7-diaryl-4,7-dihydropyrazolo[1,5-a]pyrimidine-5-carboxylic acids. Ultrason. Sonochem. **19**, 707–709 (2012)
33. Sayed, A.R., Gomha, S.M., Abd El-lateef, H.M., Abolibda, T.Z.: L-proline catalyzed green synthesis and anticancer evaluation of novel bioactive benzil bis-hydrazones under grinding technique. Green Chem. Lett. Rev. **14**, 180–189 (2021)
34. Kore, R., Uppara, P.V., Rogers, R.D.: Replacing HF or $AlCl_3$ in the acylation of isobutylbenzene with chloroaluminate ionic liquids. ACS. Sustain. Chem. Eng. **8**, 10330–10334 (2020)

35. Ratti. R.: Ionic liquids: synthesis and applications in catalysis. Adv. Chem., 1–16 (2014)
36. Mahé, O., Dez, I., Levacher, V., Brière, J.F.: Enantioselective synthesis of bio-relevant 3,5-diaryl pyrazolines. Org. Biomol. Chem. **10**, 3946 (2012)
37. Herchl, R., Waser, M.: Asymmetric cyclopropanation of chalcones using chiral phase-transfer catalysts. Tetrahedron Lett. **54**, 2472–2475 (2013)

The manufacturer's authorised representative in the EU is Springer Nature Customer Service Centre GmbH, Europaplatz 3, 69115 Heidelberg, Germany. If you have any concerns regarding our products, please contact ProductSafety@springernature.com

Printed and bound by CPI Group (UK) Ltd, Croydon, CR0 4YY
05/10/2026
02228129-0005